EINSTEIN

A Centenary Volume

The International Commission on Physics Education

Einstein in 1946

EINSTEIN

A Centenary Volume

Edited by

A. P. French

Harvard University Press
Cambridge, Massachusetts

Printed in the United States of America
Fourth printing, 1980

Library of Congress Cataloging in Publication Data

Main entry under title:

Einstein: a centenary volume.

Bibliography: p.
Includes index.
1. Einstein, Albert, 1879–1955—Bibliography—Addresses, essays, lectures. 2. Physicists—Biography—Addresses, essays, lectures. I. Einstein, Albert, 1879–1955. II. French, Anthony Philip.
QC16.E5E37 530'.092'4 [B] 78–25968

ISBN 0–674–24230–0 (cloth)
ISBN 0–674–24231–9 (paper)

Preface

At a meeting early in 1976, the International Commission on Physics Education took note of the fact that 14 March 1979 would be the centenary of the birth of Albert Einstein. Since Einstein was not only the greatest figure in twentieth-century physics but also, in many ways, a person profoundly concerned with science as part of our culture, it was felt that our Commission, involved as it is with physics education at all levels, might take some appropriate action to mark the centenary. This book is the result.

We hope that, in the spirit of our intentions, the book will appeal to a wide range of readers. In planning it, however, we had especially in mind both teachers and students of physics at secondary and undergraduate (tertiary) levels. Our main purpose has been to provide something of a picture of Einstein the man, of his scientific work and its subsequent influence, and of his role as a humanitarian and world statesman.

The preparation of this book has involved the generous cooperation of many people. First among these are the individuals who have contributed the articles and personal reminiscences that make up the main body of the text. The Commission also owes a large debt of gratitude to Unesco for its part in initiating this project and for its subsequent cooperation and financial support.

In the early stages of the planning for this book the Commission also benefited from the advice of the International Union of the History and Philosophy of Science—in particular that of its President, Professor R. Taton.

As Editor, I should like to acknowledge the invaluable help of several individuals in particular. My colleagues on the editorial committee were Peter Kennedy (University of Edinburgh), Nahum Joel (Unesco) and John L. Lewis (Malvern College). The task of selecting a wide variety of apt quotations to scatter through the book was admirably performed by Maurice Ebison (Institute of Physics). Staff members of the Niels Bohr Library and of the Center for the History of Physics at the American Institute of Physics (especially Joan Warnow and Peter Dews) were extremely helpful in identifying and supplying relevant illustrations and other material. And, in connection with the selection and use of material controlled by the Einstein Estate, I should like to thank Dr Otto Nathan and Miss Helen Dukas for their friendly and courteous cooperation.

Einstein, being a fundamentally modest man, would probably have deprecated the extent to which his first centenary has become the occasion for separate and spontaneous tributes all over the world. On the other hand, he would perhaps be gratified at the evidence that, in addition to paying homage to him as an individual, these various celebrations are an affirmation of the values, both scientific and social, that he most greatly treasured. We hope, in particular, that he would have seen this commemorative volume not as an act of idolatry, but (in the way we ourselves see it) as a modest contribution to the appreciation of science as a fascinating and fundamentally human activity of universal interest and importance.

A. P. French

The International Commission on Physics Education (ICPE) was established in 1960. It is one of a number of commissions of the International Union of Pure and Applied Physics (IUPAP). The ICPE has as its main concern the stimulation and promotion of international cooperation in physics education. Its activities consist largely in the planning and organizing of international conferences on various aspects and particular areas of physics education. The Commission has on the average about twelve members, each from a different country; the memberships come up for review every three years at a general assembly of its parent organization (IUPAP).

Contents

List of Contributors

BERGIA, SILVIO	Lecturer in Mathematical Methods of Physics, University of Bologna; Lecturer in Theoretical Physics, University of Modena, Italy.
BONDI, SIR HERMANN KCB, FRS	Chief Scientist, Department of Energy (UK); Professor of Mathematics, Kings College, London, UK.
BURGE, E. J.	Professor of Physics, Chelsea College of Science and Technology, London, UK.
COHEN, I. BERNARD	Victor S. Thomas Professor of the History of Science, Harvard University, USA.
DE BROGLIE, PRINCE LOUIS	Former Professor of Theoretical Physics, University of Paris, France; Nobel Laureate in Physics (1929); Former Permanent Secretary, French Academy of Sciences.
DORLING, GEOFFREY	Wymondham College, Wymondham, Norfolk, UK.
EBISON, MAURICE	Education Officer, Institute of Physics, London, UK; Secretary of the Committee on Education of the European Physical Society.
FRENCH, A. P.	Professor of Physics, Massachusetts Institute of Technology, Cambridge, USA; Chairman, International Commission on Physics Education.
HALSMAN, PHILIPPE	Photographer.
HOLTON, GERALD	Mallinckrodt Professor of Physics and Professor of the History of Science, Harvard University, USA.
HÖRZ, HERBERT	Professor of Philosophy, Akademie der Wissenschaften der DDR, Berlin, German Democratic Republic.
KEMENY, JOHN G.	President, Dartmouth College, Hanover, New Hampshire, USA.
KLEIN, MARTIN J.	Professor of Physics and Eugene Higgins Professor of the History of Science, Yale University, USA.
KUZNETSOV, BORIS	Professor of the History of Sciences, Institute of the History of Sciences and Technology, Moscow, USSR; Vice-President, Albert Einstein Committee of the International Union for the History and Philosophy of Science.

LORIA, ARTURO Professor of Physics and Director, Istituto di Fisica, University of Modena, Italy;
Member, International Commission on Physics Education.

PAIS, ABRAHAM Professor of Physics, Rockefeller University, New York, USA;
Visiting Senior Fellow, Princeton University (1978–).

ROGERS, ERIC M. Professor Emeritus, Princeton University, USA;
Visiting Professor, Miami University, USA;
Organiser for Nuffield Physics.

SHANKLAND, ROBERT S. Ambrose Swasey Professor Emeritus of Physics, Case Western Reserve University, Cleveland, USA.

SNOW, LORD C. P. Author and Physicist.

SPEZIALI, PIERRE Chargé de Cours d'Histoire des Sciences à la Faculté des Sciences de Genève, Switzerland;
Corresponding Member, International Academy of the History of Science.

STRAUS, ERNST G. Professor of Mathematics, University of California, Los Angeles, USA.

TAUBER, GERALD E. Professor of Physics, Tel Aviv University, Israel;
Director, Israel Center for Relativistic Astrophysics and Gravitation.

TELLER, EDWARD Hoover Institution, Stanford, USA;
Formerly Associate Director-at-Large, Lawrence Livermore Laboratory, University of California, Berkeley, USA.

WHEELER, JOHN ARCHIBALD Director, Center for Theoretical Physics, University of Texas, Austin, USA;
Former President, American Physical Society.

WIGNER, EUGENE P. Thomas D. Jones Professor Emeritus of Mathematical Physics, Princeton University, USA.

List of Illustrations

Acknowledgements

'Albert Einstein 1879–1955', reprinted from *Variety of Men* by C. P. Snow, by permission of Curtis Brown, Ltd. (New York: Charles Scribner's Sons, 1967).

'Excerpts from a Memoir' by Maurice Solovine, reprinted from *Lettres à Maurice Solovine* by Albert Einstein (Paris: Gauthier-Villars, 1956).

'Einstein's Friendship with Michele Besso', reprinted from *Albert Einstein et Michele Besso: Correspondance 1903–1955*, translated with notes and introduction by Pierre Speziali. (Paris: Collections Histoire de la Pensée 17, Editions Hermann, 1972.)

'Reminiscences of Einstein' reprinted from *Focus and Diversions* by L. L. Whyte by permission of George Braziller, Inc. (New York: Braziller, 1964).

'Anecdotes' by Philipp Frank, extracted from 'Einstein's Philosophy of Science', *Revs. mod. Phys.*, 1949, **21**, 349.

'Einstein' reprinted from *Sight and Insight* by permission of Philippe Halsman (New York: Doubleday, 1972).

'Reminiscence' reprinted from *My World Line* by George Gamow. Copyright © 1970 by the Estate of George Gamow, reprinted by permission of the Viking Press.

'A Tribute' by Pablo Casals, reprinted from *Albert Einstein: A Documentary Biography* by Carl Seelig, by permission of the Seelig Estate (London: Staples Press, Granada Publishing Ltd, 1966).

'On Albert Einstein' by J. Robert Oppenheimer, reprinted from 'Einstein's Presence' in *Science and Synthesis* by permission from Unesco (Paris: Unesco Publications, 1967).

We would also like to thank Dr Otto Nathan, trustee of the Estate of Albert Einstein, for permission to publish the many excerpts from *Einstein on Peace*, *Out of My Later Years*, and *Ideas and Opinions*.

Nathan, O., and Norden, H., *Einstein on Peace* (New York: Simon and Schuster, 1960).

Einstein, A., *Out of My Later Years* (New York: Philosophical Library, 1950).

—— *Ideas and Opinions* (New York: Dell, 1954).

SOURCES OF MARGINAL QUOTATIONS

Alfvèn, Hannes, 'Cosmology, Myth or Science?' in *Cosmology, History and Theology*, edited by W. Yourgrau and A. D. Breck (New York: Plenum Press, 1977).

Beers, Yardley, *American Journal of Physics*, 1978, **45**, 506

Bernstein, Jeremy, *Einstein* (New York: Viking, 1976).
Brillouin, Léon, *Relativity Reexamined* (New York: Academic Press, 1970).
Einstein, Albert, 'On the method of theoretical physics', 'Geometry and experience', 'On the theory of relativity', in *Ideas and Opinions* (New York: Dell, 1973).
Frank, Philipp, *Einstein: His Life and Times* (New York: Knopf, 1947).
Hoffmann, Banesh, with Dukas, Helen, *Albert Einstein: Creator and Rebel* (New York: The Viking Press, 1972). Reprinted by permission of Hart-Davis, MacGibbon and Viking Penguin Inc.
Holton, Gerald, *The Scientific Imagination: Case Studies* (Cambridge: Cambridge University Press, 1978).
Hutten, E. H., *The Language of Modern Physics* (New York: Macmillan, 1956).
Infeld, L., *Quest: The Evolution of a Scientist* (New York: 1941).
Jaki, Stanley L., *The Relevance of Physics* (Chicago: University of Chicago Press, 1966). © 1966 by the University of Chicago. All rights reserved. Published 1966.
Margenau, H. (Editor), *Integrative Principles of Modern Thought* (New York: Gordon and Breach, 1972).
Mehra, Jagdish, *The Solvay Conferences on Physics* (Dordrecht: D. Reidel, 1970).
Moszkowski, Alexander, *Conversations with Einstein*, 'Introduction' by Henry Le Roy Finch, 'Foreword' by C. P. Snow, translated by Henry L. Brose (New York: Horizon Press, 1970). Reprinted by permission of the publisher, Horizon Press, © 1970.
Newman, James R., *Science and Sensibility* (New York: Simon and Schuster, 1961), Vol. 1.
Science and Synthesis, A Unesco International Colloquium (Paris: Unesco Publications, 1967). Reproduced by permission of Unesco. © Unesco 1967.
Seelig, Carl, *Albert Einstein: A Documentary Biography* (London: Staples Press, 1956).
Synge, J. L., *Talking about Relativity* (Amsterdam: North-Holland, 1970).
Russell, B., 'Preface', in *Einstein on Peace*, edited by Otto Nathan and Heinz Norden (New York: Simon and Schuster, 1960).
Whitrow, G. J. (Editor), *Einstein: The Man and His Achievement* (London: British Broadcasting Corporation, 1967).

Chronological Biography

1879 Albert Einstein born (14 March) in Ulm (South Germany) son of Hermann Einstein and Pauline Einstein (née Koch).

1880 Family moves to Munich.

1884 The famous encounter with a pocket compass.

1884–9 Pupil at Catholic elementary school.

1889–94 Pupil at Luitpold Gymnasium.

1894 Parents move to Milan. Six months later, Einstein leaves Gymnasium and follows them.

1895–6 Einstein attends cantonal school at Aarau, Switzerland.

1896 Einstein renounces his German citizenship; enters Zürich Polytechnic (*Eidgenössische Technische Hochschule*—ETH); attends lectures by H. Minkowski (among others).

1900 Receives diploma from ETH.

1901 Acquires Swiss citizenship; completes first scientific paper ('Consequences of Capillary Phenomena').

1901–2 Tutor in a private school at Schaffhausen.

1902 Arrives in Bern (February); meets Maurice Solovine (April); obtains probationary appointment in Patent Office.

1904 Receives a definite appointment at the Patent Office.

1905 The *annus mirabilis*: papers on light quanta, Brownian motion, and special relativity (plus $E = mc^2$). Ph.D. from the University of Zürich.

1908 Appointed *Privatdozent* at the University of Bern.

1909 Leaves the Patent Office and joins the University of Zürich.

1911–12 Professor of Physics at Prague.

1912 Returns to Zürich as a professor of physics.

1913 Publishes (with Grossmann) a preliminary paper on general relativity.

1914 Moves to Berlin as professor in Prussian Academy of Sciences and Director of Kaiser Wilhelm Institute of Physics. Becomes a founding member of New Fatherland League.

1916 Publishes the general theory of relativity.

1917 Publishes paper on cosmological implications of general relativity.

1920 Becomes acquainted with Niels Bohr.

1921 Visits USA for the first time (together with Chaim Weizmann).

1922 Awarded the 1921 Nobel Prize (for his research on the theory of the photoelectric effect). Lecture tours in China and Japan.

1930–2 Three more visits to the USA.

1933 Takes up temporary residence in Belgium upon return from USA, then emigrates to USA (never to return to Europe).

1939 Letter to Roosevelt proposing atomic bomb research.

1946 Assumes chairmanship of Emergency Committee of Atomic Scientists.

1952 Is offered Presidency of Israel.

1955 Dies (18 April) at Princeton.

REMINISCENCES

Albert Einstein 1879–1955

C. P. Snow

C. P. Snow has had a distinguished career as an academic physicist, as a government scientific advisor, as a member of the British government, and as a renowned novelist. This memoir consists of excerpts from a longer article in his book, Variety of Men.

Albert Einstein, twenty-six years old, published in the *Annalen der Physik* in 1905 five papers on entirely different subjects. Three of them were among the greatest in the history of physics. One, very simple, gave the quantum explanation of the photoelectric effect—it was this work for which, sixteen years later, he was awarded the Nobel Prize. Another dealt with the phenomenon of Brownian motion, the apparently erratic movement of tiny particles suspended in a liquid: Einstein showed that these movements satisfied a clear statistical law. This was like a conjuring trick, easy when explained; before it, decent scientists could still doubt the concrete existence of atoms and molecules: this paper was as near to a direct proof of their concreteness as a theoretician could give. The third paper was the special theory of relativity, which quietly amalgamated space, time, and matter into one fundamental unity.

This last paper contains no references and quotes no authority. All of them are written in a style unlike any other theoretical physicist's. They contain very little mathematics. There is a good deal of verbal commentary. The conclusions, the bizarre conclusions, emerge as though with the greatest of ease: the reasoning is unbreakable. It looks as though he had reached the conclusions by pure thought unaided, without listening to the opinions of others. To a surprisingly large extent, that is precisely what he had done.

It is pretty safe to say that, so long as physics lasts, no one will again hack out three major breakthroughs in one year. People have complained that Einstein was not immediately recognized. That seems mildly unrealistic. Within a few months, physicists at Cracow were saying that a new Copernicus had been born. It took about four years for the top German physicists, such as Planck, Nernst, and von Laue, to begin proclaiming that he was a genius. In 1909, before he had any academic job at all, he was given an honorary degree at Geneva. Just afterward, Zürich University (not the Polytechnic) offered him a professorship. In 1911, he went to a full chair at the German University in Prague. In 1912 he was recalled to the Zürich Polytechnic, which had had, only a dozen years before, no use for him. In 1913, he was elected to the Prussian Academy of Science, at a high salary for those days, to be left free in Berlin for no duties except his research. He was by then thirty-four. He was being treated as handsomely as any scientist alive. I don't think the academic community, in particular the German-speaking academic community, comes out of that story badly.

He arrived in Berlin some months before war broke out. He was already famous in the scientific world. He was going to attract fame in the world outside such as no scientist has known before or since. He was a pacifist soon forced to watch what he regarded as German madness among, not only the crowd, but his fellow-members of the Academy. He had preserved his Swiss nationality, which was some sort of protection, when, with his habitual courage, he became an ally of Romain Rolland. But he soon came to experience the blackest unpopularity. He could shrug it off: 'Even the scientists of various countries behave as though eight months ago' (he was writing to Rolland in May 1915) 'they had had their brains amputated.'

Nevertheless, in the middle of militaristic tumult, he found both personal and creative peace. In November 1915, he wrote to Arnold Sommerfeld, himself a fine physicist, one of the classical scientific letters:

> This last month I have lived through the most exciting and the most exacting period of my life: and it would be true to say that it has also been the most fruitful. Writing letters has been out of the question. I realize that up till now my field equations of gravitation have been entirely devoid of foundation. When all my confidence in the old theory vanished, I saw clearly that a satisfactory solution could only be reached by linking it with the Riemann variations. The wonderful thing that happened then was that not only did Newton's theory result from it, *as a first approximation*, but also the perihelion motion of Mercury, *as a second approximation*. For the deviation of light by the sun I obtain twice the former amount.

Sommerfeld wrote a cautious and sceptical reply. Einstein sent him a postcard: 'You will become convinced of the general theory of relativity as soon as you have studied it. Therefore I shall not utter a word in its defence.'

It did not need defence. It was published in 1916. As soon as it reached England—across the increasing harshness of the war—scientists thought that it was almost certainly right. The greatest revolution in thought since Newton, they were saying. As a consequence of his theory,* Einstein had made a prediction. It was the prediction of an experimental effect which astronomers could test. In his paper, he asked them to do so. The English astronomers decided that this should be done. In March 1917—again across the war—they announced that on 29 May 1919 a total eclipse of the Sun would take place. The critical experiment would be set up and Einstein's theory tested.

That is an old story. The test, of course, came out as predicted, and Einstein's theory stood.

It is a strange theory. As with Rutherford, as with most scientists, if

* See page 91, 'The Story of General Relativity' below.

Einstein had never lived most of his work would soon have been done by someone else, and in much the same form. He said himself that that was true of the special theory of relativity. But, when he generalized the special theory so as to include the gravitational field, he did something that might not have been done for a generation: and, above all, might not have been done in that way. It might, some good theoreticians have suggested, have ultimately been done in a way easier for others to handle. It remains an extraordinary monolith, like a Henry Moore sculpture, which he alone could have constructed—and at which he himself hacked away, hoping to make something grander, for the rest of his scientific life.

His public life, as soon as the general theory was published (his fame had already mounted *before* the confirmation), was unlike that which any other scientist is likely to experience again. No one knows quite why, but he sprang into the public consciousness, all over the world, as the symbol of science, the master of the twentieth-century intellect, to a large extent the spokesman for human hope. It seemed that, perhaps as a release from the war, people wanted a human being to revere. It is true that they did not understand what they were revering. Never mind, they believed that here was someone of supreme, if mysterious, excellence.

Throughout the 20's he made himself the champion of good causes. He became a Zionist, though his religious thinking was quite un-Judaic: he was on the side of Zion, out of an ultimate loyalty and also, as I have said before, because the Jews were the insulted and injured of this world. He spent a lot of time trying to promote international pacifism. This sounds strange to us now, but the 20's was a period of ideals, and even Einstein, the least suggestible of men, shared them. At a later period of his life, some Americans used to call him naïve. That irritated me: he was not in the slightest naïve: what they meant was that he didn't think that the United States was always 100 per cent right, and the Soviet Union 100 per cent wrong.

He was himself Hitler's greatest public enemy. He was out of Germany when Hitler became chancellor: he was a brave man, but he knew that if he returned, he would be killed. Through most of 1933 he lived in the little Flemish seaside town of Den Haan (Coq-sur-mer).

Belgium suited him. He was more comfortable in small cosy countries (Holland was his favourite), but he wasn't safe from the Nazis. Unwillingly, he set off on his travels again, went to Princeton, and stayed there until he died.

It was a kind of exile. There is no doubt that he, who had never recognized any place as home, sometimes longed for the sounds and smells of Europe. Nevertheless, it was in America that he reached his full wisdom and his full sadness. His wife died soon after he got there. His younger son, back in Switzerland, had gone into a mental home. His merriness had finally been worn away. He was left with his duty to other men.

He was left with something else, too. He could still lose his personality, forget everything else, in speculating about the natural world. That was the deepest root of his existence: it remained strong until the night before he died. He once said in public: 'Whoever finds a thought which enables us to obtain a slightly deeper glimpse into the eternal secrets of nature, has been given great grace.' He continued—that was the grace of his solitariness—to try to find such thoughts. Quite unlike Newton, who gave up physics entirely in order to become Master of the Mint and perform textual researches on the Bible, Einstein stayed working at science long after most theoreticians, even the best, have taken to something easier. But he worked—and this was the final strangeness of his life—in a direction precisely opposite to that of his major colleagues. In the public world, against militarism, against Hitler, against cruelty and unreason, nothing had ever made him budge. In the private world of theoretical physics, with the same quiet but total intransigence, he would not budge against the combined forces of the colleagues he loved, Bohr, Born, Dirac, Heisenberg, the major intellects in his own profession.

They believed that the fundamental laws were statistical—that, when it came to quantum phenomena, in Einstein's picturesque phrase, God had to play at dice. He believed in classical determination—that, in the long run, it should be possible to frame one great field theory in which the traditional concept of causality would re-emerge. Year after year he explained and redefined his position.

To Carl Seelig: 'I differ decisively in my opinions about the fundamentals of physics from nearly all my contemporaries, and therefore I cannot allow myself to act as spokesman for theoretical physicists. In particular, I do not believe in the necessity for a statistical formulation of the laws.'

To Max Born: 'I can quite well understand why you take me for an obstinate old sinner, but I feel clearly that you do not understand how I came to travel my lonely way. It would certainly amuse you, although it would be impossible for you to appreciate my attitude. I should also have great pleasure in tearing to pieces your positivistic-philosophical viewpoint.'

To James Franck: 'I can, if the worst comes to the worst, still realize that God may have created a world in which there are no natural laws. In short, a chaos. But that there should be statistical laws with definite solutions, i.e., laws which compel God to throw the dice in each individual case, I find highly disagreeable.'

God does *not* play at dice, he kept saying. But, though he worked at it for nearly forty years, he never discovered his unified field theory. And it is true that his colleagues, who passionately venerated him, sometimes thought that he was 'an obstinate old sinner'. They believed that he had misspent half the mental lifetime of the most powerful intellect alive. They felt they had lost their natural leader.

The arguments on both sides are most beautiful and subtle. Unfortunately, they cannot be followed without some background of physics: otherwise,

Bohr's *Discussion on Epistemological Problems* and Einstein's *Reply* ought to be part of everyone's education. No more profound intellectual debate has ever been conducted—and, since they were both men of the loftiest spirit, it was conducted with noble feeling on both sides. If two men are going to disagree, on the subject of most ultimate concern to them both, then that is the way to do it.

The great debate did not reach its peak until Einstein was old, years after the war. It was never resolved. He and Bohr, with mutual admiration, drew intellectually further apart. In fact, though, when I met Einstein in 1937 he had already separated himself totally, and as it proved, finally, from the other theorists.

It was two years later that Einstein signed the well-known letter to Roosevelt about the possibility of an atomic bomb. But this event, as I mentioned before, has been wildly melodramatized. Einstein was a mythopoeic character. Some of the myths are true and significant; this myth, though factually true, is not significant.

Let me try to clear the ground. First, Einstein's work had nothing to do either with the discovery or the potential use of nuclear fission. From the moment of the Meitner–Frisch paper in January 1939 (as Niels Bohr said at the time, everyone ought to have seen the meaning of Hahn's 1938 experiments much earlier—'we were all fools'), nuclear fission was a known fact to all physicists in the field. Second, the possible use of nuclear energy had been speculated about long before Einstein produced the equation $E=mc^2$. After the fission experiments, it would have been empirically apparent if there had been no theory at all. Every nuclear physicist in the world—and many a non-nuclear physicist—was talking about the conceivability of a nuclear bomb from early 1939 onward. Third, all responsible nuclear physicists wanted to bring this news to their governments as effectively as they could. It happened in England months *before* the Einstein letter was signed. Fourth, a group of refugee scientists in America (Szilard, Wigner, Teller, Fermi) had no direct channels of communication with the White House. Very sensibly, they explained the position to Einstein. It was easy for him to understand. A letter drafted by them, signed by him, handed on by Sachs (an economist with an entry to the President) would get straight to Roosevelt. 'I served as a pillar box,' said Einstein. It was signed on Long Island on 2 August: it did not reach Roosevelt until 11 October. Fifth, if this letter had not been sent, similar messages would have been forced on Roosevelt.

It is a pity that the story of the letter has obscured the genuine moral dilemma of his later years. Which was—now that the bomb exists, what should a man do?

He couldn't find an answer which people would listen to. He campaigned for a world state: that only made him distrusted both in the Soviet Union and in the United States. He gave an eschatological warning to a mass television audience in 1950:

> And now the public has been advised that the production of the hydrogen bomb is the new goal which will probably be accomplished. An accelerated development towards this end has been solemnly proclaimed by the President. If these efforts should prove successful, radioactive poisoning of the atmosphere, and, hence, annihilation of all life on earth, will have been brought within the range of what is technically possible. *A weird aspect of this development lies in its apparently inexorable character. Each step appears as the inevitable consequence of the one that went before.* And at the end, looming ever clearer, lies general annihilation.

That speech made him more distrusted in America. As for practical results, no one listened. Incidentally, in the view of most contemporary military scientists, it would be more difficult totally to eliminate the human species than Einstein then believed. But the most interesting sentences were the ones I have italicized. They are utterly true. The more one has mixed in these horrors, the truer they seem.

He joined in other warnings, one of them signed in the last week of his life. He did not expect them to bite: he retained the hope of his strong spirit, but intellectually he seems to have had no hope at all.

He was physically a strong man. He was in spirit as strong as a man can be. He was used to being solitary. 'It is strange,' he wrote, 'to be known so universally and yet to be so lonely.' Never mind. He was isolated in his search for the unified field. And the latter was the great theme of his life. He could endure it all, impregnable, and work stoically on. He said: 'One must divide one's time between politics and equations. But our equations are much more important to me.'

From his late sixties until his death at seventy-six he was continually ill—from what appears to have been a variety of causes, an intestinal growth, a disease of the liver, finally a weakening of the aorta wall. He lived on in discomfort, and often in acute pain. He stayed cheerful, serene, detached from his own illness and the approach of death. He worked on. The end of his life was neither miserable nor pathetic. 'Here on earth I have done my job,' he said, without self-pity.

By his bedside, one Sunday night, lay some pages of manuscript. They included more equations leading to the unified field theory, which he had never found. He hoped to be enough out of pain the next day to work on them. Early in the morning the aortic blister broke, and he died.

(Taken from *Variety of Men*)

Excerpts from a memoir

Maurice Solovine

Maurice Solovine arrived in Bern from Romania in 1900 and there met Einstein for the first time. Thus began a friendship and correspondence that lasted a lifetime. This reminiscence, which was written after Einstein's death, is translated from sections of a memoir that Solovine wrote as a preface to the book, Lettres à Maurice Solovine.

Walking one day in the streets of Bern during the Easter vacation of 1902, I bought a newspaper and happened to see an announcement saying that Albert Einstein, former student at the Zürich Polytechnic, offered physics lessons for three francs an hour. I said to myself: 'Perhaps this man can introduce me to the mysteries of physics.' I found my way to the address given, went up the stairs and rang the bell. I heard a loud *Herein!*, and then Einstein appeared. As the door of his apartment opened into a dark hallway, I was struck by the extraordinary brilliance of his large eyes.

After entering his room and sitting down, I explained to him that I was studying philosophy, but also wished to deepen my knowledge of physics so as to obtain a firm knowledge of Nature. He told me that he himself, when younger, had had a keen interest in philosophy, but that its vagueness and arbitrariness had put him off, and he now confined himself to physics.

We talked for almost two hours about all sorts of questions, and it seemed that we had a community of ideas and a personal affinity. When I prepared to leave, he came with me, and we conversed in the street for about half an hour before agreeing to meet again the next day . . .

I admired his extraordinarily penetrating mind and his astonishing mastery of physical problems. He was not a brilliant speaker and he did not make use of vivid images. He explained things in a slow and even voice, but in a remarkably clear way. To make his abstract ideas more easily grasped, he would sometimes take an example from everyday experience. Although Einstein handled mathematics with incomparable dexterity, he often denounced the misuse of mathematics in physics. 'Physics,' he said, 'is essentially an intuitive and concrete science. Mathematics is only a means for expressing the laws that govern phenomena.'

Several weeks later our meetings were joined by Conrad Habicht, whom Einstein had known at Schaffhausen and who had come to Bern to complete his studies, with a view to teaching mathematics in secondary school. Einstein also had us dine together. These dinners were of an exemplary frugality. The menu was usually made up of sausage, a piece of Gruyère cheese, fruit, a small jar of honey and one or two cups of tea. But we were overflowing with good spirits . . .

At the time I met Einstein he had a probationary position at the Patent

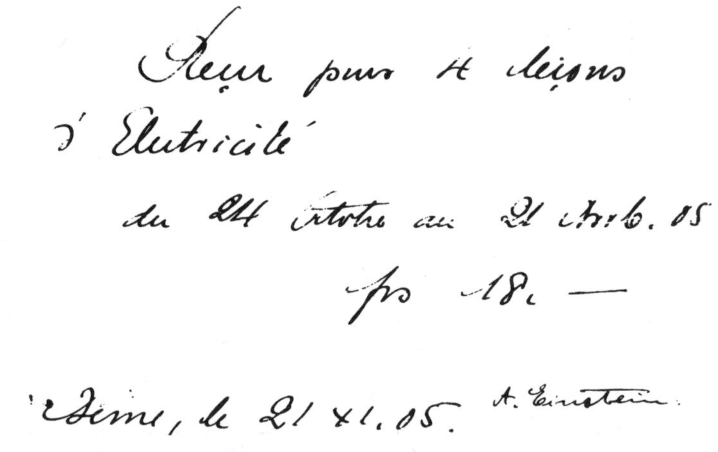

Reçu pour 4 leçons
d'Electricité
du 24 Octobre au 21 Nov. 05
frs 18.—

Berne, le 21 XI. 05. A. Einstein

Figure 1

A receipt for four lessons on electricity, given by Einstein in late 1905

Office and was impatiently awaiting a full appointment. To help support himself he had to take pupils, who were hard to find and who did not bring in much money. He said to me one day that an easier way of earning a living would be to play the violin in public places. I replied that if he decided to do that, I would learn to play the guitar to accompany him!

The end of the nineteenth century and the beginning of the twentieth was a heroic age for the fundamentals of science, and this was our main preoccupation. Einstein preferred to use the genetic method for the study of basic ideas. To clarify them, he made use of his observations of children. He also regaled us from time to time with his own work, which already exhibited the power of his mind and his great originality. It was in 1903 that he published his remarkable paper entitled *Theory of the Foundations of Thermodynamics*; in 1904, the *General Molecular Theory of Heat*; and in 1905, his great work, *On the Electrodynamics of Moving Bodies*, in which the special theory of relativity was presented. It is worth noting that, with the exception of Max Planck, nobody at the time recognized the extraordinary importance of this paper.

To give an idea of the extent to which Einstein could become absorbed in a problem that interested him, let me tell the following story.

In our walks through Bern, we used to pass a food shop in whose window were displayed various delicacies, including caviar. Seeing it reminded me how much I had enjoyed eating caviar at my parents' home in Rumania. There, it was relatively cheap, but in Bern it was prohibitively expensive. This, however, did not stop me from singing its praises before Einstein. 'Is it really so good?', he asked. 'You can't imagine how delicious it is,' I replied. And one day in February I said to Habicht: 'Let's give Einstein a special surprise, and serve him caviar for his birthday on March 14th.'

When Einstein ate something out of the ordinary he would become ecstatic and would praise it in extravagant terms. We exulted at the thought of how he would surpass himself in expressing his satisfaction on this occasion. On 14 March we went to dinner at his apartment and, in just the same way as I would have served the sausage, etc., I put the caviar on three plates on the table before joining Einstein. As chance would have it, he was talking that evening about Galileo's principle of inertia, and in so doing lost all consciousness of worldly joys and tribulations. When we sat down at the table, Einstein took mouthful after mouthful of caviar, while continuing to talk about the principle of inertia. Habicht and I furtively exchanged astonished glances, and when Einstein had finished all the caviar I said: 'Do you realize what you have been eating?' Looking at me with his big eyes he said: 'What was it, then?' 'For heaven's sake,' I cried, 'that was the celebrated caviar.' 'So *that* was caviar,' he said, in wonderment. And after a short silence he added: 'Well, if you offer gourmet foods to peasants like me, you know they won't appreciate it.'

One feature of Bern was that distinguished musicians making concert tours through Europe would always stop there to give one or two recitals, which we would make a point of attending from time to time. One day I saw a poster advertising a programme of Beethoven, Smetana, and Dvorak to be played by the then celebrated Czech Quartet. Upon coming to Einstein's apartment that evening for our usual study session, I mentioned this attractive event and said that I would plan to reserve three seats for us. '*I* think,' said Einstein, 'that we ought to forgo the concert and read David Hume.' 'All right,' I said—but when, on the day of the concert, I was walking by the concert hall, my feelings took charge, I lost my head and I bought a ticket.

As the Academy session was to take place at my lodgings that evening, I ran home to prepare dinner. Knowing that they liked hard-boiled eggs, I added four eggs to the usual menu and covered them with a sheet of paper on which I wrote (in Latin) 'To my dear friends, some hard-boiled eggs and a salutation.' I then asked the landlady to convey my excuses for having to be absent on urgent business. When they came for dinner and heard this story, they of course understood what had happened, and (after finishing their meal) knowing that I detested tobacco in any form, they proceeded to smoke furiously, Einstein with his pipe and Habicht with a big cigar. They then piled all my furniture and crockery on the bed, and pinned on the wall a sign carrying the words (likewise in Latin): 'To a dear friend, thick smoke and a salutation.'

It was my custom after attending a concert to walk for a while to let the music echo inside me and to engrave the themes and variations on my mind. I did that this evening, strolling in the streets until about one a.m. When I came home and entered my room, I was almost overwhelmed by the disgusting tobacco smoke, and thought I would suffocate. I opened the window wide, and began to remove from the bed the mound of things that reached almost to the ceiling. But when I lay down, I could not close my eyes, the pillows and bedclothes were so saturated with the abominable tobacco-smoke. It was almost morning before I got to sleep.

Next evening, when I went to Einstein's for dinner and our Academy meeting, he met me with a fierce frown and exclaimed, 'You wicked fellow, to desert a study session for a couple of fiddlers! You barbarian, if you engage in another such escapade, you will be expelled from the Academy in disgrace.'

Figure 2

The Olympia Academy: Conrad Habicht, Maurice Solovine, and Albert Einstein

Such was the rich and interesting life that we led for more than three years; I left Einstein in November 1905, to study at the University of Lyon.

I loved and admired Einstein for his profound goodness, his uniquely original mind and his indomitable moral courage. The sense of justice was developed in him to an exceptional degree. In contrast to the deplorable weakening of the moral sense among most self-styled intellectuals, he always spoke out against injustice and violence. He will live in the memory of future generations not only as a scientific genius of exceptional stature, but also as a man who embodied moral ideals in the highest degree.

(Translated from the preface to *Lettres à Maurice Solovine*)

Einstein's friendship with Michele Besso

P. Speziali

Michele Besso's lifelong friendship with Einstein is fully described by Pierre Speziali on pp. 263–9. This short anecdote is translated from the introduction to the book Einstein–Besso: Correspondance, *edited by Professor Speziali.*

For five years (1904–9) Einstein and Besso were to work side by side, becoming familiar with the procedures for patents and inventions. Freed from material worries, they enjoyed a happy existence that they recalled, much later, in their letters. Each day they shared the walk home—and sometimes went to work together in the morning. They also met during evenings and holidays. Their families got on well together, and Besso's young son, Vero, used to listen to his father's friend. This friend was always in a good mood, he was amusing and jolly, and above all he knew lots of things. One day in 1904 (or 1905?) he made for Vero a splendid kite, and they all walked into the country in the direction of a small mountain south of Bern, taking the kite with them. At the foot of the mountain one of them tried out the kite, and then put the string into Vero's hand. Was it papa's friend Albert who made the first try? That was unimportant. What Vero never forgot was that Einstein not only made the kite but could explain to him *why* it flew.

(Translated from the introduction to *Einstein–Besso: Correspondance 1903–1955*)

My meeting with Einstein at the Solvay Conference of 1927

Louis de Broglie

Louis de Broglie, world-famed theoretical physicist and one of the creators of wave mechanics, comes from a distinguished French family. This reminiscence describes his first meeting with Einstein.

Ever since I turned to the study of theoretical physics at the age of nineteen, I have been an ardent admirer of both the person and the work of Albert Einstein. I knew that at the age of twenty-five this illustrious and still youthful scholar had introduced ideas into physics that were so revolutionary in their novelty as to make him the Newton of modern science, and I studied intensely and painstakingly (for I was then merely a beginner) that elegant theory of relativity which bore his name. I also knew (and since I was immersed in the study of quantum theory, this part of his work interested me most keenly) that Einstein had produced a bold hypothesis on the subject of light, his theory of light quanta, in which he had reformulated the old corpuscular concept of radiation, abandoned since Fresnel. I knew that he had managed to interpret the laws of the photoelectric effect and to analyse the problem of energy fluctuations in black-body radiation. In short, in every aspect of my studies I encountered with growing admiration the work of this lofty thinker.

When, with increased maturity, I was able to return to my studies after the long interruption caused by World War I, it was once again the ideas of Einstein, this time on the wave-particle dualism in light, that guided me in my attempts to extend this dualism to matter and led me to propound the basic concepts of wave mechanics in the *Comptes Rendus de l'Académie des Sciences* at the end of the summer of 1923. During that period of my life I caught a glimpse of Einstein, though only from afar, while he was on a visit to Paris. He was then on a world-wide lecture tour to present his ideas on the theory of relativity which had been crowned by his discovery of general relativity in 1916. While attending his lectures at the Sorbonne and the Collège de France, I was struck by his charm and by his facial expressions which were sometimes meditative and aloof, sometimes lively and playful.

In November 1924 I had submitted to the University of Paris my doctoral thesis, in which I had put forward my new ideas on wave mechanics. Paul Langevin had sent my thesis to Einstein, who immediately evinced interest in it. A little later, in January 1925, the illustrious scientist presented a paper to the Berlin Academy of Science in which he stressed the importance of the ideas underlying my thesis and deduced a number of its consequences. That paper of Einstein's drew the attention of scientists to my work, which until then had received little notice, and for that reason alone I have always felt that I owe him a great personal debt for the encouragement that he gave me.

By 1927, the basic concepts of wave mechanics were widely acclaimed, thanks largely to the brilliant papers published by Schrödinger in 1926 and the remarkable experiments of Davisson and Germer and of G. P. Thomson on electron diffraction. In June H. A. Lorentz invited me to participate in the Fifth Solvay Congress, to be held in Brussels during October, which was to be devoted to the study of recent developments in quantum physics. I was asked to present a report on wave mechanics to that meeting. In addition to my intellectual pleasure at being asked to participate in the Congress, I was full of pleasure and curiosity at the prospect of meeting Albert Einstein and exchanging ideas with him.

My hopes were not disappointed and I did, in fact, meet the idol of my youth. During a fairly long talk he made a profound impression on me and fully confirmed my expectations. I was particularly struck by his mild and thoughtful expression, by his general kindness, by his simplicity, and by his friendliness. Sometimes, gaiety would gain the upper hand and he would strike a more personal note and even disclose some detail of his day-to-day life. Sometimes, reverting to his characteristic mood of reflection and meditation, he would launch into a profound and original discussion of a variety of scientific and other problems. I was particularly attracted by this great and sympathetic being, and have always preserved a fond memory of him.

But I must also speak of the scientific controversy in which both of us were involved during what were often the very lively arguments of an assembly of eminent physicists.

Though the concepts introduced by wave mechanics were generally accepted at the time, the details of the physical interpretation of the wave-particle dualism remained very controversial. While some of the physicists who had directly contributed to the rapid development of new forms of quantum physics, such as Schrödinger and myself, tried to present a concrete and causal interpretation, by and large in agreement with the traditional concepts of physics, others, such as Born, Bohr, Heisenberg, Pauli, and Dirac, presented a novel, purely probabilistic interpretation based on Heisenberg's recently discovered uncertainty relations.

For the previous eighteen months I had been constructing my 'theory of the double solution,' which had struck me as providing a quasi-classical interpretation of the main experimental results of wave mechanics. In my report to the Solvay Congress I presented this theory in an abbreviated form, and I waited with interest to see how it would be received by the illustrious assembly. As it turned out, the indeterminist school, whose adherents were mainly young and intransigent, met my theory with cold disapproval. On the other hand, Schrödinger was somewhat equivocal, and

A. PICCARD E. HENRIOT P. EHRENFEST Ed. HERZEN Th. DE DONDER E. SCHRÖDINGER E. VERSCHAFFELT W. PAULI W. HEISENBERG R.H. FOWLER L. BRILLOUIN

P. DEBYE M. KNUDSEN W.L. BRAGG H.A. KRAMERS P.A.M. DIRAC A.H. COMPTON L. de BROGLIE M. BORN N. BOHR

I. LANGMUIR M. PLANCK Mme CURIE H.A. LORENTZ A. EINSTEIN P. LANGEVIN Ch.E. GUYE C.T.R. WILSON O.W. RICHARDSON

Absents : Sir W.H. BRAGG, H. DESLANDRES et E. VAN AUBEL

Figure 3 The Fifth Solvay Conference, Brussels, 1927. (It is interesting to compare this photograph with that of the First Solvay Conference, 1911, on page 146.)

Lorentz, who presided over the meeting, spoke out as a convinced partisan of determinism and of the concrete images of classical physics. He did not commit himself on my theory, which, moreover, he had had little occasion to study, but presented a picture, magnificent in its precision, of what he thought a satisfactory physical theory ought to be.

But what would Einstein say? I was anxious to know. Although he had never expressed the opinion that I had found a definite solution of the problem, he had always encouraged me in private conversations to persevere on my path. But all he did do, in a very short address, was to criticize the purely probabilistic interpretation, showing through perplexing examples why he thought it incomplete.

On the return trip from Brussels to Paris with him for a commemoration of the centenary of the death of the great French physicist, Augustin Fresnel, I had a final conversation with Einstein on the arrival platform of the Gare du Nord. He told me again that he had little confidence in the indeterminist interpretation, and that he was worried about the exaggerated turn toward formalism which quantum physics was taking. Then, possibly going further than he might normally have liked to go, he told me that all physical theories, their mathematical expression apart, ought to lend themselves to so simple a description 'that even a child could understand them'. And what could be less simple than the purely statistical interpretation of wave mechanics! Outside the station he left me by saying: 'Carry on! You are on the right track!'

I was never to see him again except for a brief meeting at a lecture he gave two years later at the Institut Henri Poincaré, where I had meanwhile been appointed professor. Then there came that cruel period of trials and tribulations following the rise to power of the Hitler regime in Germany. Forced to leave his country, Einstein left for the United States, where he was received with open arms. He never returned to Europe, and my only subsequent contact with him was by letter—always cordial, but rather infrequent.

As for the controversy about the interpretation of quantum physics, it was soon settled in favour of Bohr and Heisenberg's indeterminist theory. Discouraged by the difficulties which I had met in trying to develop my theory of the double solution, I too adopted the almost unanimous view of quantum physicists in 1928, and I have ever since been teaching and expounding it. Einstein, however, stuck to his guns and continued to insist that the purely statistical interpretation of wave mechanics could not possibly be complete.

Since 1952, returning to my early ideas, I have developed afresh the causal interpretation of wave mechanics, and Einstein, during the last years of his life, was happy to learn of this development, and encouraged me to pursue my efforts in this direction.

Reminiscences of Einstein

L. L. Whyte

Lawrence Whyte was a theoretical physicist from Cambridge University who went to Germany in the late 1920s and there met Einstein who was then working in Berlin. This anecdote about Einstein and the distinguished Hungarian theorist, Cornelius Lanczos, comes from Whyte's book Focus and Diversions.

Early in 1928 Einstein was known to be looking for a mathematical assistant, someone highly skilled in the methods of relativity theory and preferably with a temperament congenial to him. One of the entrepreneurs who hover around genius suggested Cornelius (Cornel) Lanczos, a Hungarian mathematician at that time in Frankfurt-am-Main. Cornel was indeed the perfect candidate: an idealistic Jew looking like the descendant of a line of rabbis, he was an exceedingly brilliant mathematician and a profound admirer of Einstein and of his work. He was a lover of music and a sensitive person.

Einstein wrote to Lanczos, they met, and it was agreed that Lanczos would come to Berlin that October for at least a year.

One afternoon in October, Lanczos called on Einstein for the first of their weekly discussions. Einstein explained that he was interested in a certain kind of field equation E (perhaps combining gravitational and electromagnetic fields), and required a solution of the equation with certain properties, say α, β, and γ, which he described. Lanczos was to go away, think over the problem, see if he could find a suitable solution, and come back the following week.

To Lanczos this was a unique opportunity, one to be approached with religious humility and patience. He would carefully study the equation and then calmly wait for inspiration. Perhaps the divine powers would bring him the necessary insight, if he were worthy of it.

After a few days, Cornel told me, he found a solution and to his delight it had all the desired properties: α, β, and γ. He was very excited about it, but forced himself to be patient, calm, and humble, as he had been so privileged by fate.

The day came and he told Einstein in the most casual manner that he *had* been able to find a solution and that it had all three properties.

Einstein sat silent. 'Yes, very remarkable,' he commented slowly and quietly. 'Yes, you're right, it has α, β, and γ. Interesting.' And then in a burst of impatience, 'But didn't you realize, I gave you the wrong equation, didn't you see that E couldn't be the right equation, it isn't what is needed at all, it couldn't do the job.'

There was silence. It was unnecessary for anything further to be said between these two intelligent and sensitive men. Cornel knew that not even for his beloved Einstein would he allow this divine gift of the mathematical imagination to be misused. Nothing was said, and Cornel told me that Einstein silently fetched his violin, while he, Cornel, sat at the piano and they played Bach together for the rest of the hour. Perhaps they played together for several weeks. I do not know whether or not Cornel gave Einstein any further mathematical help.

Cornel told me that it was clear to him then that Einstein's judgement was no longer leading him unerringly, that his reversal of view showed that his insight was no longer as certain as it had been earlier. This is a matter on which theoretical physicists may hold contrasted views. I yield to no one in my admiration for Einstein's supreme achievements, which are those from 1905 to around 1925. But I have no doubt that the methods which he used during those twenty years when he was between twenty-six and forty-six, particularly to the restricted and the general theories of relativity, were by 1925 in this sense exhausted: they were no longer appropriate to the *new* fundamental problems which certainly had to be viewed in the light of the quantum theory of 1923–8. Perhaps Einstein unconsciously knew this, or feared that it might be so. In any case the good judgement which is one of the marks of genius began to fail him in the years from 1925 onwards, and the difficulty with Lanczos arose from an inner uncertainty which he had not known earlier. It was a symptom of something both in Einstein and in the position of physics, or in their relationship, which by 1928 was becoming disturbing.

Einstein later stated that it was part of his faith to persevere in his search for new fundamental laws of the kind that to him had always seemed appropriate. There is irony here. That faith was probably indispensable to Einstein's personality. He could not and did not face what I, and probably most quantum theorists, consider to be the fact: that from 1925 onwards radically novel methods were necessary in fundamental physics towards which Einstein, born in 1879 and maturing around 1900–5, could not be expected to be sympathetic. Would Einstein's later years, from 1935 to his death in 1955, have been any happier had he been able to recognize this? No one can say, and the question is empty. For in spite of what has been hastily written about 'the Einsteinian revolution' and in spite of his unquestioned pre-eminence in this century, he was in many ways a 'classical' mind. I mean by this that he accepted the classical ideals of Beauty, Truth, and Goodness as ultimate; that he sought for an exact mathematical harmony expressed in an invariant geometrical form; and that he knew rather little *in his own nature* of the characteristic travails of this century, of its deep moral uncertainties, its asymmetries and clashes. He suffered on account of the general condition of man, not of his own condition. He found his greatest joy in reading Dostoevsky, not science or philosophy—but for Dostoevsky's expression of men's intense yearning to realize the divine where the Christian Churches had failed, not (I believe) for

Dostoevsky's portrayal of the twists, knots, and ambiguities in human nature . . .

We may try to make a moral symbol out of Einstein, but his true place is beside Kepler and Newton.

(From *Focus and Diversions*)

Memoir

John Archibald Wheeler

John Archibald Wheeler, one of the most eminent native-born American physicists, is perhaps best known for the historic paper that he published jointly with Niels Bohr in 1939 on the liquid-drop model of nuclear fission. He was a professor at Princeton University during the years that Einstein was himself in Princeton (at the Institute for Advanced Study).

My first chance to see and hear Albert Einstein came one afternoon in the academic year 1933–4. I was in my first year of postdoctoral work with Gregory Breit in New York. He told me that there would be a quiet, small, unannounced seminar by Einstein that afternoon. We took the train to Princeton and walked to Fine Hall. Unified field theory was to be the topic, as became clear when Einstein entered the room and began to speak. His English, though a little accented, was beautifully clear and slow. His delivery was spontaneous and serious with every now and then a touch of humour. I was not familiar with his subject at that time, but I could sense that he had his doubts about the particular version of unified field theory he was then discussing. I had been accustomed before this to seminars in physics where equations were taken up one at a time or, if I may say so, dealt with in retail trade. Here for the first time I saw equations dealt with wholesale. One counted the number of unknowns and the number of supplementary conditions and compared them with the number of equations and the number of coordinate degrees of freedom. The idea was not to solve the equations, but rather to decide whether they possessed a solution and whether it was unique. It was clear on this first encounter that Einstein was following very much his own line, independent of the interest in nuclear physics then at high tide in the United States.

In 1938 I moved to Princeton and at frequent intervals called on Einstein at his house at 112 Mercer Street, climbing the stairs to his second floor study that looked out on the Graduate College. Especially vivid in my mind is a call I made in 1941 to explain the 'sum over histories' approach to quantum mechanics then being developed by Richard Feynman, whom I was fortunate enough to have as a graduate student. I had gone to see Einstein with the hope to persuade him of the naturalness of quantum theory when seen in this new light, connected so closely and so beautifully with the variation principle of classical mechanics. He listened to me patiently for twenty minutes until I finished. At the end he repeated that familiar remark of his, 'I still cannot believe that the good Lord plays dice'. And then he went on to add again in his beautifully slow, clear, well-modulated and humorous way, 'Of course I may be wrong; but perhaps I have earned the right to make my mistakes'.

I had to be away from Princeton for national reasons from 1942 to 1945 and again from 1950 to 1953; but on my return (part-time) in 1952 gave for the first time the course in general relativity in which I was to learn so much from my students over the years. In the Spring of 1953, two years before he died, Einstein was kind enough to invite me to bring the eight to ten students in the course around to his house for tea. Margot Einstein and Helen Dukas served it as we sat around the dining room table. The students asked questions about everything from the nature of electricity and unified field theory to the expanding universe and his position on quantum theory and Einstein responded at length and fascinatingly. Finally one student outdid the others in the boldness of his question: 'Professor Einstein, what will become of this house when you are no longer living?' Einstein's face took on that humorous smile and again he spoke in that beautiful, slow, slightly accented English that could have been converted immediately into printer's type: 'This house will never become a place of pilgrimage where the pilgrims come to look at the bones of the saint.' And so it is today. The tourist buses drive up. The pilgrims climb out to photograph the house—but they don't go in.

A further encounter was my last. We persuaded him to give a seminar to a restricted group. In it the quantum was a central topic. No one can forget how he expressed his discomfort about the role of the observer: 'When a mouse observes does that change the state of the universe?'

In all the history of human thought there is no greater dialogue than that which took place over the years between Niels Bohr and Albert Einstein about the meaning of the quantum. Their discussion has already been depicted in sculpture and surely will be described some day in pictures and words. Nobody can forget Einstein's letter to the young Bohr when first he met him: 'I am studying your great works and—when I get stuck anywhere—now have the pleasure of seeing your friendly young face before me smiling and explaining'. There is no greater monument to the dialogue than Bohr's summary of it in his article *Discussions with Einstein on Epistemological Problems in Atomic Physics*—written (in honour of Einstein at age seventy) for the book *Albert Einstein: Philosopher-Scientist* (edited by P. A. Schilpp).

Anecdotes

Philipp Frank

Philipp Frank was a theoretical physicist who wrote one of the best and most authoritative biographies of his close friend Einstein. The two short anecdotes given here come from the beginning of an article, 'Einstein's philosophy of science', published in Reviews of Modern Physics *in 1949.*

About ten years ago I spoke with Einstein about the astounding fact that so many ministers of various denominations are strongly interested in the theory of relativity. Einstein said that according to his estimation there are more clergymen interested in relativity than physicists. A little puzzled I asked him how he would explain this strange fact. He answered, a little smiling, 'Because clergymen are interested in the general laws of nature and physicists, very often, are not.'

Another day we spoke about a certain physicist who had very little success in his research work. Mostly he attacked problems which offered tremendous difficulties. He applied penetrating analysis and succeeded only in discovering more and more difficulties. By most of his colleagues he was not rated very highly. Einstein, however, said about him, 'I admire this type of man. I have little patience with scientists who take a board of wood, look for its thinnest part and drill a great number of holes where drilling is easy.'

(From *Einstein's Philosophy of Science*)

Memoir

Edward Teller

Edward Teller, a theoretical nuclear physicist, came to America from Hungary in the 1930s. In the reminiscence presented here he is perhaps being unduly modest, for it is reported in other sources that Teller himself was one of those who helped to draft Einstein's famous letter to President Roosevelt.

In the summer of 1939, my good and ingenious friend Leo Szilard (a business associate of Einstein since they jointly invented a novel refrigerator) was in urgent need of a chauffeur. I offered my services. I drove him out to the north end of Long Island into the neighbourhood where Einstein took his summer vacation. It was a little difficult to find Einstein. Several inquiries failed to elicit the whereabouts of this obscure personage. In the end we asked a young girl not yet ten years of age, with two fairly long braids, who responded positively to an inquiry about a nice old gentleman with plenty of white hair.

Einstein gave Szilard, and also his chauffeur, a cup of tea, and received from Szilard a letter the contents of which seemed to be quite familiar to him. The letter, addressed to President Roosevelt, predicted the atom bomb and suggested its development. Einstein signed it with hardly any comment.*

Years later Einstein wrote a second letter to President Roosevelt, proposing that the atom bomb should be demonstrated to the Japanese before it was ever used. (By that time I had lost my position as a chauffeur.) The letter was found on the day that Roosevelt died in Warm Springs, Georgia.**

* A facsimile of the letter may be found on page 191.

** Actually, the letter only expressed concern about the lack of contact between the scientists and the government with respect to the uranium project. Einstein was never party to the detailed plans.

Nr. 140217 **Klasse 108a**

SCHWEIZERISCHE EIDGENOSSENSCHAFT

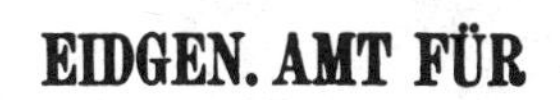

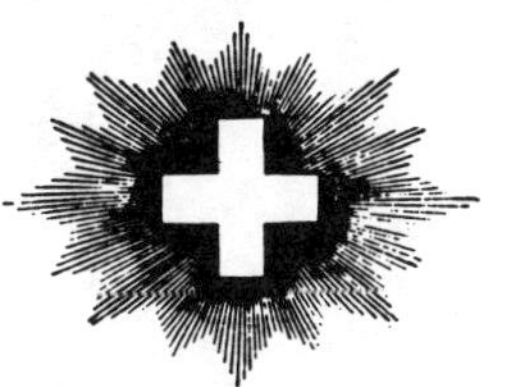

EIDGEN. AMT FÜR GEISTIGES EIGENTUM

PATENTSCHRIFT

Veröffentlicht am 16. August 1930

Gesuch eingereicht: 21. Dezember 1928, 19 Uhr. — Patent eingetragen: 31. Mai 1930.
(Prioritäten: Deutschland, 27. Dezember 1927 und 3. Dezember 1928.)

HAUPTPATENT

Prof. Dr. Albert EINSTEIN, Berlin, und Dr. Leo SZILARD, Berlin-Wilmersdorf (Deutschland).

Kältemaschine.

Die Erfindung betrifft eine Kältemaschine, bei welcher erfindungsgemäß die Betriebsenergie dadurch zugeführt wird, daß unter der Einwirkung eines Magnetfeldes, ein durch elektrischen Strom durchflossenes flüssiges Metall fortbewegt wird. Als flüssiges Metall kommen, neben Quecksilber, besonders auch Leichtmetalle in flüssigem Zustande in Frage, zum Beispiel Natrium-Kaliumlegierungen, insbesondere solche mit etwa 75% K.

Um nicht zu viele Ampèrewindungen aufwenden zu müssen, wird man dazu geführt die Einrichtung so zu treffen, daß die magnetische Kraftlinie zum größten Teil in Eisen verläuft und nur auf eine kurze Strecke das flüssige Metall durchsetzt, das heißt das flüssige Metall wird in einem Spalte, dessen Breite nur klein ist, die bewegende elektromagnetische Kraft erfahren. Da dort, wo die bewegende Wirkung auf das Metall erfolgt, die Dicke des Metalles im Verhältnis zur Breite und Länge klein ist, kann man vereinfachend auch sagen, daß das Metall auf einer Fläche sich fortbewegt, und das Vektorfeld der ponderomotorischen Kraft auf dieser Fläche (2. dimensionales Gebilde) betrachten.

Man wird nun zweckmäßigerweise die Anordnung so treffen, daß dieses Vektorfeld der ponderomotorischen Kraft wirbelfrei ist, das heißt daß für jede geschlossene, innerhalb der Fläche verlaufende (innerhalb des flüssigen Metalles im Spalt) Linie das Linienintegral der ponderomotorischen Kraft = 0 ist. Haben wir es mit einer ebenen Fläche zu tun, so ist die mathematische Bedingung hierfür

$$\frac{d\,X\,(xy)}{d\,y} - \frac{d\,Y\,(xy)}{d\,x} = o$$

wobei x und y die kartesischen Koordinaten innerhalb der Ebene und X, Y die Kraftkomponenten bedeuten.

Figure 4 First page of the refrigerator patent taken out by Albert Einstein and Leo Szilard

Figure 5 Einstein on 10 May 1947

Einstein

Philippe Halsman

Philippe Halsman is one of the best (and best-known) photographers of famous and distinguished people. This reminiscence, from his book, Sight and Insight, *describes his meeting with Einstein in 1947 when the photograph shown opposite was taken.*

I admired Albert Einstein more than anyone I ever photographed, not only as the genius who singlehandedly had changed the foundation of modern physics, but even more as a rare and idealistic human being.

Personally, I owed him an immense debt of gratitude. After the fall of France, it was through his personal intervention that my name was added to the list of artists and scientists who, in danger of being captured by the Nazis, were given emergency visas to the United States.

After my miraculous rescue I went to Princeton to thank Einstein and I remember vividly my first impression. Instead of a frail scientist I saw a deep-chested man with a resonant voice and a hearty laugh. The long hair, which in some photographs gave him the look of an old woman, framed his marvellous face with a kind of leonine mane. He wore slacks, a grey sweater with a fountain pen stuck in its collar, black leather shoes, and no socks.

On my third visit I had the courage to ask him why he did not wear any socks. His secretary, Miss Dukas, who overheard me, said, 'The professor never wears socks. Even when he was invited by Mr. Roosevelt to the White House, he did not wear any socks.' I looked with surprise at Professor Einstein.

He smiled and said, 'When I was young I found out that the big toe always ends up by making a hole in the sock. So I stopped wearing socks.' As slight as this remark was, it made an indelible impression on me. This detail seemed symbolic of Einstein's absolute and total independence of thought. It was this independence that gave him the courage when he was an unknown twenty-six-year-old patent clerk to publish a scientific paper which overthrew all the axioms held sacrosanct by the greatest physicists of his time.

The question of how to capture the essence of such a man in a portrait filled me with apprehension. Finally, in 1947, I had the courage to bring on one of my visits my Halsman camera and a few floodlights. After tea, I asked for permission to set up my lights in Einstein's study. The professor sat down and started peacefully working on his mathematical calculations. I took a few pictures. Ordinarily, Einstein did not like photographers, whom he called *Lichtaffen* (light monkeys). But he cooperated because I was his guest and, after all, he had helped to rescue me.

Suddenly, looking into my camera, he started talking. He spoke about his despair that his formula $E=mc^2$ and his letter to President Roosevelt had made the atomic bomb possible, that his scientific search had resulted in the death of so many human beings. 'Have you read,' he asked, 'that powerful voices in the United States are demanding that the bomb be dropped on Russia now, before the Russians have the time to perfect their own?' With my entire being I felt how much this infinitely good and compassionate man was suffering from the knowledge that he had helped to put in the hands of politicians a monstrous weapon of devastation and death.

He grew silent. His eyes had a look of immense sadness. There was a question and a reproach in them.

The spell of this moment almost paralysed me. Then, with an effort, I released the shutter of my camera. Einstein looked up, and I asked him, 'So you don't believe that there will ever be peace?'

'No,' he answered, 'as long as there will be man, there will be wars.'

(From *Sight and Insight*)

Reminiscence

George Gamow

George Gamow, born and raised in Russia, was famous not only as a highly inventive and original theoretical physicist, particularly in theoretical nuclear physics and cosmology, but also as an entertaining but accurate populariser of scientific ideas. This anecdote is taken from his autobiography, My World Line.

There is very little to say about my consultation work for the armed forces of the United States during World War II. It would have been, of course, natural for me to work on nuclear explosions, but I was not cleared for such work until 1948, after Hiroshima. The reason was presumably my Russian origin and the story I had told freely to my friends of having been a colonel in the field artillery of the Red Army at the age of about twenty.

Thus I was very happy when I was offered a consultantship in the Division of High Explosives in the Bureau of Ordnance of the US Navy Department.

A more interesting activity during that time was my periodic contact with Albert Einstein, who, along with other prominent experts such as John von Neumann, served as a consultant for the High Explosive Division. Accepting this consultantship, Einstein stated that because of his advanced age he would be unable to travel periodically from Princeton to Washington, D.C., and back, and that somebody must come to his home in Princeton, bringing the problems with him. Since I happened to have known Einstein earlier, on non-military grounds, I was selected to carry out this job. Thus on every other Friday I took a morning train to Princeton, carrying a briefcase tightly packed with confidential and secret Navy projects. There was a great variety of proposals, such as exploding a series of underwater mines placed along a parabolic path that would lead to the entrance of a Japanese naval base, with 'follow up' aerial bombs to be dropped on the flight decks of Japanese aircraft carriers. Einstein would meet me in his study at home, wearing one of his famous soft sweaters, and we would go through all the proposals, one by one. He approved practically all of them, saying, 'Oh yes, very interesting, very, very ingenious,' and the next day the admiral in charge of the bureau was very happy when I reported to him Einstein's comments.

After the business part of the visit was over, we had lunch either at Einstein's home or at the cafeteria of the Institute for Advanced Study, which was not far away, and the conversation would turn to the problems of astrophysics and cosmology. In Einstein's study there were always many sheets of paper scattered over his desk and on a nearby table, and I saw that they were covered with tensor formulae which seemed to pertain to the unified-field theory, but Einstein never spoke about that. However, in discussing purely physical and astronomical problems he was very refreshing, and his mind was as sharp as ever.

I remember that once, walking with him to the Institute, I mentioned Pascual Jordan's idea of how a star can be created from nothing, since at the point zero its negative gravitational mass defect is numerically equal to its positive rest mass. Einstein stopped in his tracks, and, since we were crossing a street, several cars had to stop to avoid running us down. I will never forget these visits to Princeton, during which I came to know Einstein much better than I had known him before.

(From *My World Line* by G. Gamow.

Memoir

Ernst G. Straus

Ernst Straus, an American mathematician (born in Germany in 1922), was an assistant to Einstein at the Institute for Advanced Study in Princeton from 1944 to 1948.

Since there are not many of us left who worked with Einstein, it might be good to recall the ways he described his own motivations and way of thinking.

He would say: 'All I have is the stubbornness of a mule; no, that's not quite all, I also have a nose.' This 'stubbornness of a mule' was very important because he felt that the chief task of the scientist is to find the most important question and then to pursue it without deviating from the main problem. 'You must never let yourself be seduced by any problem, no matter how difficult it is.' In that context he thought that scientific greatness was primarily a question of character, the determination not to compromise or to accept incomplete answers. Let me mention the only occasion on which he said: 'This would make a good anecdote about me.' We had finished the preparation of a paper and we were looking for a paper clip. After opening a lot of drawers we finally found one which turned out to be too badly bent for use. So we were looking for a tool to straighten it. Opening a lot more drawers we came on a whole box of unused paper clips, Einstein immediately starting to shape one of them into a tool to straighten the bent one. When I asked him what he was doing, he said, 'Once I am set on a goal, it becomes difficult to deflect me.'

His 'nose', the perception of the nature of the right direction for research and the recognition of the right answer, he would describe in many ways. 'Logical simplicity'—a term which invariably annoys my logician friends—was one of his favourite concepts. It was an aesthetic rather than a logical criterion. 'For a musical man this is convincing.' 'This is so beautiful, God could not have passed it up' or, conversely: 'This is a sin against the Holy Ghost.' When I told him that Max Planck had died, he said: 'He was one of the finest people I have ever known and one of my best friends; but, you know, he didn't really understand physics.' When I asked him how he could say such a thing against Planck, he said: 'During the eclipse of 1919, Planck stayed up all night to see if it would confirm the bending of light by the gravitational field of the Sun. If he had really understood the way the general theory of relativity explains the equivalence of inertial and gravitational mass, he would have gone to bed the way I did.'

Figure 6

A cartoon by Ippei Okamoto, autographed by Einstein: 'Albert Einstein or the nose as a thought reservoir'

In a tactless moment I once asked him how ageing had affected his thinking. His surprising answer was that he had as many new ideas as ever, but that it had become more difficult for him to decide which ones should be rejected and which ones were worth pursuing. In short, he thought that his nose had grown less certain.

He was convinced that there is an ultimate correct and aesthetically perfect physical theory, and by his famous quote: '*Raffiniert ist der Herrgott aber boshaft ist er nicht*' (God is slick, but he isn't mean) he meant that the discovery of the ultimate laws may require great mathematical and technical sophistication, but that once you overcame God's slickness he would not cheat you out of your triumph.

'If you would live a happy life, tie it to a goal, not to people or things.'

Memoir

Eugene P. Wigner

Eugene Wigner, one of the top theoretical physicists in the world, left Hungary in the 1930s and settled in the United States, at Princeton University, where he was a friend of Einstein for many years.

The personal characteristic of Einstein that is most vividly in my mind and that I like to recall most is his feeling of equality with his colleagues, his appreciation and in fact reciprocation of their friendship. My love and early admiration of physics (I studied chemical engineering) owes very much to the seminar he organized in the early twenties in Berlin on statistical mechanics. Many of the participants at the seminar, including myself, were encouraged to visit him at his home, to have personal conversations with him. We discussed, at such occasions, not only statistical mechanics, not only physics, but also personal problems, and the problems of society. His deep insights had a lasting effect on most of us, but the exchange of opinions was on an equal basis and he responded with interest to the remarks which his visitors made. In somewhat later years the subject of such conversations often turned toward politics, and his condemnation of all dictatorships, particularly Hitler's, had a great deal of influence on his friends and students. But even as far as the USSR is concerned, he wrote, when he was asked to sign a petition: 'Because of the glorification of Soviet Russia, which it includes, I cannot bring myself to sign it.'

It became more difficult for him to maintain a similarly cordial relation with his colleagues, older and younger, after moving to Princeton. Though he could speak English, he never felt at home with it. But his relations with numerous collaborators in Princeton were always cordial and, even though they were not only less widely recognized, but also considerably younger than he was, he never talked down to them, and treated them as equals. He loved to take walks, often with friends like myself, with whom the conversation was in German.

One more characteristic of Einstein which is rarely mentioned: he loved children. I recall that once my wife took some papers to his house and when Einstein asked her about our children, she had to admit that they had chickenpox, and according to local regulations, were not allowed to leave the car. Einstein said at once, 'Oh, I did have chickenpox myself, seeing them won't hurt me.' And he went down and had a nice conversation with the two. They had long remembered it (and my wife doubts very much that he knew what chickenpox was).

An Einstein anecdote

John G. Kemeny

John G. Kemeny, an American mathematician (born in Hungary in 1926), was one of the youngest of Einstein's assistants at the Institute for Advanced Study. He worked with him during the years 1948 and 1949.

I was twenty-two years old when I became Einstein's assistant. After offering me the position, he gave me time to finish writing up my Ph.D. thesis before starting work with him. I was a very excited young man when I showed up at his home one afternoon about a month later.

It was hard to believe that I was really there to work on unified field theory with Einstein. Therefore I was totally astounded when his first words were: 'Ah, Kemeny, now you must tell me about your thesis.' I protested quite vigorously. I had come to give him whatever help I was capable of providing, not to waste his time by talking about my Ph.D. thesis. Besides, my thesis was in mathematical logic, a highly abstract subject, very far from Einstein's interests. But no amount of protestation could change his intention. I had to start at the beginning, giving him detailed background on how my problem arose and explain the result that I had obtained. He listened most patiently for about half an hour, asking many questions, and when I had finished he said: 'And now let me tell you what I'm working on.'

He was Einstein at age seventy trying to find one more major breakthrough in the laws of physics. I was a twenty-two year-old unknown. But he felt that if I was going to take an interest in his work, he had to take an interest in mine.

Einstein, Newton, and success

A. Pais

Abraham Pais, an authority in modern theoretical physics, was born in Amsterdam. From 1951 to 1963 he was a professor of physics at the Princeton Institute for Advanced Study, where he came to know Einstein well.

If I had to characterize Einstein by one single word I would choose *apartness*. This was forever one of his deepest emotional needs. It was to serve him in his singleminded and singlehanded pursuits, most notably on his road to triumph from the special to the general theory of relativity. It was also to become a practical necessity to him, in order to protect his cherished privacy from a world hungry for legend and charisma. In all of Einstein's scientific career apartness was never more pronounced however than in regard to the quantum theory. This covers two disparate periods. From 1905 to 1923 he was the only one, or almost the only one, to take seriously his own light-quantum hypothesis: under certain circumstances light behaves as if it has a particulate structure. During the second period, from 1926 to the end of his life, he was the only one, or again nearly the only one, to maintain a profoundly sceptical attitude to quantum mechanics.

Yet Einstein has called 'the statistical quantum theory [i.e. quantum mechanics] . . . the most successful physical theory of our period'. Then why was he not convinced by it? I believe Einstein himself answered this indirectly in his 1933 Spencer lecture, *On the Method of Theoretical Physics* (see page 310)—perhaps the clearest and most revealing expression of his mode of thinking. The key is to be found in his remarks on Newton and classical mechanics.

In this lecture Einstein notes that 'Newton felt by no means comfortable about the concept of absolute space, . . . of absolute rest . . . [and] about the introduction of action at a distance'. Einstein then goes on to refer to the success of Newton's theory in these words: 'The enormous practical success of his theory may well have prevented him and the physicists of the eighteenth and nineteenth centuries from recognizing the fictitious character of the principles of his system.' (It is important to note that by 'fictitious', Einstein means free inventions of the human mind.) Whereupon he compares Newton's mechanics with his own work on general relativity: 'The fictitious character of the principles is made quite obvious by the fact that it is possible to exhibit two essentially different bases each of which in its consequences leads to a large measure of agreement with experience.'

Now back to the quantum theory. In the Spencer lecture Einstein mentioned the success not only of classical mechanics but also of the statistical interpretation of quantum theory. 'This conception is logically unexceptionable and has led to important successes.' But he added, 'I still believe in the possibility of giving a model of reality which shall represent events themselves and not merely the probability of their occurrence.'

From this lecture as well as from numerous discussions with him on the foundations of quantum physics I have gained the following impression. Einstein tended to compare the successes of classical mechanics with those of quantum mechanics. In his view both these theories were on a par, being successful but incomplete. For more than a decade Einstein had pondered the single question of how to extend to general motions the invariance under uniform translations. His resulting theory of 1916, general relativity, had led to only small deviations from Newton's theory. (Large deviations were discussed only much later.) He was likewise prepared to undertake the search for his own 'model of reality', no matter how long it would take, and he was also prepared for the survival of the practical successes of quantum mechanics, with perhaps only small modifications. It is quite plausible that the very success of his highest achievement, general relativity, was an added spur to Einstein's apartness. Yet it should not be forgotten that this trait characterizes his entire oeuvre and mode of life.

What did Einstein want? It is essential for the understanding of his thinking to realize that there were two sides to his attitude concerning quantum mechanics. There was Einstein the critic, never yielding in his dissent from complementarity, according to which the notion of 'physical phenomenon' *irrevocably* includes the specifics of the experimental conditions of observation. And there was Einstein the visionary, forever trying to realize an 'objectively real' world model, a deeper-lying theoretical framework which permits the description of phenomena independently of these conditions. He believed that quantum mechanics should be deducible as a limiting case of such a future theory 'just as electrostatics is deducible from the Maxwell equations of the electromagnetic field or as thermodynamics is deducible from statistical mechanics'. He did not believe that quantum mechanics itself was a useful starting point in the search for this future theory, 'just as one cannot arrive at the foundations of mechanics from thermodynamics or statistical mechanics'.

This vision which Einstein pursued can be traced back at least to 1920, well before the advent of quantum mechanics. It was a unified field theory. But by that he meant something different from what it meant and means to anyone else. He demanded that it be a local field theory, causal in the classical sense; that it unify the forces of nature; that the particles of physics shall emerge as special solutions of the general field equations; and that the quantum postulates *shall be a consequence* of these equations.

Einstein was neither saintly nor humourless in defending his solitary position on quantum mechanics, nor was he oblivious to the negative reaction to it by others. He may not have expressed all his feelings on these matters. But that was his way. 'The essential of the being of a man of my type lies precisely in *what* he thinks and *how* he thinks, not what he does or suffers.' In any event he held fast. 'Momentary success carries more power of conviction for most people than reflections on principle.'

Yet as his life drew to a close, occasional doubts on his vision arose in his mind. In the early fifties he once said to me, in essence: 'I am not sure that differential geometry is the framework for further progress, but if it is, then I believe I am on the right track.' Similar reservations are also found in his letters of that period to Max Born and to his lifelong friend Michele Besso.

Otto Stern has recalled a statement which Einstein once made to him: 'I have thought a hundred times as much about the quantum problems as I have about general relativity theory'. Einstein kept thinking about the quantum till the very end. He wrote his last autobiographical sketch in Princeton, in March 1955, about a month before his death. Its final sentences deal with the quantum theory. 'It appears dubious whether a field theory can account for the atomistic structure of matter and radiation as well as of quantum phenomena. Most physicists will reply with a convinced "No", since they believe that the quantum problem has been solved in principle by other means. However that may be, Lessing's comforting word stays with us: "The aspiration to truth is more precious than its assured possession." '

References

Einstein, A., in *Albert Einstein: Philosopher-Scientist* (Evanston: Library of Living Philosophers, 1949), p. 1.
—— 'On the method of theoretical physics', (New York: Oxford University Press, 1933); reprinted in *Phil. Sci.*, 1934, **1**, 162.
—— *J. Franklin Inst.*, 1936, **221**, 313.
—— letter to M. Besso, 24 July 1949.
—— in *Helle Zeit, Dunkle Zeit*, edited by C. Seelig (Zürich: Europa Verlag, 1956).
Jost, R., letter to A. Pais, 17 August 1977.

This paper is a fragment of a longer article entitled 'Einstein and the Quantum Theory', to appear in the October 1979 issue of *Reviews of Modern Physics*. Work supported in part by the U.S. Department of Energy under Contract Grant No. EY-76-C-02-2232B.

Conversations with Albert Einstein

Robert S. Shankland

Robert S. Shankland was a professor at the Case Institute of Technology, where Albert Michelson, together with Morley, performed experiments to search for 'ether drag'. Shankland made several visits to Princeton towards the end of Einstein's life to obtain Einstein's first-hand recollections of the circumstances surrounding the birth of special relativity.

It was my privilege to visit Albert Einstein in Princeton five times during the years 1950–54. At these meetings we discussed the experiments that had contributed to the development of the theory of relativity and especially those that Michelson, Morley, and Miller had carried out in Cleveland: work in which I had had a keen interest from my student days at Case Institute of Technology and the University of Chicago.

I was both surprised and pleased to find that Einstein had a real interest in these experiments and that he was well acquainted with the essential experimental details. His interest in experiments was also evident when we discussed Fizeau's 1851 measurement of the speed of light in moving water and the greatly refined modification of this experiment made by Michelson and Morley at Cleveland in 1886. He also talked at length about astronomical aberration, both Bradley's original discovery of 1728 and the later observations of Airy in 1871 with the water-filled telescope. He told me that he had pondered over these results and those of the moving water experiment for many years while he was working on special relativity, as they were basic for his relativistic formula for the addition of velocities and the transformation equations in general. He was also keenly interested in the latest speed of light measurements and especially the claim then being made that the value might change with time.

Our most detailed discussions were on the famous Michelson-Morley experiment of 1887 and its repetitions by Morley and Miller in 1904 and then by Miller alone at Mount Wilson (1921–26). With Einstein's encouragement and help we were able to show that Miller's extensive observations were in fact consistent with all other null results when account was taken of the temperature gradients across the interferometer. With respect to the original Michelson-Morley experiment and its influence on his own work, Einstein's statements to me varied considerably during the course of our five meetings. However, I feel confident that he was well acquainted with their result before 1905, and he was most generous in his statements to me about their work. About Michelson he stated: 'I always think of Michelson as the artist in science. His greatest joy seemed to come from the beauty of the experiment itself and the elegance of the method employed. He never considered himself a strict "professional" in science and in fact was not—but always the artist.'

He repeatedly praised H. A. Lorentz and at our last meeting he told me: 'People do not realize how great was the influence of Lorentz on the development of physics. We cannot imagine how it would have gone had not Lorentz made so many great contributions.'

Einstein talked to me about other matters, especially quantum mechanics. His well-known scepticism on this subject was clearly evident and his comments on both the subject itself and its leading proponents were often highly critical and even emotional, in contrast to his restrained and quiet explanations of relativity. Several times he expressed his conviction that the next great advances in physics will come from a fresh start beginning with general relativity, but must await major developments in mathematics so that rigorous solutions of the basic equations will be possible. He told me with complete candour that his own efforts along these lines had not satisfied him, but that in time the advances that he hoped for were sure to come.

Einstein and Newton

I. Bernard Cohen

The American historian of science, I. Bernard Cohen, a specialist in the life and work of Newton, visited Einstein at Princeton only two weeks before Einstein died. Professor Cohen based this reminiscence on his longer account of the visit, published in the Scientific American *in July 1955.*

On a Sunday morning in April, two weeks before the death of Albert Einstein, I sat and talked with him about the history of scientific thought and great men in the physics of the past.

I had arrived at the Einstein home, a small frame house with green shutters, at 10 o'clock in the morning and was greeted by Helen Dukas, Einstein's secretary and housekeeper. She conducted me to a cheerful room on the second floor at the back of the house. This was Einstein's study. It was lined on two walls with books from floor to ceiling and contained a large, low, table laden with pads of paper, pencils, trinkets, books, and a collection of well-worn pipes. There was a record-player and records. Dominating the room was a large window with a pleasant green view. On the remaining wall were portraits of the two founders of the electromagnetic theory—Michael Faraday and James Clerk Maxwell.

After a few moments Einstein entered the room and Miss Dukas introduced us. He greeted me with a warm smile, went into the adjacent bedroom and returned with his pipe filled with tobacco. He wore an open shirt, a blue sweat-shirt, grey flannel trousers, and leather slippers. There was a touch of chill in the air, and he tucked a blanket around his feet. His face was deeply lined, contemplative, tragic, and yet his sparkling eyes made him seem ageless. He spoke softly and clearly; his command of English was remarkable, though marked by a German accent. The contrast between his soft speech and his ringing laughter was enormous. He enjoyed making jokes; every time he made a point that he liked, or heard something that appealed to him, he would burst into booming laughter.

We sat side by side at the table, facing the window and the view. He appreciated that it was difficult for me to begin a conversation with him; after a few moments he turned to me as if answering my unasked questions, and said: 'There are so many unsolved problems in physics. There is so much that we do not know; our theories are far from adequate.' Our talk veered at once to the problem of how often in the history of science great questions seem to be resolved, only to reappear in new form. Einstein expressed the view that perhaps this was a characteristic of physics, and suggested that some of the fundamental problems might always be with us.

Einstein was particularly interested in the various aspects of Newton's

Figure 7

Einstein in Princeton, aged 75

personality and we discussed Newton's controversy with Hooke in the matter of priority in the inverse-square law of gravitation. Hooke wanted only 'some mention' in the preface to Newton's *Principia*, a little acknowledgment of his efforts, but Newton refused to make the gesture. Newton wrote to Halley, who was supervising the publication of the great *Principia*, that he would not give Hooke any credit; he would rather suppress the crowning glory of the treatise, the third and final 'book' dealing with the system of the world. Einstein said: 'That, alas, is vanity. You find it in so many scientists. You know, it has always hurt me to think that Galileo did not acknowledge the work of Kepler.'

Much of the time we spent together was devoted to the history of science, a subject that had long been of interest to Einstein. He had written many articles about Newton, prefaces to historical works and also biographical sketches of his contemporaries and the great men of science of the past. Thinking aloud about the nature of the historian's job, he compared history to science. Certainly, he said, history is less objective than science. For example, he explained, if two men were to study the same subject in history, each would stress the particular part of the subject which interested him or appealed to him the most. As Einstein saw it, there is an inner or intuitional history and an external or documentary history. The latter is more objective, but the former is more interesting. The use of intuition is dangerous but necessary in all kinds of historical work, especially when an attempt is made to reconstruct the thought processes of someone who is no longer alive. This kind of history, Einstein felt, is very illuminating despite its riskiness. It is important to know, he went on, what Newton thought and why he did certain things. We agreed that the challenge of such a problem should be the major motivation of a good historian of science. For instance, how and why had Newton developed his concept of the aether? Despite the success of his gravitation theory, Newton was not satisfied by the concept of the gravitational force. Einstein believed that what Newton most strongly objected to was the idea of a force being able to transmit through empty space. As Einstein saw it, Newton hoped to eliminate gravitational action at a distance by introducing an aether that could produce gravitational effects. Thus gravitational forces would be reduced to contact forces of an all-pervading aethereal matter. Here is a statement of great interest about Newton's process of thought, Einstein declared, but the question arises as to whether—or perhaps to what extent—one can document such intuition. Einstein said most emphatically that he thought the worst person to document any ideas about how discoveries are made is the discoverer. Many people, he went on, had asked him how he had come to think of this or how he had come to think of that. He had always found himself a very poor source of information concerning the genesis of his own ideas. Einstein believed that the historian is likely to have a better insight into the thought processes of a scientist than the scientist himself.

Looking back over all of Newton's ideas, Einstein said, he thought that Newton's greatest achievement was his recognition of the role of privileged systems. He repeated this statement several times and with great emphasis. This is rather puzzling, I thought to myself, because today we believe that there are no privileged systems, only inertial systems; there is no privileged frame—not even our solar system—which we can say is privileged in the sense of being fixed in space, or having special physical properties not possible in other systems. Due to Einstein's own work we no longer believe (as Newton did) in concepts of absolute space and absolute time, nor in a privileged system at rest or in motion with respect to absolute space. Newton's solution appeared to Einstein ingenious and necessary in his day. I was reminded of Einstein's statement: 'Newton, you found the only way which, in your age, was just about possible for a man of highest thought and creative power.'

A tribute

Pablo Casals

The great Spanish cellist, himself a fighter for justice and humanity, expresses here his sorrow at Einstein's death.

Although I never had the good fortune to get to know Albert Einstein personally, I developed the highest esteem for him. Certainly he was a great scholar, but beyond that he was also a pillar of the human conscience in a time when so many civilized values seemed to be tottering. I was perpetually grateful to him for his protest against the injustice to which my homeland was sacrificed. After Einstein's death it is as if the world has lost a part of its substance.

(From *Albert Einstein: A Documentary Biography*, by Carl Seelig)

On Albert Einstein

J. Robert Oppenheimer

J. Robert Oppenheimer, a brilliant American theoretical physicist and a broadly cultured scholar, was Director of the Institute for Advanced Study at Princeton from 1947 to 1966, and thus became well acquainted with Einstein. His eloquent and moving memoir was delivered in 1965, at a meeting held to commemorate the tenth anniversary of Einstein's death and that of the scientific philosopher, Teilhard de Chardin. This memoir appeared in a Unesco collection of essays entitled Science and Synthesis.

Though I knew Einstein for two or three decades, it was only in the last decade of his life that we were close colleagues and something of friends. But I thought that it might be useful, because I am sure that it is not too soon—and for our generation perhaps almost too late—to start to dispel the clouds of myth and to see the great mountain peak that these clouds hide. As always, the myth has its charms; but the truth is far more beautiful.

Late in his life, in connection with his despair over weapons and wars, Einstein said that if he had to live it over again he would be a plumber. This was a balance of seriousness and jest that no one should now attempt to disturb. Believe me, he had no idea of what it was to be a plumber, least of all in the United States, where we have a joke that the typical behaviour of this specialist is that he never brings his tools to the scene of the crisis. Einstein brought his tools to his crises: Einstein was a physicist, a natural philosopher, the greatest of our time.

What we have heard, what you all know, what is the true part of the myth is his extraordinary originality. The discovery of quanta would surely have come one way or another, but he discovered them. Deep understanding of what it means that no signal could travel faster than light would surely have come: the formal equations were already known; but this simple, brilliant understanding of the physics could well have been slow in coming, and blurred, had he not done it for us. The general theory of relativity which, even today, is not well proved experimentally, no one but he would have done for a long, long time. It is in fact only in the last decade, the last years, that one has seen how a pedestrian and hard-working physicist, or many of them, might reach that theory and understand this singular union of geometry and gravitation; and we can do even that today only because some of the *a priori* open possibilities are limited by the confirmation of Einstein's discovery that light would be deflected by gravity.

Yet there is another side besides the originality. Einstein brought to the work of originality deep elements of tradition. It is only possible to discover in part how he came by it, by following his reading, his friendships, the meagre record that we have. But of these deep-seated elements of tradition

—I will not try to enumerate them all; I do not know them all—at least three were indispensable and stayed with him.

The first is from the rather beautiful, but recondite part of physics that is the explanation of the laws of thermodynamics in terms of the mechanics of large numbers of particles, statistical mechanics. This was with Einstein all the time. It was what enabled him from Planck's discovery of the law of black body radiation to conclude that light was not only waves but particles with an energy proportional to their frequency, and momentum determined by their wave-number, the famous relations that de Broglie was to extend to all matter, to electrons first and then clearly to all matter.

It was this statistical tradition that led Einstein to the laws governing the emission and absorption of light by atomic systems. It was this that enabled him to see the connection between de Broglie's waves and the statistics of light-quanta proposed by Bose. It was this that kept him an active proponent and discoverer of the new phenomena of quantum physics up to 1925.

The second and equally deep strand—and here I think we do know where it came from—was his total love of the idea of a field: the following of physical phenomena in minute and infinitely subdividable detail in space and in time. This gave him his first great drama of trying to see how Maxwell's equations could be true. They were the first field equations of physics; they are still true today with only very minor and well-understood modifications. It is this tradition which made him know that there had to be a field theory of gravitation, long before the clues to that theory were securely in his hand.

The third tradition was less one of physics than of philosophy. It is a form of the principle of sufficient reason. It was Einstein who asked what do we mean, what can we measure, what elements in physics are conventional? He insisted that those elements that were conventional could have no part in the real predictions of physics. This also had roots: for one the mathematical invention of Riemann, who saw how very limited the geometry of the Greeks had been, how unreasonably limited. But in a more important sense, it followed from the long tradition of European philosophy, you may say starting with Descartes—if you wish you can start in the thirteenth century, because in fact it did start then—and leading through the British empiricists, and very clearly formulated, though probably without influence in Europe, by Charles Peirce: one had to ask how do we do it, what do we mean, is this just something that we can use to help ourselves in calculating, or is it something that we can actually study in nature by physical means. For the point here is that the laws of nature not only describe the results of observations, but the laws of nature delimit the scope of observations. That was the point of Einstein's understanding of the limiting character of the velocity of light: it also was the nature of the resolution in quantum theory where the quantum of action, Planck's constant, was recognized as limiting the fineness of the transaction between the system studied and the machinery used to study it, limiting

Figure 8

Einstein's blackboard at the Institute for Advanced Study, as he left it when he went into hospital in April 1955

this fineness in a form of atomicity quite different from and quite more radical than any that the Greeks had imagined or than was familiar from the atomic theory of chemistry.

In the last years of Einstein's life, the last twenty-five years, his tradition in a certain sense failed him. They were the years he spent at Princeton and this, though a source of sorrow, should not be concealed. He had a right to that failure. He spent those years first in trying to prove that the quantum theory had inconsistencies in it. No one could have been more ingenious in thinking up unexpected and clever examples; but it turned out that the inconsistencies were not there: and often their resolution could be found in earlier work of Einstein himself. When that did not work, after repeated efforts, Einstein had simply to say that he did not like the theory. He did not like the elements of indeterminacy. He did not like the abandonment of continuity or of causality. These were things that he had grown up with, saved by him, and enormously enlarged; and to see them lost, even though he had put the dagger in the hand of their assassin by his own work, was very hard on him. He fought with Bohr in a noble and furious way, and he fought with the theory which he had fathered but which he hated. It was not the first time that this has happened in science.

He also worked with a very ambitious programme, to combine the understanding of electricity and gravitation in such a way as to explain what he regarded as the semblance—the illusion—of discreteness, of particles in nature. I think that it was clear then, and believe it to be obviously clear today, that the things that this theory worked with were too meagre, left out too much that was known to physicists, but had not been known much in Einstein's student days. Thus it looked like a hopelessly limited and historically, rather accidentally conditioned approach. Although Einstein commanded the affection or more rightly the love of everyone for his determination to see through his programme, he lost more contact with the profession of physics, because there were things that had been learned which came too late in life for him to concern himself with them.

Einstein was indeed one of the friendliest of men. I had the impression that he was also, in an important sense, alone. Many very great men are lonely; yet I had the impression that although he was a deep and loyal friend, the stronger human affections played a not very deep or very central part in his life taken as a whole. He had of course incredibly many disciples, in the sense of people who, reading his work or hearing it taught by him, learned from him and had a new view of physics, of the philosophy of physics, of the nature of the world that we live in. But he did not have, in the technical jargon, a school. He did not have very many students who were his concern as apprentices and disciples. And there was an element of the lone worker in him, in sharp contrast to the teams we see today, and in sharp contrast to the highly cooperative way in which some other parts of science have developed. In later years, he had people working with him. They were typically called assistants and they had a wonderful life. Just being with him was wonderful. His secretary had a wonderful life. The sense of grandeur never left him for a minute, nor his sense of humour. The assistants did one thing which he lacked in his young days. His early papers are paralysingly beautiful, but there are many errata. Later there were none. I had the impression that, along with its miseries, his fame gave him some pleasures, not only the human pleasure of meeting people, but the extreme pleasure of music played not only with Elizabeth of Belgium, but more with Adolf Busch, for he was not that good a violinist. He loved the sea and he loved sailing and was always grateful for a ship. I remember walking home with him on his seventy-first birthday. He said, 'You know, when it's once been given to a man to do something sensible, afterward life is a little strange.'

Einstein is also, and I think rightly, known as a man of very great good will and humanity. Indeed, if I had to think of a single word for his attitude towards human problems, I would pick the Sanskrit word *Ahinsa*, not to hurt, harmlessness. He had a deep distrust of power; he did not have that convenient and natural converse with statesmen and men of power that was quite appropriate to Rutherford and to Bohr, perhaps the two physicists of this century who most nearly rivalled him in eminence. In 1915, as he made the general theory of relativity, Europe was tearing itself to pieces and half losing its past. He was always a pacifist. Only as the Nazis came into power in Germany did he have some doubts, as his famous and rather deep exchange of letters with Freud showed, and began to understand with melancholy and without true acceptance that, in addition to understanding, man sometimes has a duty to act.

After what you have heard, I need not say how luminous was his intelligence. He was almost wholly without sophistication and wholly without worldliness. I think that in England people would have said that he did not have much 'background', and in America that he lacked 'education'. This may throw some light on how these words are used. I think that this simplicity, this lack of clutter and this lack of cant, had a lot to do with his preservation throughout of a certain pure, rather Spinoza-like philosophical monism, which of course is hard to maintain if you have been 'educated' and have a 'background'. There was always with him a wonderful purity at once childlike and profoundly stubborn.

Einstein is often blamed or praised or credited with these miserable bombs. It is not in my opinion true. The special theory of relativity might not have been beautiful without Einstein; but it would have been a tool for physicists, and by 1932 the experimental evidence for the inter-convertibility of matter and energy which he had predicted was overwhelming. The feasibility of doing anything with this in such a massive way was not clear until seven years later, and then almost by accident. This was not what Einstein really was after. His part was that of creating an intellectual revolution, and discovering more than any scientist of our time how profound were the errors made by men before them. He did write a letter to Roosevelt about atomic energy. I think this was in part his agony at the evil of the Nazis, in part not wanting to harm any one in any way; but I ought to report that that letter had very little effect, and that Einstein himself is really not answerable for all that came later. I believe he so understood it himself.

His was a voice raised with very great weight against violence and cruelty wherever he saw them and, after the war, he spoke with deep emotion and I believe with great weight about the supreme violence of these atomic weapons. He said at once with great simplicity: 'Now we must make a world government.' It was very forthright, it was very abrupt, it was no doubt 'uneducated', no doubt without 'background'; still all of us in some thoughtful measure must recognize that he was right.

Without power, without calculation, with none of the profoundly political humour that characterized Gandhi, he nevertheless did move the political world. In almost the last act of his life, he joined with Lord Russell in suggesting that men of science get together and see if they could not understand one another and avert the disaster which he foresaw from the arms race. The so-called Pugwash movement, which has a longer name now, was the direct result of this appeal. I know it to be true that it had an essential part to play in the Treaty of Moscow, the limited test-ban treaty, which is a tentative, but to me very precious, declaration that reason might still prevail.

In his last years, as I knew him, Einstein was a twentieth-century Ecclesiastes, saying with unrelenting and indomitable cheerfulness, 'Vanity of vanities, all is vanity.'

EINSTEIN AND HIS WORK

I Einstein— A condensed biography

The most incomprehensible thing about the universe is that it is comprehensible.
(A.E., 'Physics and Reality')

Albert Einstein was born on 14 March 1879, in the city of Ulm in South Germany. His parents, Hermann Einstein and Pauline Einstein (née Koch), came from this region, as had their forebears through many generations.

Hermann Einstein's business failed within a year of Albert's birth and he moved to Munich to try a fresh start. Here Einstein began to grow up in a family which, though Jewish by descent, was free-thinking and little concerned with Jewish tradition—so much so that Einstein was sent to a Catholic elementary school, where he was a pupil from 1884 to 1889. From these early years the most notable incident, as recalled by Einstein himself, was when, at the age of about five, his father showed him a pocket compass: the purposeful behaviour of the isolated needle made a deep impression on him (his first experience of a force field!). However, by normal standards he was a slow starter. He was late in learning to speak, and when aged nine was still far from fluent. His parents feared that he might even be a little subnormal.

At the age of ten, in 1889, Einstein entered the Luitpold Gymnasium (secondary school). This seems to have been a typical school of that place and time, with a rigidly regimented system. Einstein disliked it intensely. It did not prevent him, at the age of twelve, from being thrilled by Euclidean plane geometry; the notion of producing concrete results by pure thought struck him as almost miraculous. However, the cumulative effect of the Gymnasium experience was to generate in him a loathing for conventional schooling, and no doubt helped to develop his lifelong antipathy towards authority.

In his early childhood he was introduced to music, and in particular to the violin, which he began to learn to play at the age of six. As he entered adolescence he began to appreciate the full power and beauty of music, and it remained a joy and a solace to him throughout his life. Also, as an adolescent, he developed a temporary but strong religiosity based on his consciousness of being a Jew.

In 1894 his father again suffered a failure in his business, and the parents, together with Einstein's younger sister Maja (born in 1881), moved to Milan. Einstein was left behind so that he might finish his schooling; instead he rejoined his family within six months, and without a diploma. There followed a period of travel and of enjoyable, undirected activity, but about a year later it was decided that Einstein should try to enter the Swiss Federal Polytechnic School at Zürich (best known in scientific circles as ETH—Eidgenössische Technische Hochschule) with a view to becoming an electrical engineer. At sixteen, in 1895, Einstein took the entrance examination, but failed. In preparation for a second attempt he became a student at a Swiss cantonal school at Aarau, about forty kilometres from Zürich. Under its director, Jost Winteler, with whose family he lived, Einstein found himself enjoying his studies; the Swiss style of schooling, which stemmed from the country's democratic tradition, was in total contrast to the Prussianism of the Munich Gymnasium. It decided Einstein to renounce his German citizenship, and for nearly six years he was officially without a country.

Einstein's move to Bern was a turning-point in his life. Although he had had no previous experience with technical inventions, he found the work in the Patent Office interesting. It was his duty to put applications into a clear form and to determine the basic idea from the often vaguely worded descriptions of the inventors. It may well have been this training that developed his remarkable faculty for seeing to the heart of a problem and quickly realizing the consequences of any hypothesis. Moreover, the work left him with ample time to pursue his own ideas. Indeed, it would seem in many ways to have been the ideal post for him at this stage of his career.

(G. J. Whitrow, *Einstein: The Man and His Achievement*)

Strengthened academically by the year of schooling at Aarau, Einstein passed the entrance examination to ETH at his second attempt, and entered the Institute in October 1896 to begin a four-year course of study for prospective science and mathematics teachers. He found little profit in the formal instruction (although one of his lecturers was Hermann Minkowski) and he relied heavily on the notes taken by his friend and fellow-student, Marcel Grossmann, to cover the many lectures that he himself did not attend. His own method of study was to read deeply into the original literature of physics by such masters as Kirchhoff, Hertz, and Maxwell. He also became acquainted with the works of the philosopher Ernst Mach, whose classic work, *The Science of Mechanics*, probed deeply into the fundamental ideas and assumptions of physics.

In 1900 Einstein graduated. The aftertaste of his formal studies was so disagreeable that he did little for a year. And even then he found it impossible to obtain a regular academic position. He stayed in Zürich and supported himself by tutoring, part-time school teaching, and such like. In 1901 he became a Swiss citizen and also wrote his first published scientific paper (on capillary phenomena). At the end of this year he applied for a position at the Swiss Patent Office in Bern; he was accepted into a probationary position there in June 1902, helped by a strong recommendation to the director from the father of his friend Marcel Grossmann.

With the move to Bern, Einstein's fortunes took an upward and happier turn. He supplemented his meagre income by tutoring, through which he met Maurice Solovine, who became a lifelong friend. With Solovine and an earlier acquaintance, Conrad Habicht, he formed what they called the

'Olympia Academy'—just the three of them, meeting in the evenings to dine and to talk about all manner of things in physics, philosophy, and other fields.*

In 1903 Einstein married a fellow member of his class at Zürich—Mileva Maric, from Hungary. They had two sons: Hans Albert, born in 1904, and Edward, born in 1910. Little is recorded about his domestic life, but it certainly did not inhibit his scientific activity. The first few years in Bern saw the full flowering of his genius as a theoretical physicist. In the process he apparently benefited greatly from numerous discussions with Michele Besso, an engineer and a colleague in the Patent Office, who also became a lifelong friend.†

The year 1905 was Einstein's '*annus mirabilis*'—a date to set beside 1543 (when Copernicus published *De Revolutionibus Orbium Coelestium*) and 1686 (when Newton completed his *Principia*). It saw not only the appearance of his special theory of relativity (in a paper entitled, 'On the Electrodynamics of Moving Bodies') but also the publication of two other major papers, one on the theory of the Brownian motion and the other, about the properties of light, introducing the fundamental concept of quantum physics—the existence of quanta of energy. The fateful equation $E = mc^2$ also made its appearance later in the year, in a paper entitled 'Does the Inertia of a Body depend on its Energy Content?'

One of the great ironies in the history of physics is that Einstein submitted his 1905 paper on special relativity to the University of Bern in support of his candidacy for an affiliation giving him the right to practise as a *Privatdozent* (entitled to offer instruction under the auspices of the University)—and it was rejected! However, his work attracted the interest of greater men, in particular Max Planck in Germany and H. A. Lorentz in Holland. In 1908 he again tried, this time successfully, to become a *Privatdozent* at Bern, and in 1909 he was appointed to a professorship at the University of Zürich. In the same year, at the annual Congress of German Scientists and Physicians, he gave a lecture on the nature and constitution of radiation. At this congress he met Max Planck. Not long before this, Minkowski, now at Göttingen, had seized upon Einstein's development of special relativity and had helped to give it prominence in connection with his own ideas of a four-dimensional space-time manifold.

The stay at Zürich was brief. In 1911 Einstein was offered a senior professorship at the German University in Prague; he accepted, but was there for only a year and a half. It was a period during which he was developing his ideas on a general theory of relativity. It was also important on a personal level, for there he met the brilliant theoretical physicist Paul Ehrenfest, who became a close friend. Einstein had by now been fully adopted into the society of the world's top physicists, and he was one of those invited to attend the First Solvay Congress at Brussels at the end of

. . . he once remarked that he never met a real physicist until he was thirty. The only person he was able to discuss his ideas with was an engineer, Michelangelo Besso, also then an employee at the patent office, whom Einstein had known since his student days in Zürich and whom he has immortalized in the last sentence of his 1905 paper: 'In conclusion I wish to say that in working at the problem here dealt with I have had the loyal assistance of my friend and colleague M. Besso, and that I am indebted to him for several valuable suggestions.'
(Jeremy Bernstein, *Einstein*)

* See page 9, 'Excerpts from a memoir' by Maurice Solovine.
† See page 13, 'Einstein's friendship with Michele Besso' by P. Speziali.

1911, along with notables of an earlier generation such as Lorentz and Madame Curie, and with younger stars such as Rutherford.

Much in demand from various quarters, he accepted an invitation to return to Zürich in 1912 on what was to have been a longterm appointment. Here, during 1912 and 1913, he worked closely with Marcel Grossmann (by then a professor of mathematics in Zürich) on the formidable mathematical problems that underlay the general theory. In 1913 they published a joint paper on this subject.

The call of the wider world again came to tempt Einstein away from Switzerland. In the summer of 1913 he was approached by Nernst and Planck with a view to persuading him to accept a chair in Berlin. Despite his negative feelings about Germany, particularly its militarism, he decided to accept, and moved there with his family in April 1914. His official appointment was as a professor in the Prussian Academy and Director of the Kaiser Wilhelm Institute of Physics. At the outbreak of war only a few months later, his wife and children returned to Zürich; it was for practical purposes the end of their marriage, although they were not formally divorced until 1919.

Einstein's great preoccupation at this time was the completion of his general theory of relativity. Two years were to pass before he reached this point, but he already had one prediction to be tested—the gravitational deflection of light by the Sun. Plans were afoot to observe it during a total solar eclipse in southern Russia in 1914; the expedition was to be led by a young Berlin Observatory astronomer, Finlay-Freundlich. The war eliminated this possibility (which was just as well, since Einstein's theory at that stage was incomplete and incorrect).

In accepting the Berlin chair, Einstein automatically qualified for German citizenship, but he still regarded himself as Swiss. In any case, in the circumstances of the war he could not feel at ease in the country of his birth, and he made known some of his pacifist and internationalist feelings. Despite the state of hostilities he was able to make visits to Switzerland and Holland; in the latter country he had discussions with Lorentz about the general theory, which was published in its final form in 1916.

Einstein almost immediately proceeded to consider the implications of the theory for the universe as a whole, and published his ideas in 1917 in a paper entitled 'Cosmological Considerations on the General Theory of Relativity'. Here he introduced the famous concept of the 'finite but unbounded' universe. Although his particular cosmological model was soon superseded, it paved the way for other 'model universes' by Willem de Sitter, Alexander Friedmann, Georges Lemaître, and others. Between 1916 and 1918 Einstein had an active correspondence with de Sitter on these matters. However, this was before Hubble and others, in the 1920s, discovered the general expansion of the universe as manifested in the red shift of light from distant galaxies—a fact that then had to be embodied as a fundamental ingredient of any viable cosmological theory.

In 1917 Einstein became seriously ill for several months. His convalescence took place under the care of his cousin Elsa, who at the time was a widow with two children (Ilse and Margot). This renewal of a friendship that had begun when they were both children in Munich culminated in their marriage in 1919.

Figure 9

Silhouettes of the Einstein family (A.E., his second wife, and his two stepdaughters), made by Albert Einstein in 1919

As the war neared its end, Einstein learned of the British plans to test his theory of the gravitational deflection of starlight at an eclipse expedition in 1919. The brilliant confirmation of his revised form of the theory must have raised his stature even further among scientists, but what was perhaps more notable was the way in which this esoteric triumph caught the public imagination and made him an almost god-like figure in the public eye. From that time on his life was never to be the same.

Although conditions in Germany after the end of the war were chaotic, and Einstein could easily have gone elsewhere, he chose to remain where he was and even officially resumed German citizenship. By staying in Germany he exposed himself to attack as a pacifist and as a Jew. Even his relativity theory was a target. Some of his German physicist colleagues came outspokenly to his defence. But in the world outside he needed no defenders; he was a hero. He became a regular visiting professor at Leiden, where in 1920 he met Niels Bohr for the first time. And in 1921 he made his first visit to America. Although the prime purpose of this visit was to help raise funds for a Hebrew University in Jerusalem, Einstein was deluged with requests to meet with many other groups in academic, political, and social circles; he received an honorary degree from Princeton University. On his way back to Germany he visited England where, despite some persistence of wartime feelings against Germany, he was widely acclaimed. Further

Figure 10

Max Planck offering the plaque of his doctoral jubilee to Albert Einstein in 1929

international travels followed—to France and to the Far East, Southeast Asia, and Palestine.

Late in 1922 Einstein was named as winner of the 1921 Nobel Prize in Physics '. . . for his services to the theory of physics, and especially for his discovery of the law of the photoelectric effect.' Profoundly important though the photoelectric result was, it may seem surprising that the prize was not given for Einstein's still greater achievements in relativity theory. It may have been because Alfred Nobel's will stated that the prizes were to be awarded for discoveries that benefited mankind—and the possible practical implications of $E=mc^2$, for either good or evil, were not then suspected.

Back in Berlin during the 1920s, Einstein worked away at his next—and last—great project, the quest for a unified theory that would bring under one roof the phenomena of both gravitation and electromagnetism. He did so in a political and social environment that became progressively more unpleasant and threatening. At the same time his fame brought innumerable demands for his time through correspondence on a multitude of topics. His position was still that of Professor of the Prussian Academy of Sciences and

Physics Research Director of the Kaiser Wilhelm Institute, the post to which he had been appointed in 1914. Among his colleagues were Nernst, Planck, and von Laue.

In physics the main focus of interest had become the quantum theory and its consequences. At the Fifth Solvay Congress in 1927, by which time wave mechanics was a fully fledged theory, Einstein had numerous discussions with Bohr based on Einstein's refusal to accept a fundamental failure of determinism in nature. The debate continued at the sixth Congress three years later. In the process Einstein began to set himself apart, intellectually, from the main body of physical thought and even of physicists.

In 1929 Einstein reached his fiftieth birthday. It was an occasion for many messages and honours, among them a medal awarded by Max Planck in his own name. But it was also a year in which the signs were mounting that Germany was no longer a good or even a safe place for Einstein to be. The Nazi party was gaining strength and organized hostility to Jews in general began to escalate. Although it was to be several more years before Einstein left Germany forever, the shape of the future was forming.

Also in 1929 he made an unexpected connection that was to last for the rest of his life. During one of his regular trips to Leiden, he received an invitation to visit the Queen of the Belgians. The King (though absent on that occasion) had an interest in science, and the Queen in music (she, like Einstein, played the violin); they both wished to make his acquaintance. After that first visit he made a number of others, and after the King died in a climbing accident in 1934 Einstein (by then permanently in America) maintained a steady correspondence with the Queen.

In 1930 Einstein made two visits to England. During one of them he met Eddington, who had done so much to bring about the first observation of the gravitational deflection of light. By a happy chance, Eddington's knighthood was announced at the time of this visit. Einstein received an honorary degree from the University of Cambridge.

The second of Einstein's visits to the United States took place the same year. This time his destination was the California Institute of Technology at Pasadena, whose President was R. A. Millikan. A matter of particular scientific interest to Einstein was the red shift of distant nebulae and its bearing on various cosmological models by himself and others. The crucial observations had been made by the astronomers at Mount Wilson, near to Pasadena (and affiliated with the California Institute). A notable and moving occasion during this visit was a dinner attended by both Michelson and Millikan—the two living experimentalists with whom Einstein's early work on special relativity and light quanta was most closely connected. Michelson, then seventy-eight years old and in poor health, died only a few months later.

Very shortly after his return from America Einstein was again in England, this time to give some lectures at Oxford University; he also received an honorary degree there. A sequel to this visit was an invitation to be a resident visiting fellow at Oxford for a short period each year thereafter.

Figure 11

Albert Einstein with Michelson and Millikan at the California Institute of Technology in 1930

At the same time, however, there were moves afoot to lure him to the United States permanently. The first overtures came from the California Institute of Technology, which Einstein was due to return to on a repeat of his previous year's visit. However, a totally new prospect soon presented itself. While Einstein was in Oxford in the spring of 1932, he was approached about the possibility of joining the faculty of a new Institute for Advanced Study at Princeton, well funded but not yet in existence. A few months later, back in Germany, he agreed to accept an appointment, at least on a part-time basis (he had not yet accepted the idea of leaving Germany permanently). He was, however, committed to one more winter visit to California, at the end of 1932.

The course of world events then proceeded to take charge. Before Einstein arrived back in Europe, Hitler had come to power in Germany (at the end of January 1933). Einstein felt that he could not return to Berlin. Quite apart from questions of personal safety, he had fundamental objections on principle: 'As long as I have any choice in the matter, I shall live only in a country where civil liberty, tolerance, and equality of all citizens before the law prevail . . . These conditions do not exist in Germany at the present time.' He proceeded to take up temporary residence in the sea-side village of Le Coq sur Mer in Belgium. Here he received various offers of appointments, including one from the Hebrew University at Jerusalem. But the die was cast, and after a last visit and lecture tour in England, Einstein sailed with his wife from Southampton, to become the first professor of the Institute for Advanced Study.

At Princeton Einstein had to establish a new life in his self-imposed exile. It was to be largely a life of isolation—as much by his own choice as anything. (He once wrote: 'I am truly a "lone traveller" and have never belonged to my country, my home, my friends or even my immediate family with my whole heart. In the face of all these ties I have never lost a sense of distance and a need for solitude—feelings that increase with the years.') He never ceased to be culturally a European, and he was never quite at home in any language other than German. After the initial stir caused by his decision to make America his home (among other things, Einstein and his wife were invited by President Roosevelt to dine and spend a night at the White House), Einstein settled down to a resumption of his researches on a unified field theory. To help him he had a succession of post-doctoral assistants, men of the highest mathematical ability as was demanded by the character of the work. His personal life, as always, was simple in the extreme; his gastronomic tastes were frugal, and he abhorred anything that could be called dressing up. He enjoyed sailing (in part because '. . . it is the sport which demands the least energy') and above all his music.

There were, of course, innumerable appeals for him to lend himself to various causes. To some of these—the preservation of peace, Zionism, the welfare of Jews in peril in Europe—he gave his best efforts, as described elsewhere in this book. Many others he justifiably turned down as frivolous, or insufficiently important, or incompatible with his own views. He could easily have let his time be completely pre-empted by matters unrelated to science, and this he would not do, for the riddle of the fundamental structure of the physical world remained, as always, the dominant preoccupation of his life.

In December 1936 his wife died, but the pattern of his life did not change significantly. He himself had nearly twenty more years to live, and to the very end he did not give up his attempt to find a valid structure for his unified field theory. Meanwhile, the developing tragedy in Europe took its course. Einstein, involved in it peripherally (though not superficially) in many ways, signed the famous letter to Roosevelt. The war came and went. With its end, in 1945, came Einstein's official retirement from the Institute. In the same year he arrived at a new formulation of his unified theory, closely related to one that he had published in Germany in 1925.

Figure 12 Einstein sailing in 1936

Naturally enough, the ending of the war brought opportunities for Einstein to move elsewhere—in particular, to Israel—but he chose to stay where he was. Age and ill health were probably the main deterrents to change. In any case, Princeton had much in its favour. The Institute and the University together formed a community of immense academic prestige to which distinguished scientists from all over the world would come. Niels Bohr made visits in 1946 and 1948; the differences between him and Einstein concerning the interpretation of quantum mechanics had done nothing to impair their great respect and regard for one another, although their fierce arguments were renewed, and Einstein continued to maintain that 'God does not play dice with the world'.

Despite his retirement from the Institute for Advanced Study, Einstein would go there every day for several hours of work. At home he had the company of his daughter Margot, of Helen Dukas, his loyal secretary of many years, and (until she died in 1951) of his sister Maja.

His seventieth birthday in 1949 was marked by a celebration at Princeton, and the year 1952 brought a singular honour—an invitation, which he immediately declined, to become President of Israel. But he felt himself to be running down. In a letter written in 1952 he said: 'As to my work, it no longer amounts to much. I don't get many results any more and have to be satisfied with playing the Elder Statesman and the Jewish Saint, mainly the latter.' There were few remaining ties to his earlier days, but in 1953 he received a postcard from Conrad Habicht and Maurice Solovine, who had met in Paris and were remembering the meetings of the 'Olympia Academy' in Bern fifty years earlier; they told him in jest, but with deep nostalgia, that they were keeping a chair ready for him.

The end came no sooner than he himself would probably have wished it. Early in 1955 he joined in a new plea to the governments of the world to settle their differences by peaceful means. Before it could be published he was taken seriously ill. At his hospital bed he worked on an address that he had agreed to prepare for Independence Day ceremonies in Israel. And, beside him, he kept his latest notes on his theoretical research. In the first hours of 18 April 1955, there came the rupture of a long-existing aortic aneurism, and, after seventy-six years, his rich and long life was over. It was perhaps symbolic of the way that science progresses that his unified field theory remained unfinished beside him when he died. It is not given to any man, not even Einstein, to write the definitive closing paragraphs in the description of nature.

A.P.F.

Some biographies of Einstein

Frank, Philipp, *Einstein: His Life and Times* (New York: Knopf, 1947).
Seelig, Carl, *Albert Einstein: A Documentary Biography* (London: Staples Press, 1956).
Kuznetsov, Boris, *Einstein* (Moscow, 1965).

Clark, Ronald W., *Einstein: The Life and Times* (New York: The World Publishing Co., 1971).
Hoffmann, Banesh (with Dukas, Helen), *Albert Einstein: Creator and Rebel* (New York: Viking Press, 1972).
Bernstein, Jeremy, *Einstein* (New York: Fontana, 1973).

To these should certainly be added Einstein's own 'Autobiographical Notes' in *Albert Einstein: Philosopher-Scientist*, edited by P. A. Schilpp (Evanston, Ill: Library of Living Philosophers, 1949) Volume I, referred to by many contributors to this book.

2 Einstein and the birth of special relativity

Silvio Bergia

Einstein rejected out of hand the idea of mankind's presence in absolute space and equally absolute intuitive time. He told us that the space and time surrounding us do not have the structure we supposed, that very structure which Kant considered so obvious he made it into one of his categories of thought. Was this not revolution indeed? Can one really assert that the time was ripe for a revolution as radical as this? It is probably the greatest mutation ever in the history of thought.
(Jean Ullmo, 'From Plurality to Unity', in *Science and Synthesis*)

INTRODUCTION

It is almost universally acknowledged that the theory of special relativity, a common inheritance nowadays of the scientific community, produced a revolution in the conceptions and laws of classical physics. A historical analysis of the process that brought about this revolution implies first of all identifying the set of problems which found their solution through the theory. It is clear that these problems arose in the course of the development of classical physics. This is the first reason why it seems desirable that a discussion of the *birth* of special relativity be preceded by an analysis of the *genesis* of the problems that confronted it. Secondly, particularly in a book in commemoration of Einstein, it seems important in view of suggestions that the theory had been anticipated by Lorentz and Poincaré that one should try to clarify whether and to what extent the theory should be credited to Einstein alone. These are the reasons why the first three sections of the present article deal with the historical developments of relativistic ideas from Galileo to Poincaré. A discussion of Einstein's motives and achievements follows. Finally, it seems expedient to examine briefly the process through which the theory, which was born with Einstein's famous paper of 1905, developed and established itself in the following years through the work of Einstein himself and of other authors, and to place particular emphasis on the historical experiments that yielded the first confirmations of the theory with respect to its fundamental aspects.

THE PRINCIPLE OF RELATIVITY FROM GALILEO TO NEWTON

In a famous passage of the *Dialogue concerning the Two Chief World Systems* (1632—imbued with good physics although not quite in the style of a modern journal!), Galileo examined what would happen to various physical phenomena occurring 'in the main cabin below decks on some large ship' moving 'with whatever velocity', provided that 'the motion be uniform and not fluctuating this way and that'; and he concluded that 'not the least change in all the effects named' should occur and that 'from no one of them could you tell whether the ship is moving or not. . . .' With this passage, Galileo aimed at refuting an objection to the idea of the motion of the earth, whose opponents maintained that things on the surface of the earth would be left behind during motion. Even if it does not lack antecedents, we may use it to mark, symbolically, the beginning of the long period of gestation of relativity theory. We say 'symbolically' because it would be rather naive to think that from the very moment that Galileo argued the validity of a principle of relativity which concerned the mechanical phenomena and laws known to him, all natural philosophers proceeded to adhere to it.

'One thing I have learned in a long life: that all our science, measured against reality, is primitive and childlike—and yet it is the most precious thing we have.'

(A.E.)

It was certainly accepted by Descartes (1644), although he took verbal precautions that permitted him not to deny the absolute rest of the earth as demanded by the Holy Scriptures. In one of his letters one finds an enunciation that stresses the relative character of the Galilean principle: '. . . of two men, one of whom moves with a ship and the other stands on the shore . . . there is nothing more positive in the motion of the former than in the rest of the latter'.

In his *Principia*, Newton devotes to the topic a single corollary: 'The relative motions of two bodies in a given space are identical whether this space is at rest or whether it moves uniformly in a straight line without circular motion.' As discussed by Mach in his *Science of Mechanics*, the corollary is important for Newton to stress that his laws of mechanics are valid in a frame of reference in uniform rectilinear motion with respect to the fixed stars. Newton's statement affirms the validity of the principle of relativity. Nevertheless, as regretted by Mach, he did not stick to the 'factual', and postulated the existence of an absolute space. However, the crucial problem of special relativity was never at the centre of his interests; his concern was to show (as with his famous experiment on the curvature of the water surface in a rotating pail) that *circular* motions were absolute. Besides the idea of an absolute space that 'in its own nature, without relation to anything external, remains always similar and immovable', Newton also codified the concept of 'absolute, true and mathematical time', which 'of itself and from its own nature, flows equably without relation to anything external, and by another name is called duration'.

These concepts of Newton's did not lack criticism even during his own time. A long debate took place on absolute space, with Huygens and Leibniz as main opponents. Oddly enough, the quarrel never concerned itself with the equivalence among systems in relative uniform rectilinear motion, but focussed on circular motions which, for Newton, reveal absolute motion through the presence of inertial forces. The keenest criticism, and one that anticipated Mach by a century as a forerunner of general relativity, was due to Bishop Berkeley.

It was, however, only in the nineteenth century that uniform rectilinear motions were considered separately and there emerged clearly their possible condition of privilege. This happened only after the question arose, via a long and tortuous path, in the field of a different discipline, viz., optics, and after a series of facts and reflections had challenged the tenets of the Newtonian heritage—tenets not even scratched by the above authors.

OPTICS AND THE QUESTION OF THE ABSOLUTE MOTION OF THE EARTH

The story had its origins within the framework of Descartes's mechanistic conception of the universe. For Descartes, during the process of evolution of the universe, three distinct forms of matter originated: '. . . all the bodies of the visible world are composed of these three forms of matter, as of three distinct elements; in fact . . . the sun and the fixed stars are formed of the first of these elements, *the interplanetary space of the second*, and the earth, with the planets and comets, of the third'. (Quoted in E. T. Whittaker, *A History of the Theories of Aether and Electricity*, hereafter to be referred to as Whittaker; emphasis added.) The matter of the second form is, for Descartes, 'the vehicle of light in interplanetary space'.

'I want to know how God created this world. I am not interested in this or that phenomenon, in the spectrum of this or that element. I want to know His thoughts, the rest are details.'

(A.E.)

The idea of a medium that permeates the interplanetary and interstellar spaces underlies the entire development that we wish to describe. This medium was given the name of 'ether', 'luminiferous ether', or 'cosmic ether'—a term borrowed from Greek science, where it joins with earth, water, air, and fire in the task of filling up the celestial regions. The physical necessity for its existence was first stated with the formulation of the wave theory of light, prefigured in the work of Robert Hooke (1667), and explicitly accomplished by Huygens in 1678. In this theory, based on the analysis of the known wave phenomena, the idea of a medium providing the support for the propagation of the waves was essential. It had previously been shown by Torricelli that light is transmitted through vacuum as through air, from which Huygens inferred that the medium, or ether, in which the propagation takes place must penetrate all matter and be present even in vacuum.

Thus was started, in a seemingly innocent way, the history of a concept that was to become crucial for the physics of the second half of the nineteenth century. It is not possible, however, to perceive this trend from the beginning. This is mainly due to the fact that the natural philosophers of Newton's generation imposed the acceptance of the corpuscular conception of light, thus showing themselves, as it were, more royalist than the king himself; Newton in fact always 'refrained from committing himself to any doctrine regarding the ultimate nature of light' (Whittaker); it should not be forgotten that some evidence had already accumulated for the existence of diffraction (Grimaldi, 1665, Hooke, 1667) and of interferometric phenomena (the 'Newton's rings', already observed by Boyle and Hooke).

However, the chief optical discovery of the first half of the eighteenth century tended to support the corpuscular theory, by which it was first, and more readily, explained. This was the discovery in 1728, by the astronomer Bradley, of the so-called stellar aberration, an apparent elliptical motion of the stars on the vault of the heaven, associated with the Earth's motion around the Sun, that cannot be explained as a result of parallax. The phenomenon can instead be readily accounted for once it is assumed that the velocity of light and the velocity of the frame of reference (the Earth) add as vectors, as if light were composed of corpuscles, in much the same way as the velocity of wind and that of a boat must be added to determine the orientation of a flag on the mast.

The fortunes of the wave theory began to brighten, as Whittaker phrases it, at the end of the century, when a new champion arose: Thomas Young. He began to write about optics in 1799; having stressed the superiority of the wave theory in explaining reflection and refraction, he stated explicitly the 'general law of the interference of light', which he used to explain Newton's rings and to interpret diffraction. Young then provided a simple explanation of the stellar aberration in terms of the wave theory: if we suppose the ether surrounding the Earth to be at rest and unaffected by the Earth's motion, the light waves will not partake of the motion of the telescope and the image of the star will therefore be displaced by a distance equal to that which the Earth describes while the light is travelling through the telescope, in agreement with what is actually observed. One should note that this hypothesis implied that even the ether within the telescope is unaffected by the motion of the matter that constitutes it; Young was therefore led to the belief 'that the luminiferous aether pervades the substance of all material bodies with little or no resistance, as freely as the wind passes through a grove of trees' (see Whittaker, p. 115) —a point of view that was not generally shared.

A number of other problems had meanwhile arisen. The discovery by Malus (1808) that one could obtain polarized light by refraction drew attention to the phenomenon and had the effect of encouraging the adherents of the corpuscular doctrine; in fact, the wave-theorists, misled by the analogy of light with sound, were unable to give any account of polarization. Fresnel and Arago (1816) performed a key experiment when they tried to obtain interference by using two rays polarized at right angles. From the negative result, Fresnel and Young independently drew the conclusion that light vibrations must be transverse. This conclusion was in time going to pose many problems to the ether theorists.

In the meantime an experiment performed by Arago (1810), had raised a new question which was indirectly to exert a marked influence on the future developments and to impress a turning point on the whole matter. The question whether rays coming from the stars are refracted differently from rays originating in terrestrial sources had been raised originally by John Mitchell at Cambridge (1784). Arago saw how the matter could be submitted to the test of experiment. From the corpuscular point of view, the velocity with respect to the earth of the light projectiles emitted by a star depends upon the direction of the earth's motion. Such rays of different velocities should undergo different deviations in going through a prism. Thus upon observing, say, at 6 a.m., a star aligned with the earth's motion and at 6 p.m., a star lying in the opposite direction, one should observe a difference in the angle of deviation of the order of the ratio, v/c, of the

Figure 13

Einstein at his desk in the Patent Office, Bern, in the early 1900s

velocity of the Earth to the velocity of light. This was, at least, the principle of Arago's experiment. It should be recalled that the velocity of light had been measured by Roemer (1675) on the basis of the delays observed in the eclipses of the satellites of Jupiter, and the value had been made more precise by the value of the aberration, another phenomenon depending linearly on v/c. On the basis of a value of v of the order of magnitude of the velocity of the orbital motion of the Earth, one could expect v/c to be of the order of 10^{-4}. The effect to be looked for was therefore a small one; Arago found, however, no effect at all. Fresnel, asked by Arago whether he could explain the null result from the point of view of the wave theory (see Rosser, *An Introduction to the Theory of Relativity*), furnished an interpretation which was due to last a long time. It should be noted that the result cannot be explained in terms of Young's hypothesis; in fact, if material bodies pass through the ether without dragging it at all, the velocity of light, as measured in the two directions, must be different, and an effect should result. Fresnel was led to formulate the hypothesis of a partial dragging of the ether by bodies, such as Arago's prism, having index of refraction larger than that of the vacuum. Fresnel's assumptions were that the ethereal density in any body is proportional to the square of the refractive index n, and that, when a body is in motion, it carries along part of the ether within it, that part which constitutes the excess of its density over the density of ether in vacuum. From these hypotheses he deduced a 'drag coefficient':

'If you want to find out anything from the theoretical physicists about the methods they use, I advise you to stick closely to one principle: don't listen to their words, fix your attention on their deeds. To him who is a discoverer in this field, the products of his imagination appear so necessary and natural that he regards them, and would like to have them regarded by others, not as creations of thought but as given realities.'
(A.E., 'On the Method of Theoretical Physics')

$$f = 1 - 1/n^2$$

Fresnel was also able to infer that the aberration would be unaffected if observed with a telescope filled with water (an experiment suggested by Boscovich already in 1776 and performed eventually by Airy in 1871, resulting in a confirmation of Fresnel's prediction). In 1818, Fresnel published an investigation of the influence of the Earth's motion on light, which was a study on relativity; he showed in fact that the apparent positions of terrestrial objects, carried along with the observer, are not displaced by the Earth's motion, and that experiments on refraction and interference are not influenced by any motion which is common to source, apparatus, and observer.

More generally the following theorem can be shown to hold: if one takes into account Fresnel's dragging coefficient and neglects terms of the order of $(v/c)^2$, optical phenomena on Earth due to terrestrial sources are independent of the Earth's motion. It is perhaps worthwhile to stress parenthetically the meaning of this result. If the theorem did *not* hold, classical optical experiments should reveal the effects of the Earth's motion through the ether and, in principle, allow a determination of the velocity of this motion. By virtue of Fresnel's hypothesis, this possibility seems to disappear. It may be said, however, that from the moment it was formulated the attention of the physicists focussed on the problem, as it was realized that at stake was the possibility of revealing the *absolute* motion of the Earth.

Ether was becoming a central problem for other reasons too. Fresnel had already pointed out that, in order to be able to transmit transverse vibrations, the ether should behave in some way as an elastic solid. At that time general mathematical methods for studying the properties of elastic bodies had not been developed. But under the stimulus of Fresnel's ideas, 'some of the best intellects of the age were attracted by the subject' (Whittaker); among them Navier (1821), Cauchy (1828, 1830, 1836), Poisson (1828), Green (1837), Neumann (1837). The idea underlying all these investigations was to treat ether as an elastic solid, in which light waves propagate similarly to sound waves in material bodies. A first objection to the elastic ether arises from the necessity of attributing to it a sufficiently high rigidity to explain the high velocity of the waves. How is it, then, that the planets are able to journey through it without encountering any perceptible resistance? Stokes (1845) tried to answer this objection by recalling the existence of substances, such as pitch, that are so rigid as to be capable of elastic vibration and yet sufficiently plastic to permit other bodies to pass slowly through them. The ether 'may have this combination of qualities in an extreme degree', 'behaving like an elastic solid for vibrations so rapid as those of light', like a fluid for 'the much slower progressive motions of the planets' (Whittaker).

Finding it extremely hard to accept Fresnel's hypothesis, Stokes formulated an aberration hypothesis alternative to that of Young. He thought that the ether might be dragged by the earth much in the same way as layers of a fluid are dragged due to friction, by a body in motion through it. (If this were so, the theorem on 'optical relativity' previously stated would be automatically guaranteed, insofar as in terrestrial laboratories the ether would be at rest.) As far as aberration is concerned, one would tend to draw the conclusion that no effect should arise in these conditions. Stokes was nevertheless able to show that, if the ethereal motion is irrotational, the observed effect could be exactly reproduced. Much later (1886) Lorentz raised a major objection to Stokes' theory by showing the incompatibility of the hypothesis of the irrotationality of the motion with the one that should take place in the vicinity of the Earth.

He was convinced that his ideas were fundamentally very simple despite their very heavy mathematical mechanisms. He had a firm conviction, which I do not think was justified, that he could explain it to everybody. For instance, as I remember quite clearly, we were working on something in unified field theory and he came down rather cheerfully and said, 'I explained it this morning to my sister and she also thinks that it is a very good idea.'

(E. Straus, in G. J. Whitrow, *Einstein: The Man and His Achievement*)

But in the meantime, already in 1851, Fizeau had performed his famous experiment (repeated by Michelson and Morley in 1886) confirming that the speed of light in flowing water was changed just as one would expect from Fresnel's theoretical drag coefficient. Other important events had occurred at about the same time. After the measurements of the velocities of light in air with terrestrial sources (Fizeau, 1849; Foucault, 1862), Foucault measured the velocity of light in water, confirming the prediction of the wave theory that it should be less than in vacuum, whereas the corpuscular theory would have required it to be greater; this, to most physicists, appeared as the most convincing proof of the wave theory.

Parallel to the development that we have thus far summarized, concerning the 'luminiferous ether', another one had been going on, starting from the first studies of Faraday (1831), concerning the theory of the electric medium, i.e., the theory of how electric and magnetic influences are transmitted through space. To W. Thomson (1844, 1847) is due the credit of having properly initiated the theory of the electric medium: he suggested that 'the propagation of electric or magnetic force' might 'take place in somewhat the same way as changes in the elastic displacement are transmitted through an elastic solid' (Whittaker). Maxwell's early investigations on the subject may be regarded as an attempt to connect the ideas of Faraday with the mathematical analogies that had been devised by Thomson. In the years between 1855 and 1862, he completed his construction of a mechanical theory of the propagation of the fields through the electric medium. The question then presented itself whether one need suppose, for these phenomena, an ether distinct from the light medium; but by identifying the velocity of propagation of electromagnetic disturbances with the velocity of light, Maxwell was able to conclude: 'We can scarcely avoid the inference that light consists in the transverse modulations of the same medium which is the cause of electric and magnetic phenomena (Whittaker).

Like Newton, Einstein could and did concentrate on individual problems for *years* at a time. The special theory of relativity required, from all accounts, nearly a decade of preparatory thought, although, as he later remembered it, the final formulation and the writing of the manuscript took only five or six weeks. The general theory of relativity and gravitation took about seven years to complete allowing for all of the false starts, and Einstein worked constantly on the unified field theory—an attempt to unite gravitation and electromagnetism —despite the critical opposition of most of his contemporaries, who were as convinced that he was on the wrong track as he was convinced that he was not, for more than three decades.
(Keynes, *Newton the Man*)

The present-day reader, accustomed to imagining without difficulty an electromagnetic field oscillating in empty space, may wonder why there should be need of an ether at all. As a matter of fact, although the mechanical model of the ether had been an essential tool for the building of the theory, already in 1864 Maxwell had presented an account of his theory in which the 'architecture of his system was displayed, stripped of the scaffolding by aid of which it had been first erected' (Whittaker). Before the process was complete, however, some more steps had to be made (we shall come to them later on); towards the end of the seventies ether was still a solid reality.

In 1879 Maxwell wrote an acknowledgement of some astronomical tables he had received from D. P. Todd of the U.S. Nautical Almanac Office in Washington. In his letter Maxwell suggested an experiment which would have allowed a determination of the velocity of the solar system through the ether; the experiment was a repetition of the early Roemer experiment and implied 'comparing the values of the velocity of light deduced from the observation of the eclipses of Jupiter's satellites when Jupiter is seen from the earth at nearly opposite points of the ecliptic' (as explained by Maxwell himself in an *Encyclopaedia Britannica* article on 'Ether'; see J. C. Maxwell, *Scientific Papers*, (New York: Dover Publications, 1952); the next two quotations from Maxwell are from the same article). It must be noted that the experiment escaped the trap of Fresnel's 'relativity' theorem. Maxwell inquired in his letter whether astronomical measurements had an accuracy sufficient to detect the effect. (By the way, we may quote from Max Born (1920, 1962), that the standard requested had still not been reached in 1920.) Maxwell thought that his was 'the only practicable method of determining directly the relative velocity of the ether with respect to the solar system'; attention should be drawn to the adjective *practicable*. Indeed he was aware of the fact that terrestrial measurements of the velocity of light, such as Fizeau's and

Foucault's, are capable, in principle, of showing the effect of the motion. Let us see why. We recall that, by Fresnel's theory, the air of terrestrial laboratories does not drag the ether in an appreciable way; therefore, within the laboratories, the ether will flow in the direction opposite to the earth's motion. This situation has been picturesquely described in terms of an 'ether wind' 'blowing' through the laboratories, giving resultant light velocities between a minimum of $c-v$ (c is the light velocity in the medium at rest, v is the ether wind velocity) and a maximum of $c+v$. We may remark parenthetically that the above reasoning attaches an operational meaning to the privileged frame of reference at rest in the ether; it is the only one in which propagation of light is isotropic. The above discussion shows that indeed the effects of the ether wind should manifest themselves in the measurements of the velocity of light. In discussing the problem, however, Maxwell makes a remark whose importance for the future developments can hardly be overestimated; to give it in his words: 'All methods . . . by which it is practicable to determine the velocity of light from terrestrial experiments depend on the measurement of the time required for the double journey from one station to the other and back again, and the increase of this time on account of a relative velocity of the ether equal to that of the Earth would be only about one hundred millionth part of the whole time of transmission, and would therefore be quite insensible'.*

The relevance of Maxwell's remark stems from the implicit challenge it gave to experimenters. This challenge was soon met. While working, like Todd, at the Nautical Almanac Office, the American physicist, A. A. Michelson, had the opportunity of studying Maxwell's letter (as reported by Shankland). Only two years later, during a stay in Berlin, he set up an experiment where the seemingly insurmountable difficulty due to the smallness of the foreseen effect was overcome. The basic idea of the experiment is of comparing the times (or velocities) along two paths at right angles, one of which, to be definite, may be thought of as aligned with the earth's motion. The velocity along the perpendicular direction (according to Michelson's ideas) should not be affected by the motion and

On the fireplace over the Professors' room at the Institute there is an inscription which Einstein contributed at the time the building was being constructed. It says, '*Raffiniert ist der Herr Gott aber boshaft ist Er nicht.*' (God is cunning but He is not malicious.) In other words the world has been put together in a very complicated and subtle way, but still the Lord gives us a chance to find out how it is done.

(J. A. Wheeler, in G. J. Whitrow, *Einstein: The Man and His Achievement*)

* This can be checked at once in the simplified case of a measurement of the velocity of light, say by Fizeau's method, along a base line of length d, which will be supposed aligned with the direction of the earth's motion. The distance d will be covered at the velocities $c-v$ ('against the wind') and $c+v$ ('with the wind'), and the total time taken by the light will be:

$$t=\frac{d}{c+v}+\frac{d}{c-v}=\frac{2d}{c}\cdot\frac{1}{1-v^2/c^2}=t_0\frac{1}{1-v^2/c^2},$$

where $t_0=2d/c$ is the time the light would take to travel the same distance in absence of ether wind.

For $v \ll c$, we can approximate:

$$t \simeq t_0\,(1+v^2/c^2)$$

Hence $(t-t_0)/t_0$ is of the order of v^2/c^2, that is of *second order* in this ratio. With $v \simeq 30$ km s^{-1}, $c \simeq 3\times 10^5$ km s^{-1}, Maxwell's conclusion follows.

Figure 14 *The Michelson–Morley experiment*

(a) A sketch of the apparatus.
(b) Plan view of the optical system.
(c) Variation of fringe position during one rotation of the apparatus.
(Adapted from A.A. Michelson, *Studies in Optics*, University of Chicago Press.)
(d) The fringes during two complete revolutions of the apparatus in a later repeat of the experiment by G. Joos in 1930 (from G. Joos, *Lehrbuch der Theoretischen Physik*, Akademische Verlagsgesellschaft, Leipzig).

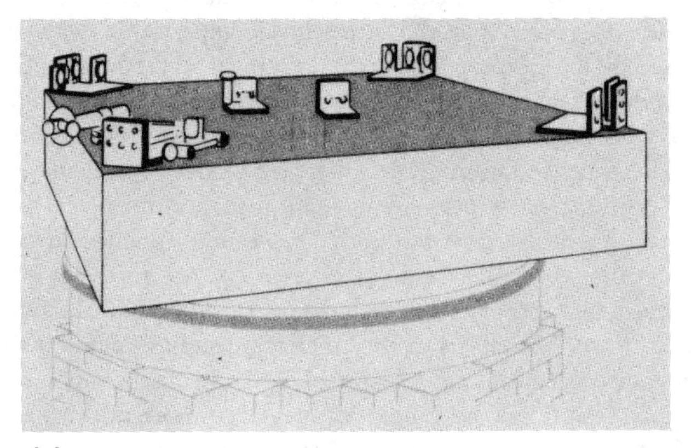

(a)

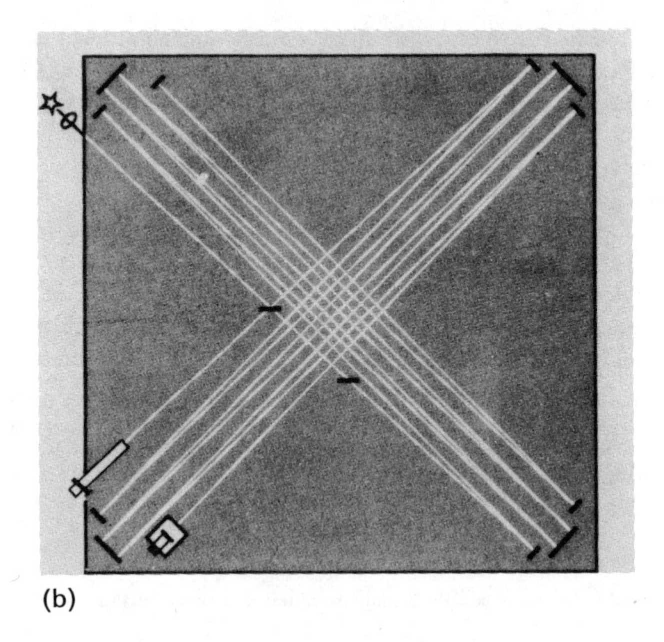

(b)

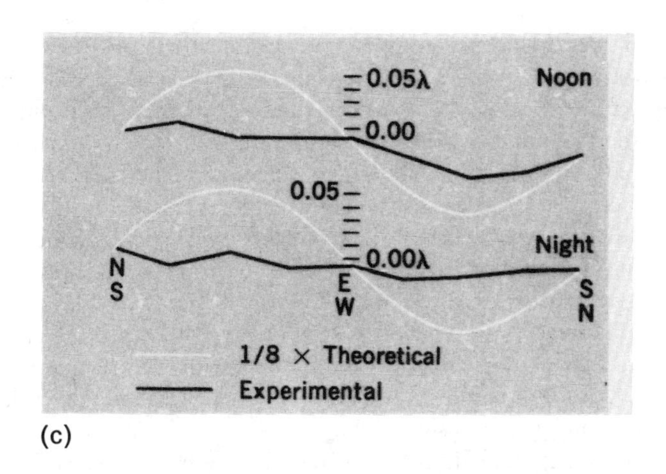

(c)

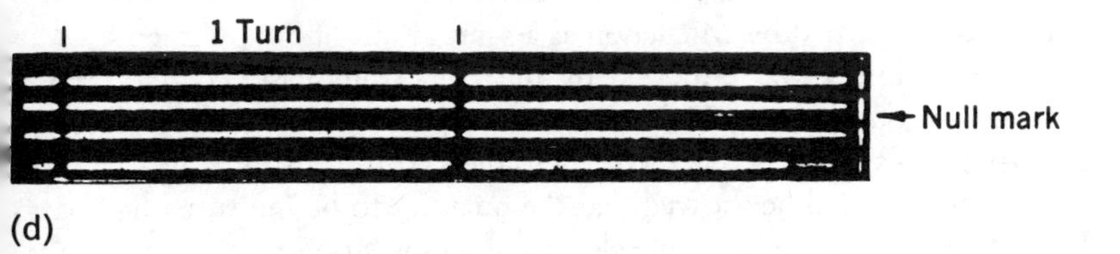

(d)

Trials of the Michelson–Morley Experiment*

Observer; year	*l, cm*	δ_{calc}	δ_{obs} *(upper limit)*	*Ratio*
Michelson; 1881	120	0.04	0.02	2
Michelson and Morley; 1887	1100	0.40	0.01	40
Morley and Miller; 1902–1904	3220	1.13	0.015	80
Miller, 1921	3220	1.12	0.08	15
Miller; 1923–1924	3220	1.12	0.03	40
Miller (sunlight); 1924	3220	1.12	0.014	80
Tomaschek (starlight); 1924	860	0.3	0.02	15
Miller; 1925–1926	3200	1.12	0.08	13
Kennedy; 1926	200	0.07	0.002	35
Illingworth; 1927	200	0.07	0.0004	175
Piccard and Stahel; 1927	280	0.13	0.006	20
Michelson *et al.*; 1929	2590	0.9	0.01	90
Joos; 1930	2100	0.75	0.002	375

* From a review by Shankland *et al.*, *Rev. Mod. Phys.*, 27, 167 (1955).

the comparison should make possible a determination of v. The problem was: how are the two times to be actually compared? Michelson's answer was: by letting the two light rays in question interfere. The difference between the two times is converted into a difference of optical path and this may have an observable influence on the interference pattern. Michelson's interferometer was an instrument capable of revealing differences in optical paths between coherent rays which travel equal distances along perpendicular paths. When the apparatus is set up in the laboratory with any given orientation, one does not know the direction of the ether wind through it. However, if the velocity of light is not isotropic in a terrestrial laboratory, the two times should vary anyway with the orientation of the apparatus. Upon rotating it through 90°, so as to exchange their roles, the interference fringes should shift.

Often the lecturers, sometimes very good scientists and occasionally post-graduate students, were somewhat obscure. Einstein would rise after the lecture and ask whether he might put a question. He would then go to the blackboard and begin to explain in simple terms what the lecturer had been talking about. 'I wasn't quite certain I understood you correctly,' he would say with great gentleness and then he would make clear what the lecturer had been unable to convey.

(E. H. Hutten, in G. J. Whitrow, *Einstein: The Man and His Achievement*)

Michelson's experiment, performed in Potsdam in 1881, had a null result. 'There is no displacement of the interference bands', he concluded. 'The result of the hypothesis of a stationary ether is thus shown to be incorrect.' Michelson's last sentence refers to Fresnel's theory; his result was indeed regarded by Michelson as a vindication of Stokes' theory.

The experiment was criticized by the Dutch physicist, H. A. Lorentz (1886), who pointed out, in the first place, that the velocity along the arm by hypothesis aligned perpendicularly to the Earth's velocity is also affected by the motion; and he showed that the quantity to be measured had only half the value supposed by Michelson. Lorentz went on to suggest that the null result of the experiment might be explained by a combination of Fresnel's and Stokes' theories, whereby the ether near the Earth is moving irrotationally (as in Stokes' theory), but at the surface of the Earth the ethereal velocity is not necessarily the same as that of ponderable matter.

Lorentz's criticisms were among the reasons that pushed Michelson to an improved repetition of the experiment, which was carried out by him and Morley at Cleveland in 1887. A moment's reflection should convince one that the shift of the interference pattern must turn out proportional to the change Δt in the difference of the times the light takes to travel along the two arms in the two settings of the apparatus. One should have

$$\Delta t \simeq \frac{2d}{c}\frac{v^2}{c^2}$$

where d is the length of either of the arms. The experiment was a very delicate one: nothing was to perturb the interference pattern, especially during rotation of the apparatus, which was therefore mounted on a stone slab floating on mercury. Multiple reflection yielded $d \simeq 11$ m. The order of magnitude for the fringe shift, $\Delta\lambda$, may be computed by multiplying Δt by the velocity of light, c. One finds (taking $v^2/c^2 \simeq 10^{-8}$)

$$\Delta\lambda \simeq 2 \times 11 \times 10^{-8}\text{m} = 2200\text{Å}$$

ompared with the wave length of the yellow light used, where $\lambda \simeq 5500$Å. Michelson and Morley therefore expected a shift of the order of four tenths of a fringe, which they would have been able to detect and measure. The enewed null result led Michelson and Morley to the conclusion 'that if here be any relative motion between the Earth and the luminiferous ether, t must be small.' Indeed 'quite small enough to refute Fresnel's explanation of aberration', i.e., we should say, Fresnel's theory altogether. After quoting orentz's criticism of the Stokes theory (see above) and Lorentz's own heory, Michelson and Morley conclude: 'If now it were legitimate to onclude from the present work that the ether is at rest with regard to the arth's surface, according to Lorentz there could not be a velocity potential nd his own theory also fails.'

[Einstein's] unconventionality and his courage are attested not only by his scientific work, but also by his more incidental utterances. Common sense, he said, is merely the layer of prejudices which our early training in science has left in our minds. Following him, other scientists of this period took the view that common sense is no infallible and innate *lumen naturale* but merely the residue which advancing science has left in its wake and which has penetrated popular thinking. They heeded D'Alembert's admonition, 'Allez en avant, la foi vous viendra!'
(H. Margenau, in *Integrative Principles of Modern Thought*)

HE PRE-RELATIVISTIC THEORIES

According to a legend deeply rooted among physicists, the entire scientific ommunity stood astonished in front of that 'unreasonable, apparently nexplicable experimental fact', to quote Millikan's words. Particular credit s given to the story that Michelson and Morley, in view of the consideration hat the earth might have been by chance at rest in the ether at the moment he experiment was carried out, had it repeated after six months, i.e. with he Earth's velocity pointing in the opposite direction along its orbit. ndeed they had planned to repeat their observations at regular intervals hroughout the year, but in fact made no further trial, as 'they soon became nvolved in new lines of research that absorbed all their interests' (R. S. hankland (1964)). This does not seem to indicate, at least on their part, a eeling of dismay. The second part of the story is that, again to quote Millikan, 'for almost twenty years after this fact came to light, physicists vandered in the wilderness in the disheartening effort to make it seem easonable'. This wandering did not occur in full darkness. Indeed, already n 1882, the Irish physicist Fitzgerald, and, independently five months later, orentz had produced an explanation of the result. This was based on the ypothesis that bodies in motion through the ether undergo a contraction of heir dimension aligned with the motion's direction by a factor $\sqrt{1-v^2/c^2}$. This factor would cause a contraction of the length of the arm of the nterferometer aligned with the direction of the earth's motion; it is then a imple exercise to verify that the time difference Δt vanishes.

As remembered by Whittaker, after Lorentz's communication, the ypothesis 'won favour in a gradually widening circle, until eventually it ame to be generally taken as the basis of all theoretical investigations on he motion of ponderable bodies through the aether'.

Lorentz's hypothesis was inserted in the research on ether that he had ackled in another paper of that same year. His research programme was ntended to supplement the scheme of the Maxwell's equations, taken as escribing the ether's behaviour, with a 'granular' point of view on the ature of electricity, i.e. a point of view that attributes electromagnetic phenomena to the action of electric charges in motion. One of the successes of Lorentz's theory was a reinterpretation of Fresnel's results which was achieved on the basis of the supposition that the polarized molecules of the dielectric increase the dielectric constant and 'that it is (so to speak) this augmentation of the dielectric constant which travels with the moving matter' (Whittaker). This removed an objection that had been moved to Fresnel's theory, namely that it required a different relative velocity of ether and matter for light of different colours; Lorentz's theory only required in fact that the dielectric constant be different for light of different colours.

Einstein's total energies were always turned in the direction of unremitting questioning, and the answers were actually won inch by inch over periods of many years as the questions themselves underwent change after change. The precondition for this obstinacy, this struggle, its life-long cost, was an inner isolation and solitude which few men could have endured, but which had become the very air Einstein breathed and which had to be that if his single-minded devotion to scientific thought was to be possible.

(Henry Le Roy Finch, in *Conversations with Einstein*)

A short treatise of Lorentz's of 1895 brings us fully into a pre-relativistic climate. In this work, Lorentz tackles the problem of the influence of the Earth's motion on electric and optical phenomena: the most general approach to the problem consists of transforming the fundamental equations of the theory, which are supposed to be valid in the frame of reference of the ether, to a frame of reference in motion with the Earth. Now, if one requires that experiments carried on in a terrestrial laboratory should not reveal any effect of the motion, the equations of the theory must keep their forms unaltered when going from the frame of reference S of the ether to the terrestrial one S', at least as long as terms of the order of $(v/c)^2$ are neglected. To get this result, Lorentz was compelled to introduce a transformation for the time of the form

$$t' = t - \frac{v}{c^2}x \; \star$$

(here t denotes the time in S, t' in S'; the relative motion is supposed to take place along the x axis). To t' Lorentz gave the name of *Ortzeit* (local time), without attributing to it anything but a purely formal significance. Much later (1927), Lorentz was to give the following testimony about this point: 'A transformation of the time was necessary, so I introduced the conception of local time which is different for different systems of reference which are in motion relative to each other. But I never thought that this had anything to do with real time. This real time for me was still represented by the older classical notion of an absolute time, which is independent of any reference to special frames of coordinates. There existed for me only one true time. I considered my time-transformation only as a heuristic working hypothesis, so the theory of relativity is really solely Einstein's work' (see Rosser, page 67). In the last chapter of the treatise Lorentz discussed the experimental results unexplained by the preceding hypotheses, among them Michelson and Morley's experiment which required the separate hypothesis of the contraction.

As early as 1895 the outstanding French mathematician and mathematical physicist Poincaré had expressed criticism of the method followed in approaching the ether problem; the contraction hypothesis struck him

⋆ One should note that this result is obtained, as a first order approximation, from the relativistic formula. It is then clarified why Lorentz's *Ansatz* was able to account for the negative result of the experiments to first order in v/c.

as 'a fudge factor (*coup de pouce*) provided by nature to avoid the movement of the earth being revealed by optical phenomena' (as quoted by Goldberg). In 1900 the English physicist J. Larmor had formally obtained, anticipating Lorentz, the transformation rules for space and time under which Maxwell's equations remain exactly invariant; this revealed a remarkable connection between the equations of transformation and the Lorentz-Fitzgerald contraction.

Pushed either by Poincaré's criticism or by Larmor's results, or both, in 1904 Lorentz successfully set himself the problem of determining the rules of transformations which left invariant in form 'for any velocity of translation smaller than *c*' the equations of the electron theory when passing from the frame of reference of the Earth to that of the ether, with no need for a separate assumption on contraction. They were the transformations which, with good reason, are known nowadays as the Lorentz transformations.*

Temporally and conceptually we are thus at the threshold of relativity. But if we want to complete the picture of the background from which the Einsteinian edifice stood out, we must briefly analyse the development of another field of studies: the one concerning the foundations of mechanics. The second half of the nineteenth century was characterized by a rediscussion of the fundamental concepts of Newtonian mechanics: from mass to force, from absolute space to absolute time. Kirchhoff (1876) and Hertz (1894) tried to build their mechanics prescinding from the concept of force. Ludwig Lange (1885), in the attempt to find a way out of the paradoxical situation arising from the adherence to the concept of absolute space, on the one hand, and its absence from practical physics on the other hand, pointed out that the essential, or, as we would rather say, *operational* content of the law of inertia was preserved if the idea of absolute space was substituted by the concept of inertial system, a concept that is essential in the theory of special relativity. In his treatise, *The Science of Mechanics* (first editions in 1883, 1888, 1897), the Austrian physicist and philosopher, Ernst Mach, besides giving a fascinating reconstruction of the historical development of mechanics, also carried out a critical examination of it which basically follows the principle of expunging from physics anything that may smell of metaphysics and retaining only that which expresses relations among observable quantities. Thus, on the basis of a critique (reminiscent of Berkeley's and prefiguring general relativity) of Newton's 'rotating pail' experiment, absolute space is refuted as a 'conceptual monstrosity'. As far as absolute time is concerned, Mach stresses that 'it can be measured by comparison with no motion; and it has therefore neither a practical nor a scientific value; and no one is justified in saying that he knows aught about

What I remember most clearly was that when I put down a suggestion that seemed to me cogent and reasonable, he did not in the least contest this, but he only said, 'Oh, how ugly.' As soon as an equation seemed to him to be ugly, he really rather lost interest in it and could not understand why somebody else was willing to spend much time on it. He was quite convinced that beauty was a guiding principle in the search for important results in theoretical physics.

(H. Bondi, in G. J. Whitrow, *Einstein: The Man and His Achievement*)

* Actually the transformations are left 'indeterminate to a certain extent' by the above requirement. A factor, l, differing from 1 by terms of the order of v^2/c^2, 'may appear' in all the 'primed' coordinates. Furthermore, Maxwell's equations are not completely covariant, due to a spurious term left over in the expression for div**D**, as Lorentz later frankly acknowledged.

it. It is an idle metaphysical concept . . .' (quoted in *Relativity Theory: Its Origins and Impact on Modern Thought*, edited by L. Pearce Williams).

With Poincaré, the process of revision of the space–time concepts makes a further step to what has been called a transitional stage to the fully relativistic theory. Between 1895 and 1904, he developed an increasing conviction about the impossibility of detecting the absolute motion of the Earth by any experimental means, either mechanical or electromagnetic. In 1904, in an address delivered at the International Congress of Arts and Sciences in St Louis, he codified what he called the 'principle of relativity', according to which '. . . the laws of physical phenomena must be the same for a "fixed" observer or for an observer who has a uniform motion of translation relative to him: so that we have not, and cannot possibly have, any means of discerning whether we are, or are not, carried along in such a motion' (Whittaker). This paper of Poincaré's was in many respects prophetic as he went so far as to foretell the advent of 'an entirely new mechanics', which would be, above all, characterized by the fact 'that no velocity could surpass that of light', while 'inertia would become infinite when one approached the velocity of light'. Furthermore he was able to interpret Lorentz's 'most ingenious idea' of local time: this is the time marked by watches synchronized by light signals in a way substantially equivalent to what was to be Einstein's procedure. The watches adjusted in this manner do not mark, however, 'the true time' if the frame of reference is moving, and this is so because the velocity of light is not isotropic. However, this will lead to no contradiction, as an observer there has no means of detecting the anistropy and thereby of ascertaining a difference between true and local time. Poincaré took up the argument again in a subsequent paper where he stressed that self-consistency is guaranteed experimentally by the Michelson–Morley result and theoretically by the Fitzgerald–Lorentz contraction. On the basis of the above result, it has been claimed that Poincaré anticipated Einstein and that, in particular, '. . . all that one would have to do to bring the foregoing into accord with Einstein's general definition of time as it appears in his 1905 paper would be to eliminate Poincaré's non-relativistic references to 'fixed' and 'moving' systems which reveal his retention of the ether as a physically meaningful concept' (Scribner, 1964). However, it is apparent from Poincaré's reasoning that 'the velocity of light is not really a universal constant in all inertial frames of reference; it only *appears* (emphasis added) to be a constant in those frames moving with respect to the ether', and that this fact is compensated by the Lorentz–Fitzgerald contraction (Goldberg, 1967). Is it only a matter of formal differences? We think not. To cite one single point: the Lorentz–Fitzgerald contraction is a one-way effect, which has no reciprocal when passing from the ether to a moving system; moreover, it is in principle unobservable (as measuring rules are contracted in the same ratio); Einstein's time dilation is reciprocal and observable. This is not a minor point. There is an all-important difference of method between Poincaré and Einstein.

ANNALEN

DER

PHYSIK.

BEGRÜNDET UND FORTGEFÜHRT DURCH

F. A. C. GREN, L. W. GILBERT, J. C. POGGENDORFF, G. UND E. WIEDEMANN.

VIERTE FOLGE.

BAND 17.

DER GANZEN REIHE 322. BAND.

KURATORIUM:

F. KOHLRAUSCH, M. PLANCK, G. QUINCKE,
W. C. RÖNTGEN, E. WARBURG.

UNTER MITWIRKUNG

DER DEUTSCHEN PHYSIKALISCHEN GESELLSCHAFT

UND INSBESONDERE VON

M. PLANCK

HERAUSGEGEBEN VON

PAUL DRUDE.

MIT FÜNF FIGURENTAFELN.

LEIPZIG, 1905.

VERLAG VON JOHANN AMBROSIUS BARTH.

Figure 15

The title page of *Annalen der Physik*, Volume 17, 1905, in which Einstein's first relativity paper appeared

Poincaré's principles were 'distilled' from experience, or, as it has been said, his approach was that of 'induction to first principles'. His is a theory which explains the existent (Goldberg, 1967). Einstein's method is 'deduction *from* first principles'. His is a theory that predicts the not-yet existent. The theory of relativity was in fact built by Einstein and, later, by other authors such as Lewis and Tolman, by exploiting the heuristic value of its two basic postulates or of the clash between them.

Thus Whittaker's account of Einstein's work as if he had but 'set forth the relativity theory of Poincaré and Lorentz with some amplification' is incorrect not only because of the *formal* fact that the latter was not a theory of relativity, but a theory of the ether, but also because of the *substantial* fact that Einstein's theory had, by its structure, a far greater predictive value, a point which was particularly stressed by Karl Popper (for instance in *The Logic of Scientific Discovery*).

'Whoever finds a thought which enables us to obtain a slightly deeper glimpse into the eternal secrets of nature, has been given great grace. But the man who, in addition to this, experiences the recognition, sympathy and encouragement of the best of his age, has been given almost more happiness than a man can bear.'
(A.E., on receiving the 1925 Gold Medal of the Royal Astronomical Society)

However, even though Poincaré did not invent special relativity, he obtained results which, like the Lorentz transformations themselves, could be taken over into relativity. In a paper of 1906 published in *Rendiconti del Circolo Matematico di Palermo* (a partial commented translation has been carried out by Schwartz), his mastery of all fields of mathematics led him to demonstrate that the Lorentz transformations formed a group,* to anticipate the four-vector calculus of Minkowski and to interpret the Lorentz transformation as a rotation about a fixed origin in four-dimensional space.

THE EINSTEINIAN REVOLUTION

In his 1905 paper, the twenty-six-year-old Einstein gave a decisive turn to the process outlined in the last section. Aware of the experimental situation and of the status of Lorentz's theory up to 1895, but not of the 1904 results (convincing pieces of internal evidence in Einstein's paper that he had not read Lorentz's paper of 1904 have been produced by G. Holton, 1960), he gave an answer completely at variance with that of his predecessors to the problems presented by this field of physics. As everyone knows, he based his construction on the principle of relativity, extended to all physical phenomena and expressed as the equivalence of all frames of reference 'for which the equations of mechanics hold good', i.e. inertial systems. Besides this he stated another postulate, 'which is only apparently irreconcilable with the former, namely, that light is always propagated in empty space with a definite velocity c which is independent of the state of motion of the emitting body'. In the expression 'only apparently irreconcilable' lies the clue to grasp the logic of the entire paper. What Einstein means is that the second postulate can actually be reconciled with the principle of relativity, *taken in its physical content*, which states the equivalence of all inertial systems *independently of specific rules of transformations*. What must be discarded are

* Better stated: he showed that the requirement that they formed a group fixed the parameter l of Lorentz (see footnote on page 79) with no need of a separate assumption.

the rules, i.e. the Galilean transformations, and, with them, the prevailing ideas on space and time that underlie them.

The entire question of the ether, more than two centuries old, was dismissed in a single sentence: 'The introduction of a "luminiferous ether" will prove to be superfluous in as much as the view here to be developed will not require an "absolutely stationary space" provided with special properties. . . .'

How was Einstein led to take this bold attitude? A widespread opinion, linked to the belief that induction is the key to the formation of scientific theories, tends to favour the thesis that Einstein was predominantly or exclusively concerned with an explanation of the null result of Michelson's experiment. This ignores the essential fact that the experiment had already been 'explained', at the time of Einstein's paper, thirteen years earlier, in terms of the Lorentz–Fitzgerald contraction. The 'ether-drift' experiments were known to Einstein, who refers to them as 'the unsuccessful attempts to discover any motion of the earth relatively to the "light medium." ' As documented in particular by Holton (1969), however, Michelson's experiment did not provide a primary motivation of his reflections. Not only does Einstein make no direct reference to Michelson's work, but, questioned about it, he supplied some significant though not unambiguous testimonies. Among them we quote a passage from a letter written in 1954 in answer to a question on whether Michelson had influenced his thinking and perhaps helped him to work out his theory of relativity: 'In my own development Michelson's result has not had a considerable influence. I even do not remember if I knew of it at all when I wrote my first paper on the subject (1905). The explanation is that I was, for general reasons, firmly convinced that there does not exist absolute motion, and my problem was only how this could be reconciled with our knowledge of electrodynamics. One can then understand why in my personal struggle Michelson's experiment played no role or at least no decisive role' (Holton, 1969).

Einstein was, apparently, not a particularly good mathematical calculator and he did not find his results as a consequence of long calculations. (He often claimed that his memory was bad, and prodigious calculational ability is usually combined with an exceptional memory.) He found his results by a phenomenal intuitive instinct as to what the results *should* be.

(Jeremy Bernstein, *Einstein*)

But by what means had Einstein convinced himself that 'there does not exist absolute motion'? It is almost universally acknowledged that, as we shall discuss later on concerning this as well as other similar problems, what was decisive was Einstein's reading of Mach's treatise while a student; by direct testimony of Einstein himself (see *Autobiographical Notes*), 'it was Mach who, in his *The Science of Mechanics*, shook' his 'dogmatic faith' 'in mechanics as the final basis of all physical thinking', so that Einstein had for a long time shared in particular the conviction that absolute space was a meaningless concept, though he could not have found in Mach, as previously discussed, a clearcut indication of privilege of the inertial system. Concerning the possibility, indicated by Einstein, of reconciling this idea with Maxwell's electrodynamics 'as usually understood at the present time', as he quotes in his paper, it must be recalled that in Maxwell's theory, *expressed in pre-relativistic language*, a different description is given, to take Einstein's example, of the situation in which a magnet is moving with respect to an electric circuit or the other way round. If the only thing that matters, as confirmed on the other hand by experiment, is *relative* motion, then electrodynamics must be reformulated so as to make the description depend only on relative motion. This is, by the way, one of the achievements of Einstein's paper, and possibly the one that concerned him most.

In his autobiographical notes, Einstein tells us, 'By and by I despaired of the possibility of discovering the true laws by means of constructive efforts based on known facts. The longer and the more despairingly I tried, the more I came to the conviction that only the discovery of a universal formal principle could lead us to assured results. The example I saw before me was thermodynamics. The general principle was there given in the theorem: the laws of nature are such that it is impossible to construct a *perpetuum mobile* (of the first and second kind). How, then, could such a universal principle be found? After ten years of reflection such a principle resulted from a paradox upon which I had already hit at the age of sixteen: If I pursue a beam of light with the velocity c (velocity of light in a vacuum), I should observe such a beam of light as a spatially oscillatory electromagnetic field at rest. However, there seems to be no such thing, whether on the basis of experience or according to Maxwell's equations. From the very beginning it appeared to me intuitively clear that, judged from the standpoint of such an observer, everything would have to happen according to the same laws as for an observer who, relative to the Earth, was at rest. For how, otherwise, should the first observer know, that is, be able to determine, that he is in a state of fast uniform motion?' We deliberately restrain ourselves from touching upon the point of logical internal consistency in this passage (Grunbaum); we rather call attention to the 'intuitive' conclusion it leads to: as a spatially oscillatory field at rest 'does not make sense', no observer, i.e. no material body, can reach the velocity c, which is therefore a limiting velocity: as such it must be the same for all inertial observers. One can thus agree with Einstein on the fact that 'in this paradox, the germ of the special relativity theory is already contained'.

'Men like Einstein or Niels Bohr grope their way in the dark toward their conceptions of general relativity or atomic structure by another type of experience and imagination than those of the mathematician, although no doubt mathematics is an essential ingredient.'
(H. Weyl, quoted in Stanley L. Jaki, *The Relevance of Physics*)

Having thus reviewed the background of the two postulates, let us go back to the question of their apparent irreconcilability, clearly exhibited in the paradoxical situation described above. 'Today everyone knows, of course,'—continues Einstein—'that all attempts to clarify this paradox satisfactorily were condemned to failure as long as the axiom of the absolute character of time, viz. of simultaneity, unrecognizedly was anchored in the unconscious. Clearly to recognize this axiom and its arbitrary character really implies already the solution of the problem. The type of critical reasoning which was required for the discovery of this central point was decisively furthered, in my case, especially by the reading of David Hume's and Ernst Mach's philosophical writings.' Concerning this point two comments are in order. In the first place one must stress the role that is played in the creative mental processes of science by the stage where the mere existence of a problem is recognized; in this case it was of foremost importance to realize that one *could set oneself the question of the nature of time*. Only after this stage did it become possible to acknowledge the arbitrariness of the concept of absolute time. Once the nature of the problem was identified, an analysis of the space–time concepts permitting a check of where the prejudice had crept in became necessary. It is at this stage that an operational analysis of the concepts becomes essential. As stressed by Bridgman in *The Logic of Modern Physics*, this analysis is not very revolutionary in itself. The point is that, before Einstein, no one had made a constructive use of it. This leads us to our second comment. As acknowledged by Einstein himself, he owes something to Mach concerning this point too. At the same time, we see his unique capacity to make the next step, the most important one from the strictly scientific point of view. Therefore, a comment by Philipp Frank, a physicist and biographer of Einstein's, seems particularly pertinent. According to him, the criticism of the mechanistic philosophy had 'ploughed the ground where Einstein could cast his seed'.

After discussing the relativity of lengths and times, in Section 3 of his paper Einstein proceeds to derive the equations of transformation between two frames of reference in a state of relative uniform rectilinear motion, under the hypothesis of the validity of the two postulates of the theory; he thus obtained independently the same formal conclusion as Lorentz. We need only recall the different basis (and method) upon which this derivation is achieved and the overall different meaning they have in the two theories: for Lorentz they are equations of transformations which make the equations of the electron theory covariant; for Einstein, expressions of the general properties of space and time.

Many people probably felt relieved by being told that the true nature of the physical world could not be understood except by Einstein and a few other geniuses. Paradoxically enough, Einstein may have been hailed by the general public not because he was a great thinker, but because he saved everybody from the duty of having to think.

(Hannes Alfvèn, 'Cosmology: Myth or Science?', in *Cosmology, History and Theology*)

In Einstein's 1905 paper, two short kinematical sections follow. In the first one, a 'peculiar consequence' of the Lorentz transformations is discussed, namely the observable time lag introduced by motion between two synchronized clocks, i.e. what came later to be known as the 'twin paradox'. In the second one the addition theorem for velocities is derived; while he might explain with great ease the results of Fizeau's experiment he omits however to do so. In the following two sections he deals with the relativistic covariance of Maxwell's equations and the relative nature of electric and magnetic forces, and with the Doppler effect and aberration. Here again he does not point out that Fresnel's theory on the observation of aberration may be immediately re-interpreted; neither does he explicitly stress that a transverse Doppler effect arises. In Section 8 he obtains the transformation of light energy; concerning the result obtained on this point he makes what has been called (by A. I. Miller) 'one of the great understatements in the history of science': 'It is remarkable that the energy and the frequency of a light complex vary with the state of motion of the observer in accordance with the same law', a result which is clearly related to his earlier work on light quanta. In the same section he derives an equation for the pressure of radiation exerted on a reflector, 'in agreement with experiment and with other theories'; he does not state that he has results concerning the long discussed problem of reflection by a moving mirror, but just concludes the section by pointing out that 'all problems in the optics of moving bodies will be reduced to a series of problems in the optics of stationary bodies'. The final section is devoted to the dynamics of the electron. Here an explicit prediction is made as to the trajectory of an electron in a uniform magnetic field. The section initiates relativistic dynamics.

There has been much discussion concerning the apparent carelessness of the paper on the points discussed above. Whether it be due to 'lack of serious concern with the messy details of experimental physics', instrinsic need of elegance, or sheer intellectual arrogance (G. Holton, 1969), we cannot tell; we simply remark that it was to take years for the scientific community to work out the detailed consequences, both experimental and theoretical, of this densely packed paper.

In June 1933 he delivered the Herbert Spencer lecture at Oxford in which he attempted to analyse what he called 'the Method of Theoretical Physics.' The title is somewhat misleading because what he really describes is *his* method of doing theoretical physics, which was, by this time, almost completely distinct from that of any of his contemporaries. In fact in some bizarre sense his 'method' has more in common with the philosophical attitudes of Plato, with the Platonic emphasis on perfect shapes and forms, than with any physicist one can think of since and including Newton.
(Jeremy Bernstein, *Einstein*)

RELATIVITY'S PROGRESS

Contrary to a widespread opinion, the full body of relativity was not shaped at once with Einstein's first paper. To begin with, Einstein himself made in that same 1905 another contribution of paramount importance. Basing himself on his previous results, he came to the conclusion, in a subsequent paper, that 'if a body emits the energy *E in the form of radiation*, its mass decreases by E/c^2'. Emphasis is added to stress that the statement had not yet attained its full generality. The full meaning of $E=mc^2$ was discussed in a comprehensive paper that Einstein published in 1907 in *Jahrbuch der Radioaktivität und Elektronik*. In the meanwhile some notice had been taken of the existence of Einstein's theory. Planck, one of its early defenders and propagandists, had shown (1907) that the equations of motion of special relativity could be derived from the principle of least action in terms of a Lagrangian $L=-mc^2(1-v^2/c^2)^{1/2}$; he had thus removed an ambiguity left over by Einstein concerning the most suitable definition of force. In the same year von Laue had applied relativistic kinematics to a derivation of Fresnel's drag coefficient and an interpretation of Fizeau's experiment. Fresnel's theory had been thus successfully re-interpreted for the second time. Einstein's paper aimed at presenting a systematic survey of the entire field. But he went further. Apart from the already quoted result on $E=mc^2$, he posed the physical basis of general relativity, that he was to develop during the following nine years.

Both theory and experiment made notable steps in 1909. The American physicists Lewis and Tolman applied consistently the heuristic power of the principle of relativity foreshadowed by Einstein to give a solid physical basis to relativistic dynamics. The method makes use of the condition that the laws of physics be invariant in form when going from one inertial system to another one by using Lorentz transformations: it is then clear that the laws of classical mechanics must be modified. Lewis and Tolman showed that the validity of a covariant principle of the conservation of momentum for an isolated system could be obtained only if the momentum p had the expression $p=mu$ in terms of the velocity u and the 'relativistic mass' m:

$$m=\frac{m_0}{\sqrt{1-u^2/c^2}}$$

(m_0=rest mass). In the same year the mathematician Minkowski obtained the elegant four-dimensional formulation of the theory which is nowadays followed in formal treatments. Also in 1909, the German physicist Bucherer confirmed (contradicting an earlier result by Kaufmann) the dynamical behaviour of electrons as foreseen by the theory.

Concerning relativity's progress during its early years, some general comments are in order. In the first place it should be noted that experiment did not, and could not, discriminate between relativity and Lorentz's theory: indeed the latter predicted also an increase of the mass according to the above formula (the fact that, for Lorentz, u had to be interpreted as velocity with respect to the ether had no practical relevance), and in fact Kaufmann had stated that his measurements were 'not compatible with the Lorentz–Einstein fundamental assumption'.

Nevertheless, even in the absence of an unambiguous experimental confirmation, the theory of relativity gained new adherents by degrees, after a period in which it had been ignored by the majority of physicists. Like old soldiers, Lorentz's theory never died, it only faded away; as von Laue put it in his text on relativity (1911), the first to be written about the subject: '. . . a really experimental decision between the theory of Lorentz and the theory of relativity is indeed not to be gained; and that the former, in spite of this, has receded into the background, is chiefly due to the fact that, close as it comes to the theory of relativity, it still lacks the great simple universal principle, the possession of which lends the theory of relativity from the start an imposing appearance.' Thus relativity's success seems to correspond more closely to a gestaltic re-orientation, in the sense of Kuhn, than to a fair competition between research programmes as in Lakatos' theory (1970) of the growth of knowledge.

A turning point in the process was marked by the appearance of Minkowski's paper, which won favour for the theory by the mere strength of its formal elegance. The final success of relativity principles both within and outside the scientific community was sealed by the experimental confirmation of one of the effects foreseen by the theory of *general* relativity: the deviation of light rays by a strong gravitational field.

Experimental confirmations of relativistic dynamics began to pile up quickly after the years 1924–5, when Bothe and Geiger, and Compton and Simon, verified, through the Compton effect, both Einstein's hypothesis of light quanta and the relativistic laws of the conservation of energy and momentum. As far as the $E=mc^2$ is concerned, it had already been applied in 1913 by the French physicist, Langevin, to nuclear physics to explain the deviations of the atomic weights from integral values. However, the first experimental confirmation of the $\Delta E = c^2 \Delta m$ was obtained by Cockcroft and Walton in 1932 for a nuclear reaction. The classical check of the $E = mc^2$ in the creation of electron–positron pairs by gamma rays and the annihilation of such pairs into photons was obtained by Blackett and Occhialini the next year. Finally, it was only in 1938 that the German physicist Hahn, and others, discovered in nuclear fission the process in which the mass–energy equivalence could be exploited for practical purposes. Two further emblematic dates concerning $E = mc^2$ are Fermi's 'pile' (1942) and Hiroshima (1945).

Experimental checks of the kinematic effects were very late. It was only in 1938 that Ives and Stilwell were able to verify experimentally the relativistic prediction on the transverse Doppler effect by using canal rays as a moving source. This was an indirect confirmation of the time dilation. The first direct confirmation of the effect dates from 1941 and was due to Rossi and Hall with an experiment on the lifetime of muons.

Bibliography

Born, Max, *Einstein's Theory of Relativity* (London: Methuen, 1924; New York: Dover, 1962).

Bridgman, P. W., *The Logic of Modern Physics* (New York: Macmillan, 1927).

Einstein, A., 'Autobiographical Notes', in *Albert Einstein: Philosopher-Scientist* (Evanston, Illinois: The Library of Living Philosophers, 1949).

Galileo, G., *Dialogue Concerning the Two Chief World Systems* (1632). Translated by Stillman Drake (Berkeley: University of California Press, 1953).

Goldberg, S., 'Henri Poincaré and Einstein's Theory of Relativity', *Am. J. Phys.*, 1967, **35**, 934.

Grunbaum, A., 'The Special Theory of Relativity', in *An Introduction to the Theory of Relativity*, edited by W. G. V. Rosser (London: Butterworths, 1964).

Holton, G., *Am. J. Phys.*, 1960, **28**, 627.

—— *Am. J. Phys.*, 1969, **37**, 968.

—— *Isis*, 1969, **60**, 133–97.

Kuhn, T. S., *The Structure of Scientific Revolutions*, 2nd edition (University of Chicago Press, 1970).

Latakos, I., 'Criticism and the Growth of Knowledge', in *Criticism and the Methodology of Scientific Research Programmes* (Cambridge: Cambridge University Press, 1970).

Mach, E., *The Science of Mechanics*, 9th edition, 1933 (La Salle, Illinois: Open Court Publishing Company, 1960).

Maxwell, J. C., *Collected Scientific Papers* (New York: Dover, 1952).

Miller, A. I., *Am. J. Phys.*, 1976, **44**, 912.

Millikan, R. A., 'Albert Einstein on his Seventieth Birthday', *Revs. Mod. Phys.*, 1949, **21**, 343.

Pearce Williams, L. (Editor), *Relativity Theory: Its Origins and Impact on Modern Thought* (New York: John Wiley, 1968).

Popper, Karl, *The Logic of Scientific Discovery* (New York: Harper and Row, 1965).
Rosser, W. G. V., *An Introduction to the Theory of Relativity* (London: Butterworths, 1964).
Schwartz, H. M., *Am. J. Phys.*, 1971, **39**, 1287.
Scribner, G. Jr., *Am. J. Phys.*, 1964, **32**, 672.
Shankland, R. S., 'The Michelson–Morley Experiment', *Am. J. Phys.*, 1964, **32**, 16.
Whittaker, E. T., *A History of the Theories of Aether and Electricity* (2 volumes) (London: Thomas Nelson, 1951 and 1953; New York: Harper Torchbooks, 1960).

3 The story of general relativity

A. P. French

1. THE DEVELOPMENT OF THE THEORY

This one man changed human thinking about the world as only Newton and Darwin had changed it.

(*New York Times*)

It has been remarked that, in the years just before 1905, the results and ideas necessary for the development of special relativity theory had attained a wide currency—special relativity was 'in the air', so to speak, and if Einstein had not crystallized it, someone else would have done so before very long. Whether or not this is true (see the article by Silvio Bergia on page 65), it is certain that Einstein, in creating the general theory of relativity, took a further step that was uniquely his, and which might not have been made for decades if he had not been there to show the way.

It is a remarkable thing that Einstein was far from being content with his triumph of reconciling mechanics and electromagnetism, through special relativity, via his assertion that all inertial frames are equivalent with respect to all physical processes. Instead, he almost immediately tried to broaden the scope of his synthesis, and to bring within it the mystery of gravitation.

In 1905, only a few months after the paper that created special relativity, Einstein published in *Annalen der Physik* another paper with the title 'Does the Inertia of a Body Depend on Its Energy Content?' It was in this paper that he proposed the relation $E=mc^2$ as a general connection between energy and inertial mass. In his *Autobiographical Notes* (1946), he looks back at his thinking of forty years earlier, and remarks: 'That the special theory of relativity is only the first step of a necessary development became completely clear to me only in my efforts to represent gravitation in the framework of this theory.' He proceeds to explain how he recognized that a satisfactory theory had to encompass the following results:

(a) The special theory clearly requires that the inertial mass of a body depends on its total energy (and therefore, for example, increases as the kinetic energy increases).

(b) Very accurate experiments (particularly the delicate torsion-balance experiments of Eötvös) show (as Newton had shown to about 1 part in 1000 through pendulum experiments) that the gravitational mass of a body is exactly proportional to its inertial mass.

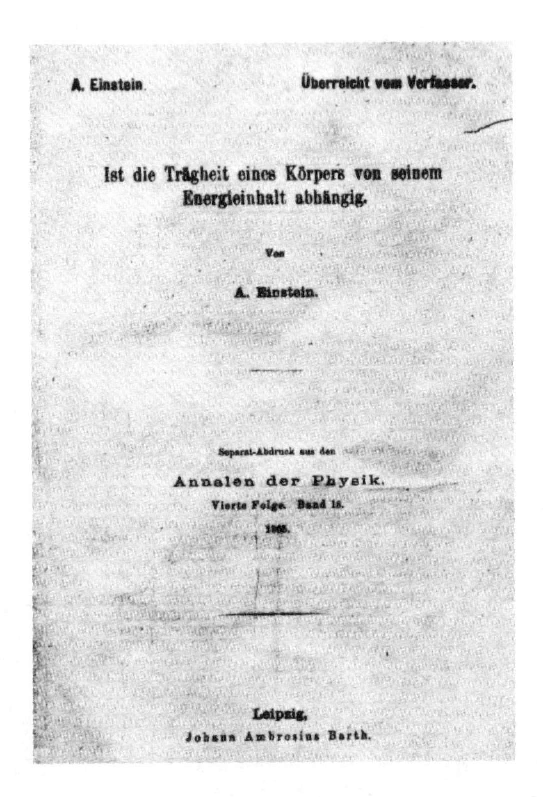

A. Einstein. Überreicht vom Verfasser.

Ist die Trägheit eines Körpers von seinem Energieinhalt abhängig.

Von

A. Einstein.

Separat-Abdruck aus den

Annalen der Physik.

Vierte Folge. Band 18.

1905.

Leipzig,

Johann Ambrosius Barth.

Figure 16

The cover of a reprint of the article which first introduced the equation $E=mc^2$

By putting together these two results, it follows that the weight of a body depends in a precisely determined way upon its total energy. Einstein says, 'If the theory did not accomplish this or could not do it naturally, it was to be rejected.' No such connection was forthcoming from the special theory of relativity—but then, Einstein recalls, the crucial idea came to him:

> The fact of the equality of inertial and gravitational mass, i.e., the fact that the gravitational acceleration is independent of the nature of the falling substance, may be expressed as follows: In a gravitational field (of small spatial extent) things behave as they do in a space free of gravitation, if one introduces, in place of an 'inertial system', a reference system that is accelerated relative to an inertial system.

Thus was born the famous Principle of Equivalence, according to which a gravitational field of force is precisely equivalent, over a limited region, to an artificial field of force associated with a general acceleration of the reference frame. Einstein published this principle in a lengthy paper in *Jahrbuch der Radioaktivität* in 1907.

In the process of developing his ideas, Einstein in 1911 published a paper on the effect of gravity upon the propagation of light. Its main result was a prediction that light rays passing close by the Sun's surface would be changed in direction by about 0.85″ of arc. This result (only half as large as the result from Einstein's later, fully developed theory of gravitation) corresponds precisely to what one would infer from a simple Newtonian analysis, treating light as being made up of particles of mass m travelling at the speed c. The theoretical deflection is easily calculated on this model to be equal to $2GM/c^2R$, where M and R are the mass and radius of the Sun, and G is the constant of universal gravitation. (Such a calculation had in fact been carried out more than a century earlier, by J. Soldner in 1801.) However, Einstein seems to have been aware that more was needed to give proper expression to his vision of a fully relativistic theory. In his *Autobiographical Notes* he comments on the fact that it took so long for him to proceed from his first ideas to the final form of the theory (achieved during the period 1914–16): 'Why were another seven years required for the construction of the general theory of relativity? The main reason lies in the fact that it is not so easy to free oneself from the idea that coordinates must have an immediate metrical meaning.' Underlying this remark was Einstein's realization that space is not merely the stage on which material objects move and interact, but that the fundamental geometry of space is contingent on the presence and distribution of matter, as expressed in the statement: 'Gravity is due to a change in the curvature of space-time, produced by the presence of matter' (Whittaker, 1953).

Einstein showed (see also the following article by Hermann Bondi) that the minimum formal structure needed to give quantitative expression to these ideas was in terms of ten parameters (gravitational potentials) which determine the *metric** of space–time and define the minimal paths (geodesics) along which objects will tend to move.

Einstein's development of the theory took him deeply into tensor analysis, and he was greatly aided in this research by his friend and former fellow-student at Zürich, the geometer Marcel Grossmann. This collaboration took place during the period 1912–13, and resulted in a joint paper entitled, 'Outline of a Generalised Theory of Relativity and a Theory of Gravitation'. In 1914, however, Einstein left Zürich to take up a professorship in Berlin, and the partnership ended. In a letter to a friend at this

* The metric for a 'space' of any given number of dimensions is simply the explicit mathematical connection between the length of an elementary displacement in the space and the component displacements in terms of which it can be analysed. The characterization of a type of space through its metric must, of course, be independent of any particular choice of coordinate system.

time he wrote: 'The Germans are betting on me as a prize hen; I am myself not sure whether I am going to lay another egg.' He need not have doubted, for during 1915 he began extracting the observable consequences of the theory, and presented his findings at three consecutive sessions of the Prussian Academy of Sciences in November 1915. In 1916 he finally published, in *Annalen der Physik*, a full account of the general theory. The style was similar to that of his 1905 paper introducing the special theory—a thorough and self-contained development of another epoch-making intellectual achievement.

2. THE THREE CLASSIC TESTS OF GENERAL RELATIVITY

'I am sending you some of my papers. You will see that once more I have toppled my house of cards and built another; at least the middle structure is new. The explanation of the shift in Mercury's perihelion, which is empirically confirmed beyond a doubt, causes me great joy, but no less the fact that the general covariance of the law of gravitation has after all been carried to a successful conclusion.
(A.E. to Wladyslaw Natanson, 15 December 1915)

At the end of his great paper, Einstein described three observational implications of the new theory; these were the so-called 'precession of the perihelion' of planetary orbits, the slowing down of clock rates in a gravitational field, and the deflection of light by a massive body. Although the reasoning involved in the development of the theory is subtle and complex, its final expression in these observable phenomena is relatively simple. Each of these effects was found to conform to the theory; let us consider them in some detail.

(a) *Precession of the perihelion of Mercury*

The greatest triumph of Newton's theory of gravitation was to explain in detail the elliptic paths of the planets around the Sun. One of the most remarkable features of his theory (insufficiently emphasized in most discussions) is that, under a pure inverse square force from a central body, the elliptic path of an orbiting object *closes upon itself* and retraces itself indefinitely; in other words, the orbit is a closed curve, fixed in space. If the orbit is elliptic rather than circular, the major axis of the ellipse points always in the same direction in the frame of reference defined by the fixed stars.

This simple result does not quite hold for the planets because, in addition to the main force provided by the Sun, their mutual gravitational interactions disturb the orbits slightly in a way that can be calculated, again on the basis of Newton's law of gravitation. It was, of course, the analysis of such 'perturbations' in the case of Uranus that led to the discovery of Neptune in 1846 and provided a further brilliant vindication of Newton's gravitational theory.

A more general consequence of the interplanetary interactions is that the Keplerian ellipses do not in fact stay fixed in space; they rotate very slowly in the plane of the solar system (i.e. to all intents and purposes in their own plane). This rotation is described in terms of what is called the 'precession of the perihelion', i.e. a progressive change in direction of the line from the Sun to the point, at one end of the major axis, that represents the planet's point of closest approach to the Sun (Diagram 3.1).

According to Newtonian theory, if the effect of the interplanetary perturbations could be subtracted away we should be left once again with stationary, closed orbits. However, the marvellously precise results of astronomical observations showed that this was not so. In the most striking case, that of the planet Mercury, the calculated precession of the perihelion is an angle of about 8.85 minutes of arc per century relative to the fixed stars; the observed amount is about 9.55′, leaving an unaccounted discrepancy of about 0.7′ or 42″ of arc per 100 years.

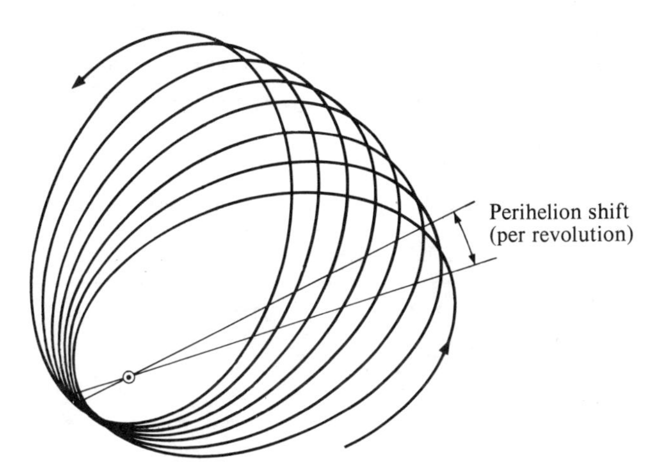

Diagram 3.1
Precession of the perihelion of an orbit (greatly exaggerated).

Such a residual precession of the orbit would come about if the gravitational force due to the Sun were not *exactly* an inverse-square force. Now Einstein's theory, as applied to the Kepler problem, contained just this feature. The curvature of space–time in the vicinity of a gravitating mass expresses itself, to a very good approximation, as a small additional term in the force law, corresponding to an extra attractive force varying as $1/r^4$ (i.e. a $1/r^3$ correction to the gravitational potential). The formulation of the Einstein theory actually involves modifications of the scales of both time and radial distance in a gravitational field, but the net result can be represented by a modification of the force law in ordinary (Euclidean, non-relativistic) space–time. The actual predicted amount of precession per revolution is given by the formula

$$\Delta\phi = \frac{6\pi GM}{c^2(1-\epsilon^2)a} \qquad (1)$$

where M is the Sun's mass, and a and ϵ are the semi-major axis and the eccentricity of the planetary orbit respectively.

By virtue of its small orbit and large eccentricity ($\epsilon = 0.206$) Mercury would be expected to show by far the largest relativistic precession of any of the planets. Substituting the appropriate numerical values into equation

(1) (which, it may be noted, contains no adjustable parameters) gives a calculated precessional rate agreeing almost perfectly with the observed value. When Einstein discovered this agreement, towards the end of 1915, he was understandably elated, and in a letter to Arnold Sommerfeld he wrote: 'This last month I have lived through the most exciting and the most exacting period of my life; and it would be true to say that it has also been the most fruitful. . . . The wonderful thing that happened was that not only did Newton's theory result from it [General Relativity] *as a first approximation*, but also the perihelion motion of Mercury, *as a second approximation*.' Table 3.1 shows the observed and calculated precessional rates (per century), not only for Mercury, but also for Venus, Earth, and the asteroid Icarus.

Table 3.1 Perihelion precession rates (measured in arc seconds per century)

Body	*Observed rate*	*General relativity rate*
Mercury	43.11 ± 0.45	43.03
Venus	8.4 ± 4.8	8.6
Earth	5.0 ± 1.2	3.8
Icarus	9.8 ± 0.8	10.3

It has been suggested that the orbital precession might be tested using artificial satellites of the Earth, but uncertainties caused by the non-sphericity and uneven mass distribution of the Earth would make the observations difficult to interpret.

(b) The gravitational red-shift

According to the general theory, the rate of a clock is slowed down when it is in the vicinity of a large gravitating mass—i.e. if it is in a region of negative gravitational potential. Since the characteristic frequencies of atomic transitions are, in effect, clocks, one has the result that the frequency of such a transition occurring, say, on the surface of the Sun should be lowered by comparison with a similar transition observed in a terrestrial laboratory; it manifests itself as a gravitational red shift in the wavelengths of spectral lines.

The phenomenon can be regarded as a direct application of the Equivalence Principle. Suppose, to take a particularly simple case, that light from a certain atomic transition is emitted at point A in a uniform gravitational field, g, and detected at point B, at a distance h higher up (Diagram 3.2a). According to the equivalence principle, one could replace the gravitational field by a general acceleration of both A and B in the upward direction (Diagram 3.2b). However, this would mean that, in the time t ($=h/c$) which the light takes to travel from A to B, B acquires an upward velocity v of magnitude gh/c, and the frequency of the received light would be Doppler-shifted downward, by a fractional amount equal to v/c or gh/c^2. This is equal to the change $\Delta\phi$ of gravitational potential (gravitational energy per unit mass) between A and B, divided by c^2.

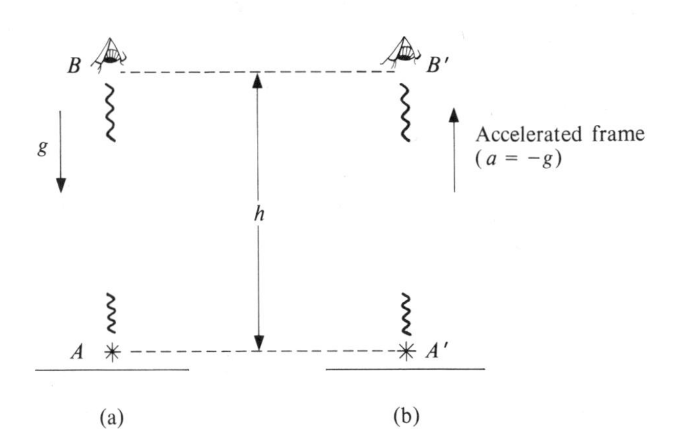

Diagram 3.2
(a) Red-shift of light travelling upward from A to B in a gravitational field.
(b) Equivalent process in an upwardly accelerated frame.

More generally, the change of gravitational potential from the surface of a spherical gravitating mass to an infinitely distant point is given by

$$\Delta\phi=\int_R^\infty \frac{GM}{r^2}\,\mathrm{d}r=\frac{GM}{R} \qquad (2)$$

The factor $f(R)$ by which a clock rate is different at $r=R$ and $r=\infty$ is then given by

$$f(R)=1-\frac{GM}{c^2R} \qquad (3)$$

Although Einstein suggested in his 1916 paper that this result might be tested by observations on spectral lines reaching us from the surface of large stars, it proved extremely hard to verify the phenomenon in the presence of other disturbing effects, such as Doppler effect due to convective motion of the radiating atoms in the stellar atmosphere. However, various measurements made during the past 25 years have given good evidence for the gravitational red-shift for the Sun and other stars—notably by Snider (1971) who verified the correctness of equation (2), as applied to the Sun, with a stated accuracy of about 6 per cent.

What is perhaps even more impressive is the detection of the far smaller red-shift due to the gravitational field of the Earth. Indeed, by far the most accurate measurement of this consequence of general relativity was made by observing the minute shift resulting from a vertical displacement of only about 22.5 metres at the surface of the Earth. The experiment was made possible through the Mössbauer effect (recoilless emission of low-energy gamma rays) in certain crystalline structures, which results in extremely narrow gamma ray lines from certain emitters. Using this method, Pound and Snider in 1964 (refining an earlier experiment by Pound and Rebka, 1960) verified the fractional red-shift gh/c^2 (equal in their experiment to only 2.45×10^{-15}) with an accuracy of 1 per cent.

Still more recently (1971) Haefele and Keating made a comparison of actual clock rates at different altitudes, using caesium atomic clocks. A reference clock was kept on the ground, and other clocks were carried around the world at an altitude of about 10 000 metres on commercial jet aircraft. The change of clock rate involves not only the effect due to the gravitational potential, but also the kinematic time dilation of special relativity, due to the velocity relative to an inertial frame. Since this velocity is the combination of the aircraft's ground speed with the rotational motion of the Earth, the kinematic effect is different for east-west and west-east flights. Table 3.2 shows the results obtained, and the comparison of the observed and theoretical gravitational components of the over-all time dilation.

Table 3.2 Time dilation with atomic clocks

	Time gain, W to E (s)	*Time gain, E to W (s)*
Observed difference	$(-59 \pm 10) \times 10^{-9}$	$(273 \pm 7) \times 10^{-9}$
Kinematic correction	$(-184 \pm 18) \times 10^{-9}$	$(96 \pm 10) \times 10^{-9}$
Remainder	$(125 \pm 21) \times 10^{-9}$	$(177 \pm 12) \times 10^{-9}$
Gravitational effect (theory)	$(144 \pm 14) \times 10^{-9}$	$(179 \pm 18) \times 10^{-9}$

(*c*) *The bending of starlight by the sun*

It was this third prediction of general relativity that provided the most famous and dramatic test of the theory. Although the effect itself was so very small, and had no practical implications, the observation of it seized hold of the public imagination, and cemented Einstein's reputation as the magician who had probed and mastered the deepest mysteries of the universe—which indeed he had.

Before recounting the story, let us recall its theoretical basis. In special relativity, a space-time interval between two events is described, in polar coordinates (r,θ) by the expression:

$$ds^2 = c^2 dt^2 - dr^2 - r^2 d\theta^2 \tag{4}$$

What Einstein showed was that, in the vicinity of a gravitating mass M, this relation was slightly modified, and became:

$$ds^2 = \gamma(r) c^2 dt^2 - \frac{1}{\gamma(r)} dr^2 - r^2 d\theta^2 \tag{5}$$

where

$$\gamma(r) = 1 - \frac{2GM}{c^2 r} \tag{6}$$

Here r is the distance from the centre of the (spherical) mass M.

The gravitational deflection of light can be regarded as a process of refraction. In the vicinity of a mass M, the speed of light is reduced (see equation (5)), to $c\gamma(r)$, where $\gamma(r)$ is given by equation (6). The result is a bending of the wave-fronts towards the mass, just as sound waves are bent towards the Earth's surface if the air temperature increases with height. By integrating this refractive effect over the complete trajectory of a light ray passing by a massive object, one finds a net change of direction given by

$$\alpha = \frac{4GM}{c^2 r_0} \tag{7}$$

where r_0 is the shortest distance between the light path and the centre of M (see Diagram 3.3). This value, as noted earlier, is just twice what would be predicted by Newtonian theory, and for light rays passing as close as possible to the Sun the deflection is 1.75 seconds of arc.

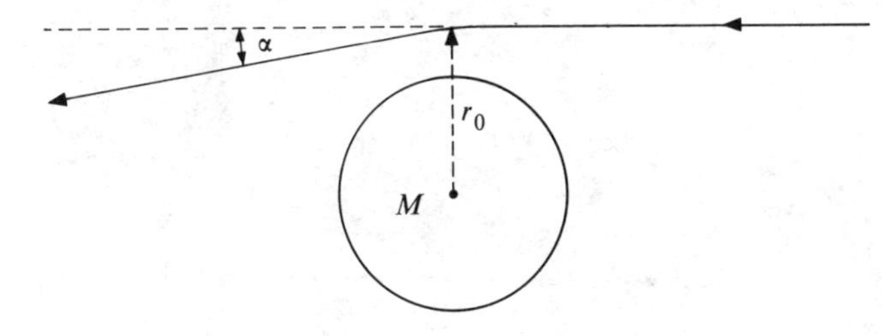

Diagram 3.3
Deflection of a ray of light by a star (greatly exaggerated).

When Einstein, then in Berlin, first arrived at this result in 1915, England and Germany were at war. Nevertheless, the proposal and the initial planning for an astronomical expedition to test the theory occurred in England, after copies of Einstein's paper had been smuggled to Arthur Eddington via the Dutch astronomer Willem de Sitter. Paradoxically, the state of war actually helped to promote this development, according to a fascinating account by S. Chandrasekhar (1975). Eddington, as a Quaker, was a conscientious objector, and the authorities approved the project to avoid the embarrassment of putting such a distinguished scientist into an internment camp.

It was Sir Frank Dyson, then Astronomer Royal, who in 1917 drew attention to the fact that 29 May 1919 was going to be an exceptionally propitious date to test the theory. A total solar eclipse was necessary in order to make observations on stars whose light passed close to the Sun on the

Figure 17

Einstein with Eddington at Cambridge in 1930, taken by Eddington's sister

way to the Earth, and the 1919 eclipse was to occur when the Sun was in a region of the sky (the Hyades) rich in bright stars that would be visible against the solar corona.

The path of the total eclipse ran across South America and Africa within a few degrees of the Equator, and it was decided to set up two observation stations, one at Principe Island in the Gulf of Guinea and the other at Sobral in Brazil (as shown on the map). Feverish work to prepare the necessary instruments began as soon as the war ended.

On the day of the eclipse the weather at Sobral was perfect; that at Principe (where Eddington went) was overcast, but cleared slightly at the

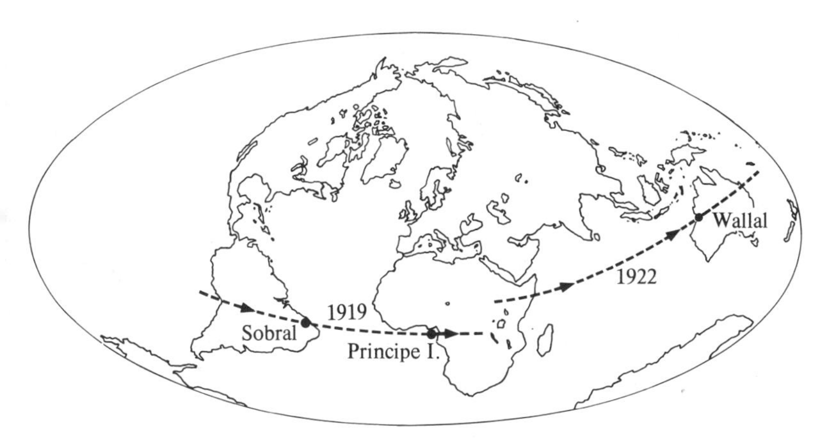

crucial moment. Both expeditions were able to obtain sets of photographs showing the apparent positions of the stars as modified by the gravitational deflection. These had to be compared with the positions observed when the Sun was at a quite different part of the sky. Such comparison photographs were already available, but the full analysis of the data took several months. When it was completed, the results gave strong support to Einstein's

Figure 18

Eclipse instruments used by Eddington at Sobral

theory. As against his predicted deflection of 1.75″ for $r_0 = R$ (R = Sun's radius) the observations yielded the following values (when reduced to $r_0 = R$):

Sobral	$1.98 \pm 0.12''$
Principe	$1.61 \pm 0.30''$

(See also Diagram 3.4.)

The definite news of the success reached Einstein at the end of September 1919, in a telegram from his Dutch physicist friend H. A. Lorentz.

Only three years later another quite favourable eclipse occurred, with an eclipse path crossing Australia (see map). Observations (see Diagram 3.5) were made at Wallal, Western Australia by Campbell and Trumpler, and gave a mean deflection (for $r_0 = R$) of $1.72 \pm 0.11''$. Since that time a number of other eclipse observations have been made, but without significantly improving on the results of the first two expeditions.

Figure 19

A postcard from Albert Einstein to his mother, dated 27 September 1919. The first sentences read 'Good news today. H. A. Lorentz has telegraphed me that the British expeditions have definitely confirmed the deflection of light by the Sun'

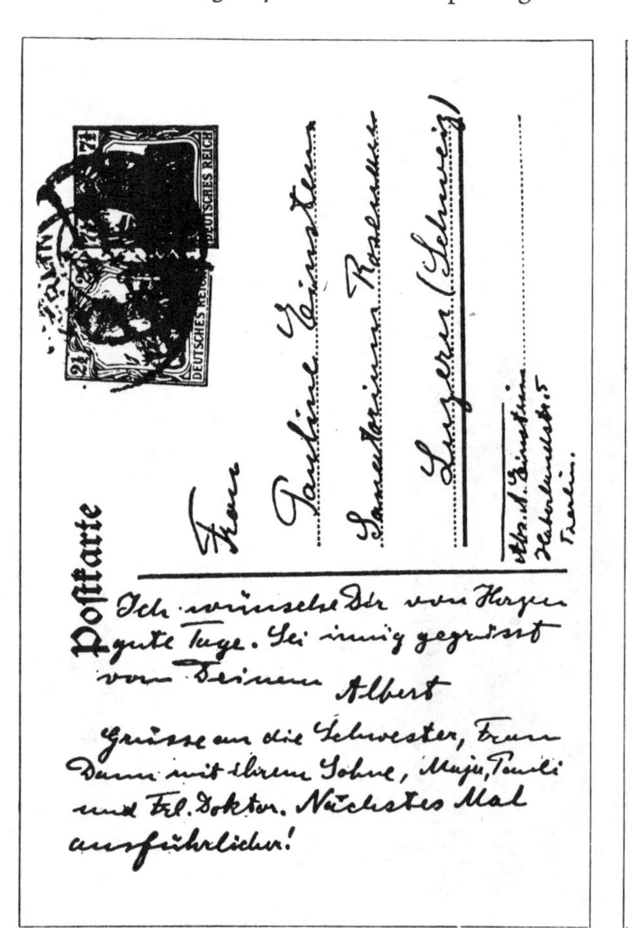

Postkarte

DEUTSCHES REICH

Frau Pauline Einstein
Sanatorium Rosenau
Luzern (Schweiz)

Abs. A. Einstein Haberlandstr. 5 Berlin.

Ich wünsche Dir von Herzen gute Tage. Sei innig gegrüsst von Deinem Albert

Grüsse an die Schwester, Frau Dunn mit ihrem Sohne, Maja, Pauli und Frl. Doktor. Nächstes Mal ausführlicher!

27. IX. 19

Liebe Mutter!

Heute eine freudige Nachricht. H. A. Lorentz hat mir telegraphiert, dass die englischen Expeditionen die Lichtablenkung an der Sonne wirklich bewiesen haben. Maja schreibt mir leider, dass Du nicht nur viel Schmerzen hast, sondern dass Du Dir auch noch trübe Gedanken machst. Wie gern würde ich Dir wieder Gesellschaft leisten, dass Du nicht dem hässlichen Grübeln überlassen wärest! Aber eine Weile werde ich doch hier bleiben müssen und arbeiten. Auch nach Holland werde ich für einige Tage fahren, um mich Ehrenfest dankbar zu erweisen, obwohl der Zeitverlust recht schmerzlich ist.

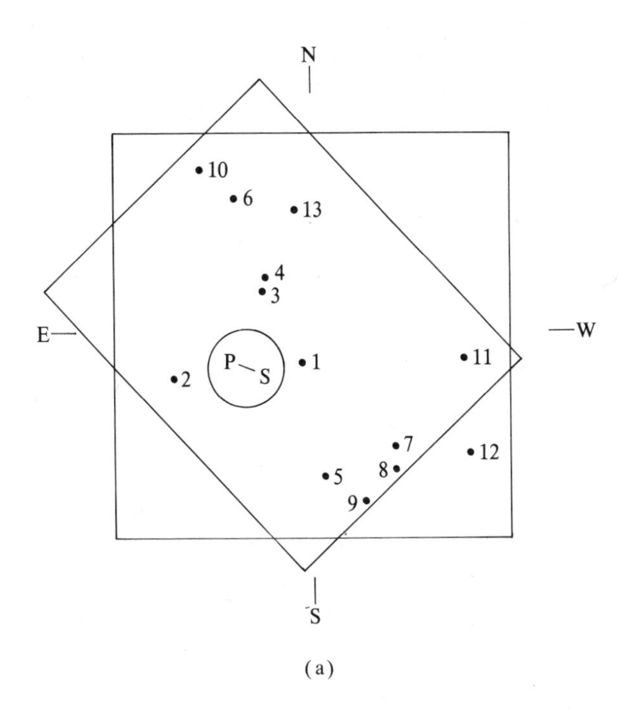

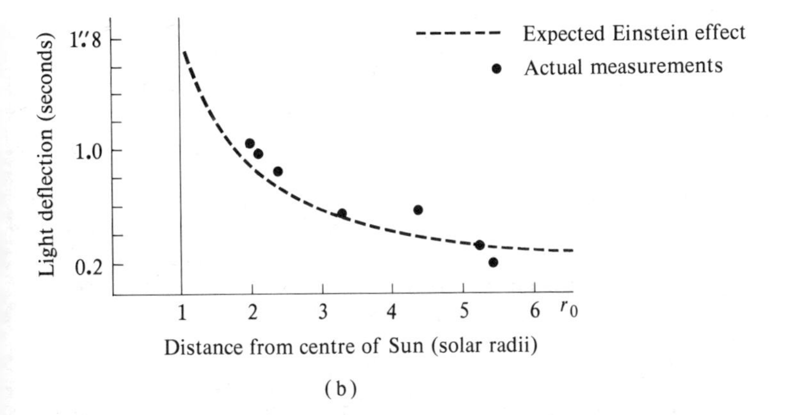

Diagram 3.4
The 1919 eclipse expedition.
(a) The square shows the extent of the star field at Principe, the rectangle that at Sobral. The centre of the sun moved from S to P during the time between total eclipse conditions at the two stations.
(b) A test of the relation between the angle of deflection and the distance between the light ray and the centre of the Sun (theoretically $\alpha \sim 1/r_0$).

It has been remarked by Einstein's biographer Banesh Hoffmann that the impact of Einstein's theory would have been vastly less if (as had been planned by a German astronomer) his prediction in 1911 of the 'half-deflection' (0.85″) had been tested first. Then, as Hoffmann says, 'Imagine

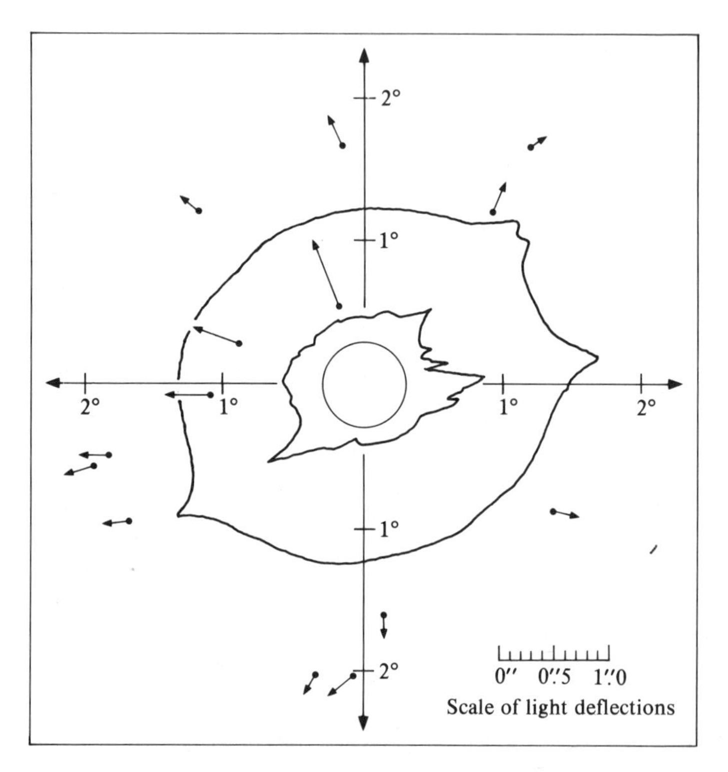

Diagram 3.5
Displacements of the images of the best-measured stars in the 1922 eclipse. Note that the displacements, indicated by arrows, are drawn on a greatly enlarged scale compared to the scale for the relative angular *positions* of the stars. The small circle shows the Sun's disc; the irregular lines surrounding it show contours of the Sun's corona at two different levels of intensity.

how tame Einstein's 1915 calculation of 1.7 seconds of arc would have seemed. . . . He would have been belatedly changing the value after the event, having first been shown to have been wrong.' But the eclipse at which this test was to have been made took place in Russia in 1914, and the war intervened. Thus, the theory of General Relativity was vindicated through a bold and novel prediction for which there was no prior observational basis of any kind.

The story does not quite end there. As long ago as 1913, Einstein had written to the American astronomer, George Hale, asking if there was any

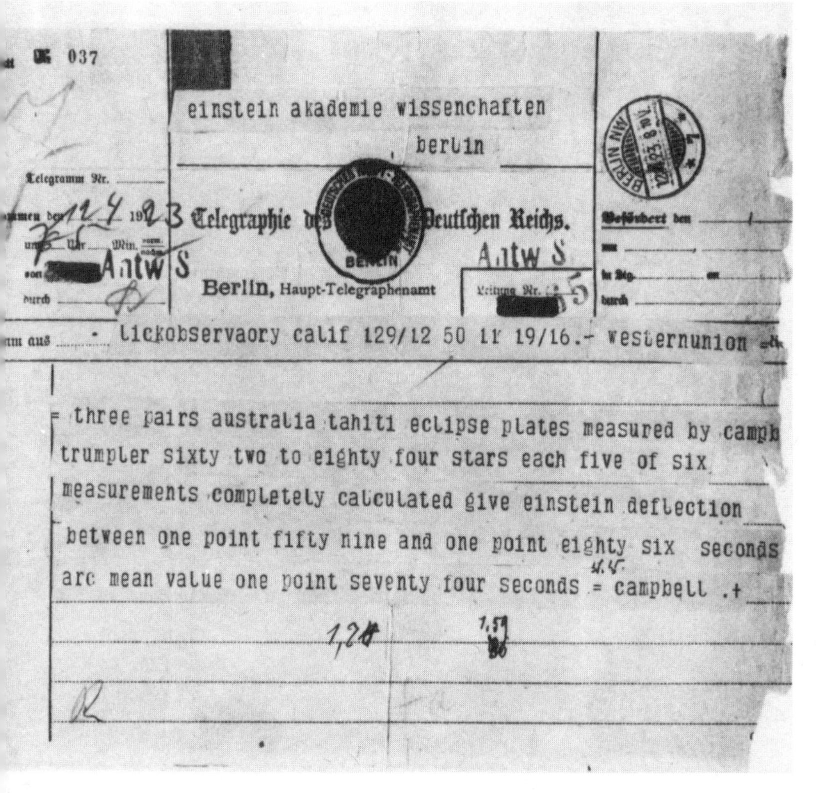

037

einstein akademie wissenchaften
berlin

Telegraphie des Deutschen Reichs.
Berlin, Haupt-Telegraphenamt

lickobservaory calif 129/12 50 11 19/16.- westernunion

three pairs australia tahiti eclipse plates measured by campb
trumpler sixty two to eighty four stars each five of six
measurements completely calculated give einstein deflection
between one point fifty nine and one point eighty six seconds
arc mean value one point seventy four seconds = campbell .+

Figure 20

The telegram sent to Einstein by W. W. Campbell to confirm the results of the 1922 eclipse measurements

possibility of detecting the gravitational deflection of starlight at times other than total eclipses. The verdict was negative, but the development of radio-astronomy has transcended this difficulty. The powerful radio emissions from certain quasars can be detected under any conditions; it is just a matter of choosing a time when the Sun's rim is close to the line of sight. Using this method, measurements at a number of observatories have confirmed the Einstein prediction with an accuracy as high as 1 per cent (Fomalont and Sramek, 1975).

3. A FOURTH TEST OF GENERAL RELATIVITY

As mentioned in Section 2(c) above, the bending of light by a massive body is linked to a reduction in the speed of light in its vicinity. Thus the time taken for a signal to pass from one point to another in space is slightly lengthened if the path passes close to a massive object such as the Sun.

With the development of sophisticated radar ranging techniques, it became feasible to measure such time delays for radar signals sent out from Earth and reflected from other planets. This possibility was proposed by Shapiro (1964) and led to a highly successful series of observations on radar

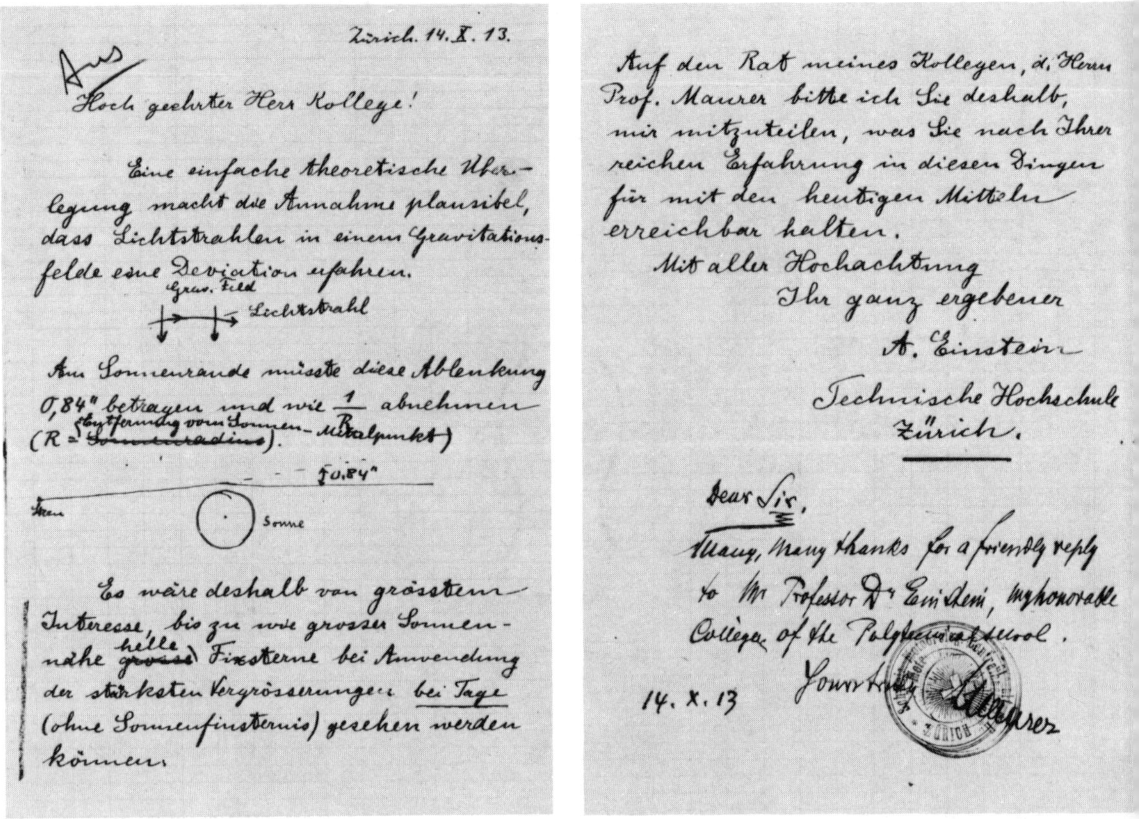

Zürich. 14. X. 13.

Aus

Hoch geehrter Herr Kollege!

Eine einfache theoretische Überlegung macht die Annahme plausibel, dass Lichtstrahlen in einem Gravitationsfelde eine Deviation erfahren.

Grav. Feld

Lichtstrahl

Am Sonnenrande müsste diese Ablenkung 0,84" betragen und wie $\frac{1}{R}$ abnehmen (R = ~~Sonnenradius~~ Entfernung vom Sonnen-Mittelpunkt).

0,84"

Stern

Sonne

Es wäre deshalb von grösstem Interesse, bis zu wie grosser Sonnennähe ~~grosse~~ helle Fixsterne bei Anwendung der stärksten Vergrösserungen bei Tage (ohne Sonnenfinsternis) gesehen werden können.

Auf den Rat meines Kollegen, d. Herrn Prof. Maurer bitte ich Sie deshalb, mir mitzuteilen, was Sie nach Ihrer reichen Erfahrung in diesen Dingen für mit den heutigen Mitteln erreichbar halten.

Mit aller Hochachtung

Ihr ganz ergebener

A. Einstein

Technische Hochschule

Zürich.

Dear Sir,

Many, many thanks for a friendly reply to Mr Professor Dr Einstein, my honorable College of the Polytechnical School.

14. X. 13 Your truly Maurer

Figure 21

The letter from Einstein to George Hale (14 October 1913), in which he asks whether the gravitational deflection of starlight could be detected when the Sun is not eclipsed. Note that the letter quotes Einstein's first, incorrect value (0.84″) for the theoretical deflection

echoes from Mercury, Venus, and Mars. Such measurements call for a fa more precise knowledge of the planets—their orbital dimensions, topo graphy, etc.—than had been previously available. Thus a great deal o effort went into exploring these details so as to be able to extract th relativistic delay with accuracy.

Diagram 3.6 shows the results of a set of observations reported b Shapiro *et al.* (1971) for radar echoes from Venus as a function of time The peak of the curve corresponds to a date when Venus and Earth were a opposite ends of the line passing through the Sun (i.e. a so-called 'superio conjunction'). This maximum delay amounts to about 200 microsecond in a total travel time of about half an hour—i.e. fractionally about on part in ten million. To determine such delays with an uncertainty of les than 20 microseconds requires a knowledge of the relevant distances t within a few kilometres—an impressive achievement.

4. GRAVITATIONAL RADIATION

Implicit in any field theory of gravitation is the possibility (first explore by Einstein in 1916) of gravitational waves. To some extent this questio can be considered by analogy with electromagnetism. If an electric charg

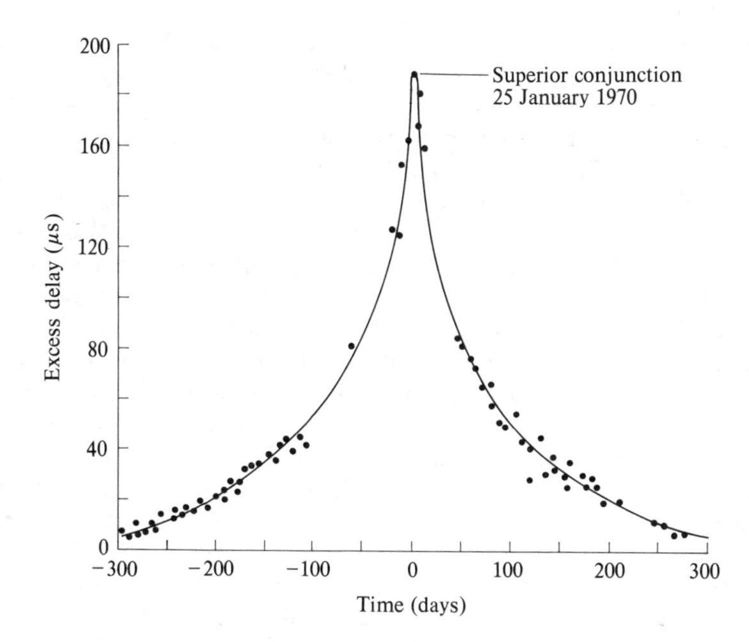

Diagram 3.6
The fourth test of general relativity: time delay of radar echoes from Venus, showing maximum delay when the Sun's edge touches the line between Earth and Venus (after Shapiro *et al.*, 1971).

undergoes a sudden change of motion (as measured by an acceleration) the information that this change has taken place is not communicated instantaneously to distant points; the message is carried in a pulse of electromagnetic radiation travelling at the speed of light, c. Moreover, in contrast to the inverse-square dependence of the electrostatic field, the radiation field strength falls off only as $1/r$, so that its influence can be carried to large distances (and, indeed, represents a net energy flow that is the same through any closed surface, however far from the charge).

In very much the same way an accelerated mass can be expected to produce at distant points a gravitational radiation field whose strength is proportional to the acceleration and inversely proportional to r. Like light, the gravitational field travels at the speed c. The measure of this field is the force per unit mass that it causes in distant objects; hence dimensionally it is an acceleration. If one could use a complete analogy to the electromagnetic case, the theoretical magnitude of the gravitational radiation field would be given by:

$$f(r) = \frac{GMa}{c^2 r} \tag{8}$$

where a is the acceleration of the mass M producing the radiation.

However, the analogy begins to fail at this point, because of a fundamental difference between gravity and electromagnetism. In electricity we have charges of opposite signs, and the basic radiating system is a dipole made of equal and opposite charges oscillating in antiphase, whereas in gravitation the gravitational 'charge' (i.e. mass) has only one sign, and the basic radiating system is a quadrupole (such as one would have in electromagnetism if two *like* charges were made to oscillate in antiphase). The consequence is that the theory of gravitational radiation diverges widely from that of electromagnetic waves. Whereas the quantum of the electromagnetic field (the photon) is a particle of angular momentum $h/2\pi$, the quantum of the gravitational field (the graviton) has twice this amount.

As a practical matter, gravitational waves are exceedingly hard to detect, and indeed are still far from being detected, even though their existence is confidently believed in. A great deal of work over the past two or three decades has gone into designing suitable detectors and identifying promising sources. The universe is full of cataclysmic gravitational events—supernovae, collapsing stars, etc.—and the amounts of energy released in such events are enormous, but only relatively small amounts go into gravitational radiation. Moreover, the coupling of this radiation to a detector is extremely weak. Just as the prototype source of gravitational radiation is a pair of masses oscillating in antiphase along a straight line, so the prototype detector or gravitational antenna is a pair of masses whose separation l is changed by some amount Δl when a gravitational wave passes by. Instead of separate masses, a solid bar of length l has until now been the standard detector. Estimates of the effect of various possible cosmic sources of gravitational radiation give calculated values of $\Delta l/l$ of the order of about 10^{-17} at most—equivalent to a change of separation of about one nuclear diameter between masses one kilometre apart! Despite such dismayingly unfavourable estimates, there are hopes of achieving such sensitivities before the end of this century.

His powers of imagination are closely related to reality. He told me that he visualizes the gravitation waves with the help of an elastic body, and at the same time he made a movement with his fingers as though he were pressing an indiarubber ball. For students he is a very cosy man as long as you understand how to interest him, and by your question to make him forget the time. Then he gets under way on his own. I had to admire the clarity and the penetrating power of his thought. He is never in doubt and wherever doubts exist they are lucid doubts.
(From the diary of R. J. Humm, quoted in Carl Seelig, *Albert Einstein: A Documentary Biography*)

A quite different approach to the problem is to infer the existence of the gravitational radiation from the change in the radiating system as it loses energy. If, for example, one has a binary star system, the loss of energy by gravitational radiation requires that the distance between the two stars should decrease as time goes on, with an accompanying shortening of the orbital period. If the system consists of two stars of equal mass M in orbits of radius R (i.e. separated by a distance $2R$) the characteristic time τ for the decrease of orbital period is given, in rough order of magnitude, by

$$\tau = \frac{c^5 R^4}{(GM)^3} \tag{9}$$

The radius R is, of course, related to the orbital period T through Kepler's third law.

For binary systems that are resolvable as such telescopically, the value of τ is of the order of 10^{23} years, i.e. effectively infinite. However, for a system of two neutron stars (each of about one solar mass) separated by a relatively short distance so that the period T is about one day, the theoretical value of τ is of the order of 10^9 or 10^{10} years. This is still immensely long, but it means that over the space of ten years the period of such a system would decrease by about one part in 10^8 or 10^9. Systems of this type are known (revealed through the periodicity in their electromagnetic radiation). Given the extreme precision possible for time or frequency measurements, there is reason to hope that this slow change of period can be measured, and (very important) can be ascribed unambiguously to gravitational radiation rather than to tidal effects that also dissipate energy.

5. BLACK HOLES

As we have seen, the observational tests of general relativity were based on small and subtle effects. Although in conceptual terms the theory represented a profound and sweeping change in our picture of the physical world, its practical consequences appeared to be slight. But then, in about the middle of this century, there began a major expansion of our knowledge of the universe. Astronomy, which had until then been confined to the visible or near-visible spectrum, began to study the information carried to us via radiations of all kinds, from gamma rays to long-wavelength radio waves. Aided by new and sophisticated techniques of experimental physics, this search showed the universe to be an even richer and stranger place than we had imagined. And perhaps the most bizarre development of all was the emergence of observational and theoretical bases for the existence of the objects called 'black holes'.

It was actually Laplace, in 1796, who first conceived the possibility that a sufficiently massive object might, through its gravity, prevent the escape of light. In his treatise, *Exposition of the System of the World*, he wrote:

> A luminous star, of the same density of the Earth, and whose diameter would be two hundred and fifty times larger than that of the sun, would not, in consequence of its attraction, allow any of its rays to escape to us; it is therefore possible that the largest luminous bodies in the universe may, through this cause, be invisible.

The basis of this estimate was a model, such as Soldner used a few years later in calculating the gravitational deflection of light, in which light was treated as Newtonian particles emitted with the speed c. The escape speed for a particle of mass m in starting out from the surface of a spherical body of mass M and radius R is given (in Newtonian mechanics) by

$$\tfrac{1}{2}mv_0^2=\frac{GMm}{R} \qquad (10)$$

Hence, if we put $v_0=c$, we have the result

$$\frac{2GM}{c^2R}=1 \qquad (11)$$

I had my first glimpse of Einstein in June 1921 when during the general excitement following the reported confirmation of his general relativity theory he came to England and lectured at King's College, London. I doubt if any scientific advance, not excluding the space explorations of more recent times, has ever roused the general public to such a pitch of enthusiasm as that which was then experienced. The idea that our most elementary notions of space and time had been found erroneous caught the imagination of the public, and 'Space Caught Bending' became the most prominent headline in a leading newspaper.

(H. Dingle, in G. J. Whitrow, *Einstein: The Man and His Achievement*)

It is easy to verify that Laplace's calculation conforms to this equation. However, the general theory of relativity provides a very different (and much firmer) basis for the theoretical result expressed by equation (11), which is indeed correct.

The essence of the relativistic analysis is Einstein's fundamental idea that the geometry of space–time is modified by matter, as described by equations (5) and (6). One can see that for $2GM/c^2r=1$ the equation for the metric develops a singularity. This critical condition corresponds to a closing of curved space upon itself. For any given M, there is a radius—the so-called *Schwarzschild radius*, R_S, equal to $2GM/c^2$—defining a volume from which no radiation or information of any kind can escape, i.e. a black hole. For the Sun ($M=2\times 10^{30}$ kg) this radius is three kilometres; that is, if all the mass of the Sun were confined within a radius of three kilometres or less, it would act as a black hole. The mean density corresponding to this mass and radius would be of the order of a hundred times that of nuclear matter.

Fundamentally, Einstein created a new physical belief. He was the Martin Luther of physics. He created a new mode of thinking, a new scientific culture. We all are nourished by him still.

(S. Müller-Markus, in H. Margenau, *Integrative Principles of Modern Thought*)

For a long time the notion that matter in bulk might exist with a density as high as that of atomic nuclei, or greater, was not seriously considered by most physicists. But then, in 1967, came the discovery of the first pulsars—objects emitting short radio-bursts with clock-like regularity. It was soon accepted that these were rotating neutron stars, having masses of the order of one solar mass and radii of the order of ten kilometres. It was known theoretically that such objects might come into being as a result of gravitational collapse after a normal star had used up all its nuclear fuel.

After this discovery, it was not such a great step to envisage the possibility that a somewhat more massive star (perhaps of the order of ten solar masses) might be forced through gravitational contraction below the Schwarzschild radius. At this instant it would become a black hole; any knowledge of its subsequent evolution—whether it contracts to a point or arrives at some limiting configuration—is then beyond the reach of observation. What *can* be observed, however, is any process taking place down to the bounding surface defined by the Schwarzschild radius, and it is on such evidence that the search for black holes depends.

A very general indication of catastrophic gravitational collapse is the emission of violent bursts of light and radio waves. Numerous candidates for identification as black holes can be recognized in this way, but in most cases alternative explanations are possible. A more selective test (although still not definitive) is possible if a putative black hole is a member of a binary star system, in which the other partner is a normal star. The black hole, through its intense gravitational field, can capture material from the other star, and in the process intense X-ray emissions will occur; the details of this X-ray emission, coupled to other observational and theoretical evidence, can constitute a kind of 'signature' through which a fairly convincing identification is possible. Most (but not all) astrophysicists believe that the necessary conditions are met by an X-ray source in the constellation Cygnus.

The formation of a black hole should be accompanied by a large burst of gravitational waves, and such events are believed to represent one of the most promising sources for detectable gravitational radiation.

Although a mass of the order of ten solar masses appears to be a minimum for leading to the degree of collapse necessary to form a black hole, there is no natural maximum. Thus it has been suggested that, in addition to black holes being a rather frequent product from the collapse of individual stars, there also exist monstrous black holes (in terms of mass) formed from thousands or millions of stars condensing together at the cores of galaxies. Perhaps quasars, with their huge energy output, are systems of this kind.

6. CONCLUDING REMARKS

In the general theory of relativity we see one of the most marvellous products of speculative but disciplined thinking about the physical world. It can be said to have begun with a question so simple yet so profound that most people would not think to ask it, or would be content with a superficial explanation: 'Why do all objects, whatever their nature, fall under gravity with the same acceleration?' Einstein, by concentrating on this question, created for the first time a genuine theory of gravitation. (Newton, it must be remembered, did not claim to have provided an explanation of gravity—*Hypotheses non fingo*.)

The question has been raised whether Einstein should be considered the sole author of general relativity. The reason for this doubt is that the great mathematician, David Hilbert, became deeply interested in Einstein's geometrical approach to gravitation. Working at Göttingen, he eagerly followed the development of Einstein's ideas, and in November 1915, simultaneously with Einstein's first presentation of general relativity in Berlin, Hilbert presented a note on 'The Foundations of Physics' to the Royal Society of Science in Göttingen. In it he incorporated the geometry of curved space–time, for which he had derived the ten requisite metric coefficients in a more elegant way than Einstein did. Yet (as with Poincaré and Einstein in the case of special relativity) it was Einstein the physicist who provided the crucial insights. Hilbert himself, on numerous occasions, made it clear where the credit lay. With a certain degree of exaggeration he once remarked: 'Every boy in the streets of Göttingen understands more about four-dimensional geometry than Einstein. Yet, in spite of that, Einstein did the work and not the mathematicians' (quoted in Constance Reid's biography of Hilbert, 1970). And of all Einstein's great scientific achievements, the general theory of relativity is perhaps supreme in its originality and intellectual grandeur.

We have refrained from entering into a discussion of general relativity as applied to cosmology, since this is a huge field in itself and since the principal development of this subject was carried out by others, subsequent to Einstein's own first paper (1917) on relativistic cosmology—after which his own interests turned mainly to the development of his unified field theory. An excellent account of cosmologies may be found in G. J. Whitrow, *The Structure and Evolution of the Universe* (New York: Harper Torchbooks, 1959). More recent discussions may be found in the books by Berry, Davies, and Ohanian listed below.

Suggested Reading

Bergmann, Peter G., *The Riddle of Gravitation* (New York: Charles Scribner's Sons, 1968).

Berry, Michael, *Principles of Cosmology and Gravitation* (Cambridge: Cambridge University Press, 1976).

Born, Max, *Einstein's Theory of Relativity* (New York: Dover, 1962).

Davies, P. C. W., *Space and Time in the Modern Universe* (Cambridge: Cambridge University Press, 1977).

Eddington, Sir Arthur, *Space, Time and Gravitation* (New York: Harper Torchbooks, 1959).

Einstein, A., *Relativity: The Special and General Theory* (New York: Crown, 1961).

Einstein, A., and Infeld, L., *The Evolution of Physics* (Cambridge: Cambridge University Press, 1938).

Mehra, Jagdish, *Einstein, Hilbert and the Theory of Gravitation* (Dordrecht: D. Reidel, 1974).

Ohanian, Hans C., *Gravitation and Spacetime* (New York: W. W. Norton, 1976).

Sciama, D. W., *The Physical Foundations of General Relativity* (New York: Doubleday, 1969).

Tonnelat, M. A., *Histoire du Principe de Relativité* (Paris: 1971).

Whittaker, E. T., *A History of the Theories of Aether and Electricity* (London: Nelson, 1951, 1953; New York: Harper Torchbooks, 1960).

4 Relativity theory and gravitation

Hermann Bondi

'Mathematics are all well and good but Nature keeps dragging us around by the nose.'
(A.E. to Hermann Weyl, 1923)

1. NEWTONIAN GRAVITATION AND OBSERVATION

1.1 The essential characteristic feature of gravitation was discovered by Galileo, namely that all bodies fall (i.e. accelerate) equally fast at a given place. 'Galileo's principle', as we may call it, has been tested with very high precision. Notably early in this century Eötvös (1908) established its validity with an accuracy of one part in 10^8, and more recently Dicke (1962) has driven the precision to the astonishing level of one part in 10^{11}. Thus it is entirely reasonable to try to establish the consequences of the assumption that Galileo's principle holds exactly.

The contrast with other forces is profound. In every other case there is a property, that bodies may or may not have, that determines whether a force does or does not act on them. Thus an electric field acts only on bodies that have electrical characteristics (charge, dipole moment, etc.). Remove these, and the force disappears; strengthen them, and the force increases. (It is true that on an atomic scale matter is necessarily electrical, but if we confine our attention to bodies no smaller than specks of dust these complexities disappear.) In much the same way the response of a body to a magnetostatic field is wholly determined by its magnetic characteristics. With most materials there is little difficulty in reducing their magnetic response to very low levels indeed.

Gravitation is unique in acting not on any abolishable property of a body, like its charge or magnetic moment, but on its inalienable feature of inertia. For inertia (or mass) is, by Newton's second law, that by which force has to be divided to yield acceleration. If all bodies have the same acceleration, the forces on them must be proportional to their inertial masses. Thus inertia or mass, the very feature by which we recognize a body as a body, is also that which responds to gravitation.

1.2 At first sight the equality of the response of all bodies to gravitation appears to be a simplifying feature. But in fact the opposite is the case when we try to *observe* gravitation. Perhaps an analogy will help. Imagine a world in which all materials had the same coefficient of thermal expansion. How would you then construct a liquid-in-glass thermometer?

Of course we are all aware of gravitation; standing makes our legs tired, we can measure our weight on scales, etc. But these are all means more or less confined to the surface of the Earth where we happen to live. Gravitation as a *universal* force (Newton and the motion of the Moon!) must be measurable *everywhere*, and our position on the surface of a massive body—the Earth—is highly atypical of the universe, most of which is empty. How does one observe gravitation in empty space? Since everything falls the same way, nothing measurable seems to be left. We are nowadays familiar with the weightlessness of astronauts in an orbiting spacecraft: we know very well that they cannot measure their masses by stepping on scales, and that their soup floats around in drops. So, to all appearances, there is nothing of gravitation left measurable in space. Are we thus talking about a pseudo-force, one which can be observed if one has solid ground under one's feet but not otherwise, e.g. in space?

A closer analysis shows this pessimism to be misplaced. Though all bodies fall equally fast, this common acceleration varies with position. Consider a spacecraft in orbit near the Earth (Diagram 4.1) and remember that it is of finite size, small though it is compared to the scale of its orbit. The acceleration of free fall at the point of the spacecraft closest to the Earth is higher than in its middle, where it is in turn higher than at the point of the spacecraft furthest from the Earth. There will therefore be a stress on the spacecraft trying to elongate it along the line joining it to the centre of the Earth. While the structure of the spacecraft will easily be strong enough to resist

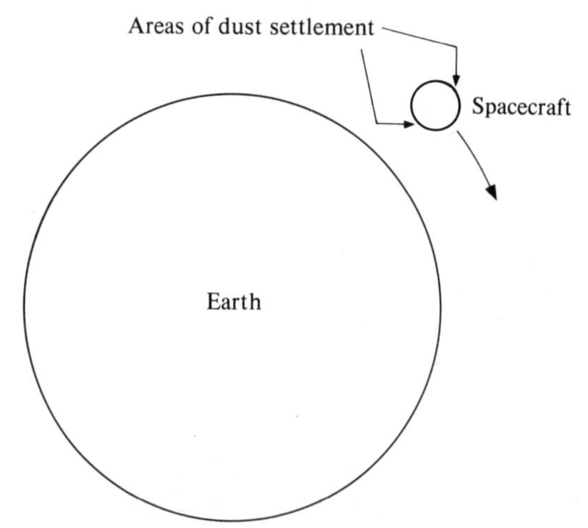

Diagram 4.1
Residual gravity in a spacecraft in close orbit.

this stress, specks of dust near the part of the spacecraft furthest from the Earth will tend to drift further that way, for they will fall with the local acceleration which will be marginally less than the 'compromise' acceleration adopted by the spacecraft as a whole. Similarly, dust near the part of the spacecraft closest to the Earth will fall a little faster than the spacecraft. Thus the astronaut will observe dust settling in the two portions of the spacecraft furthest from and nearest to the Earth. From this fact he will be able to infer that he is in a gravitational field. (Indeed, this effect has been used to engineer a 'gravity gradient' stabilization for certain spacecraft.)

We can easily transfer this consideration to 'spacecraft Earth' in its orbit round the Sun. While we cannot feel directly the enormous gravitational pull of the Sun, since we and the Earth under us are all falling towards the Sun equally, the 'softest' parts of the Earth—the oceans—respond to the effect discussed by elongating the sphere of water both towards and directly away from the Sun, producing the *solar tides*. (The somewhat larger lunar tide is produced in just the same way, but the different ratios of distances and masses render the visualization a little harder). Thus, although there is no direct observable effect of the Sun's field, the tides are a plain demonstration of its non-uniformity and we could deduce from them the existence of the Sun (and the Moon) even if we could not see them.

Thus everywhere an observable of gravitation may be measured through the fact that, although different particles fall equally fast if they are in the same place, there is a difference in their accelerations if they are in different places, even if these are close by. *So the universal observable of gravitation is the relative acceleration of neighbouring particles.*

1.3 Since this relative acceleration will be small if the particles are near to each other, and tends to zero as they tend towards coincidence, it is reasonable to suppose that this relative acceleration depends linearly on the separation. Since both the separation and the acceleration have direction as well as magnitude, we are talking about a rather complex linear relation, in which different directions are by no means equivalent. (What would have happened if instead of choosing the parts of the spacecraft nearest and furthest from the Earth, we had chosen the leading and trailing parts?) The important point, however, is not this complexity, but the fact that it is *only* this non-uniformity of the field that is observable everywhere. A field in which the acceleration is the same throughout, both in magnitude and in direction, is *unobservable* and should thus not be regarded as a field. Hence we arrive at the following conclusion: since in physics we always define quantities by how we measure them, *a gravitational field is a relative acceleration of neighbouring particles.* If this relative acceleration vanishes we do not have a field. (The reader should note that in other presentations he may find the concept of a 'uniform gravitational field', i.e. one where the acceleration is the same throughout the field. According to the analysis here this would be described as a zero gravitational field.)

While our definition is universally applicable, it may leave the reader wondering whether this relative acceleration is what makes his feet tired when he stands around too long. What in fact happens with a solid Earth (as with the rigid spacecraft earlier) is that it integrates these small relative accelerations through its body, leading to a substantial difference in acceleration ($2g$) at opposite ends of a diameter of the Earth, or g between its surface and its centre. The Earth as a whole, though massively compressed by all these effects, moves with its centre effectively in free fall (towards the Sun). This integrated difference, g, is thus what we feel. (Of course, there is no appreciable change in its magnitude between our head and our toes, but we are conscious of it because the ground prevents us from falling freely.)

Talking about words, what about the words 'theory of relativity'? I believe that they possess a strong emotive attraction. Take the word *theory*. It is a dignified word, but it carries the implication that this is only a theory, a speculation, not a fact, and keeps us guessing whether it is true or not. As for *relativity*, it has a polysyllabic mystery in it, and makes one wonder what is relative to what. I often find myself wishing Einstein had chosen some other title, for the theory is fact and you have sometimes to search hard to find what is relative to what.
(J. L. Synge, *Talking About Relativity*)

1.4 *Reciprocity* is a universal feature of physics and in dynamics is described by Newton's third law, relating the equality of action and reaction. Since mass is what gravitation acts on, mass must also be that which *produces* gravitation. Thus mass generates gravitation, just as electric charge generates an electric field. Neither mass nor electric charge can vary arbitrarily; both satisfy a *law of conservation*. Important though this constraint is in the electrical case, it is even more significant in the gravitational case because although electric charges of both signs exist, *we know of no negative mass*. Thus we may have an electric field generated by separated equal and opposite charges and, by bringing them to coincidence, we can wipe out both source and field. The absence of negative mass makes this impossible in the gravitational case, resulting in a remarkable permanence of the sources and therefore of the fields. Moreover, the law of conservation of momentum (which has no parallel in the electric case) further constrains the motion of the sources.

Why is there no negative mass? Before attempting to answer this question it may be worthwhile to distinguish three kinds of mass, each defined, as any physical quantity should be, by the method of measuring it:

(i) Inertial mass, measured by the acceleration produced by a known force, or the velocity by a known impulse (e.g. response of a ping-pong ball to the bat).
(ii) Passive gravitational mass, i.e. that property of matter that the gravitational field hooks on to. This may be measured by the force produced in a known gravitational field, for example by weighing a body on a spring balance at the Earth's surface.
(iii) Active gravitational mass which *produces* a field, measured through observing the orbit of a body in its field (e.g. the mass of the Sun can be deduced from the motion of the Earth and a knowledge of the Sun–Earth distance).

By Galileo's principle, (i) ≡ (ii) and by Newton's third law (action equals reaction), (ii) ≡ (iii). Thus if any mass is negative, all are negative. A negative inertial mass would be strange, for such a body would come towards you if you pushed it away, and move away if you pulled it towards you. Not perhaps inconceivable, but we may be pleased at not having discovered any such material.

1.5* To put the results of the end of Section 1.2 and the beginning of Section 1.3 into mathematical form, we shall use tensor notation (in three dimensions). Thus the relative acceleration vector δf^i will be linearly connected with the relative position vector δx^i. Such a linear connection can only be given by a tensor

$$\delta f^i = a^i{}_j \, \delta x^j \qquad (1)$$

Accordingly the gravitational field is fully described by the nine observables $a^i{}_j$.

(We are using here the standard simplified notation of tensor analysis, in which it is understood that a summation is to be made over any repeated index. Thus equation (1) must be read as

$$\delta f^i = \sum_j a^i{}_j \delta x_j \qquad (j = 1, 2, 3)$$

We shall also be using raised and lowered indices, in the accepted conventions of tensor analysis, to distinguish between so-called contravariant and covariant quantities. The reader should consult a mathematics text that deals with tensor analysis if he is unfamiliar with such matters and wishes to follow this analysis in detail.)

Next, consider a small sphere (a spherically symmetrical body). If the accelerations described by equation (1) were to result in an angular acceleration (i.e. through the force acting as a couple) then there would be nothing to stop the field making the sphere spin faster and faster, gaining more and more kinetic energy. There is nothing to suggest that this spin could weaken or otherwise alter the field, so, since energy must be conserved, there can never be any such couple. Calculation shows that this implies that a_{ij} (i.e. the tensor with its i suffix lowered) is symmetric, so that

$$a_{ij} = a_{ji} \qquad (2)$$

Hence there are six free components of the observable which thus describe the field.

Next, consider the relative acceleration f^i of particles P and Q, a finite distance apart. This quantity is itself observable and is evidently given by

$$f^i = \int_P^Q a^i{}_j \delta x^j \qquad (3)$$

Since f^i is observable, it cannot depend on the route taken to link P and Q. Thus the line integral is route independent and $a^i{}_j$ can be written:

$$a^i{}_j = \partial W^i / \partial x^j \qquad (4)$$

Frequently, I remember, if I brought up a mathematical argument that seemed to him unduly abstract, he would say, 'I am convicted but not convinced', that is to say, he could no longer get out of agreeing that it was correct, but he did not yet feel that he had understood why it was so. For, in order to convince himself that something was so, he had to reduce it to a certain simplicity of concept.
(E. Straus, in G. J. Whitrow, *Einstein: The Man and His Achievement*)

* Sections involving the mathematics (tensor analysis) needed for the formal development of general relativity are marked with an asterisk. Such sections can be passed over if desired. The essential features of our account do not depend upon them.

Combining equations (2) and (4) (and assuming suitable conditions of smoothness) it follows that

$$a_{ij} = -\frac{\partial^2 V}{\partial x^i \partial x^j} \tag{5}$$

where the minus sign is conventional and V is now the usual Newtonian gravitational potential, completing the link between the presentation given here and earlier ones.

Finally the connection between the field and its sources is given by Poisson's equation which takes the form

$$-\nabla^2 V = a^i{}_i = -4\pi G\rho \tag{6}$$

where ρ is the density of matter and G the constant of gravitation. Note that $\nabla^2 V$ appears very simply as the trace of the tensor, so that the density is proportional to a linear combination of the tensor components: the sum of its diagonal components.

Notice that the whole of Newtonian theory is based on equations (5) and (6) which have thus been expressed in terms of our observable $a^i{}_j$. It is clear from the derivation that V and its gradient cannot themselves be observable.

. . . it is very likely that future generations will refer to the first half of the twentieth century as the age of Einstein, just as historians think of the latter half of the seventeenth century as the age of Newton. The irony of this is that Einstein's work is understood by such a small percentage of the people whose lives and intellectual outlook have been, often unwittingly, influenced by it.

(Jeremy Bernstein, *Einstein*)

2. RELATIVITY

2.1 Newtonian mechanics, however excellent at describing velocities small compared with that of light, ceases to be logically or experimentally tenable at high velocities. In particular it is quite impossible to make the gravitational theory just described cover the motion of light in any credible sense.

Special relativity is known to describe perfectly, in the absence of gravitation, both mechanics at all speeds and the propagation of light. It is reasonable to assume that when weight is abolished, as in an orbiting spacecraft, at least the main principles of special relativity will still apply within this modest volume. However, a difficulty immediately presents itself if we then try to describe gravitation through the Newtonian observable. If the relative acceleration of neighbouring particles were independent of velocity, it would be possible to arrange the particles and their velocities so that one of them was accelerated from below to above the speed of light, which is forbidden by special relativity. Thus the relative acceleration of neighbouring particles must depend on their velocities, but to get agreement with Newtonian theory for small velocities this dependence must be negligible for such small speeds. Although these requirements for making the notion of our observable fit with special relativity are of a somewhat formal nature, they turn out to be crucial in formulating the equations of the theory.

2.2 Of more immediate physical significance is an ideal experiment, first discussed by Einstein, which reveals the deep connection between light, gravitation, and time that emerges when well established features of relativistic and quantum physics are linked to Galileo's principle. These features are:

(i) Atoms of one species have a well defined set of states, each identifiable and of a particular energy, the state of lowest energy being called the ground state. For our purposes it is sufficient to focus attention on this and on one other state (called the excited state).
(ii) Light of any given frequency (i.e. colour) exists only in units (photons) whose energy is a universal constant times their frequency.
(iii) When an atom makes a transition from the excited to the ground state, the energy lost by it is radiated as a photon of this energy (and therefore of the appropriate frequency). Conversely, light of this frequency (and therefore energy) can be absorbed by an atom in the ground state, thereby putting it into the excited state. Although, in general, some blurring of the sharpness of the frequency concerned occurs due to momentum and other effects, suitable choices can make this blurring very small and then the excellent definition of frequency is used for our best time measuring devices (caesium and ammonia clocks). Indeed the elasticity of the balance spring of a watch is controlled by interatomic forces of just the same nature as the atomic forces defining frequencies. Equally, it is also feasible to use nuclear transitions for measuring time.
(iv) Light reflected by a moving mirror shows a shift of frequency (Doppler shift). This is towards the blue (higher frequencies) for an approaching mirror and towards the red for a receding mirror.
(v) Like other physical quantities, time is defined by the means used to measure it, i.e. clocks.
(vi) On a mirror light exerts a well defined pressure which, while small in laboratory circumstances, is nonetheless readily measurable.
(vii) Energy has mass, by Einstein's well-known equation $E=mc^2$. This is a thoroughly tested relation. Although in our case the difference is too small for measurement, there is no doubt that the mass of an excited atom exceeds that of an atom in the ground state by precisely the amount corresponding to its extra energy. (For certain nuclear transitions the mass difference is actually measurable.)

'For me, Einstein is not only a leading research worker who has acquired the right to lay aside the daily work of other physicists, but also a man of terrific strength of character. He does not shrink from spending fifteen years on a task which finally turns out to be fruitless. Just as serene as he was when originally convinced of its success, he can say at the close: "I've once more turned my back on it." '

(Hermann Weyl, quoted in Carl Seelig, *Albert Einstein: A Documentary Biography*)

With this preamble, we now consider a tower on the Earth, with an endless chain of buckets linking a wheel at the top to one at the bottom (Diagram 4.2). The buckets, all equal, are filled with equal numbers of atoms of one and the same species, but all the buckets on the side labelled *G* are filled with atoms in the ground state, while all those on side *E* contain atoms in the excited state. Since the excited atoms have more energy (available for release as light) than the atoms in the ground state, they have more mass and thus, by Galileo's principle, more weight. Accordingly side E is heavier than side G and, with sufficiently freely rotating wheels, side E will begin to move downwards, with side G going upwards.

We now induce the excited atoms, as they arrive at the bottom, to make the transition to the ground state, emitting light of the appropriate frequency in the process. Thus as the buckets arrive at side G, the atoms in them will

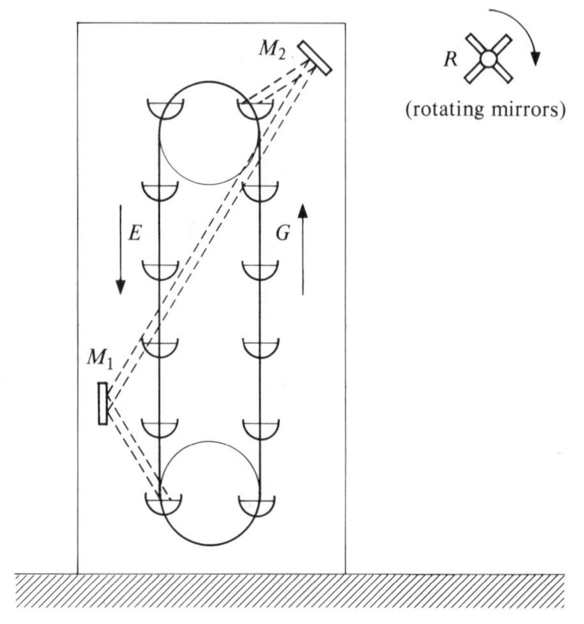

Diagram 4.2
The ideal experiment establishing gravitational red shift. Buckets full of excited atoms (E) descend on the left; buckets of lighter atoms in the ground state (G) ascend on the right. Radiation is transferred from bottom to top via mirrors M_1 and M_2. M_2 is later replaced by a set of rotating mirrors (R).

be in the ground state, like the atoms already on side G. The light emitted at the bottom is caught by a suitable arrangement of fixed mirrors, producing a beam travelling to the top of the tower which is there focussed on the atoms in the G buckets arriving at the top. Since (by (iii) above) the frequency emitted by an atom making the transition from the excited to the ground state is just that needed by an atom in the ground state to make it excited, our arrangement will ensure that the situation remains as in Diagram 4.2. Thus side E will always contain excited atoms and side G atoms in the ground state; hence the chain will keep moving, driving the wheels and yielding energy, with no recognizable reaction on the field or on its sources, the Earth. We have thus devised a *perpetuum mobile*, generating energy from nothing. As this is known to be impossible, there must be a mistake in the argument. But where? Every step seems sound and tested by experiment, directly or indirectly (see (i), (iii), (vii) above). So how can a contradiction have arisen?

2.3 The one possible chink lies in the reciprocity of (iii). Though we know that an atom changing from the excited to the ground state emits light of just the right frequency to be absorbed by an atom in the ground state, thereby making it excited, this has only been established *with the atoms side by side*. Perhaps this does not work if the emitting atom is at the bottom of the tower and the absorbing one on top. If the frequency of the light on arriving at the top were too low (i.e. if it were too red) the photons would have insufficient energy to excite the atoms there and thus our *perpetuum mobile* would not work. (If the frequency on arrival at the top were too high, the light could excite the atoms easily and the problem would persist.) How can we test this ingenious way out? If the only reason the system does not work lies in the light being too red on arrival, giving it an appropriate blue shift could return it to the right frequency and thus would make the system work again. Since reflection from an approaching mirror causes such a blue shift, we fit a wheel of mirrors at the top of the tower (Diagram 4.2) and spin it so that the incoming light is blue-shifted by this reflection. At the right speed of rotation, the reflected light should now have the correct frequency to excite the atoms arriving at the top, so that the system can work and deliver energy. However, we need energy to keep turning the wheel of mirrors against the pressure exerted on the mirrors by the light. Thus the answer is clear. The red shift of the light is such that in moving the mirrors to compensate for it, the wheel of mirrors uses up exactly the energy produced by the chain, since energy can be neither created nor destroyed.

Thus we can calculate this gravitational red-shift* or Einstein shift, which turns out to be a fractional lowering of frequency by $\Delta V/c^2$ where ΔV is the difference in Newtonian potential between top and bottom and c is the speed of light. For a tower of height h on the surface of the Earth, we can put $\Delta V = gh$, so that the fractional shift is equal to gh/c^2. With a 27 m high tower, for example, this is 3×10^{-15}, a very small number indeed. Generalizing our formula it turns out that for light emitted on the surface of the Sun and received here on Earth the shift is about 2×10^{-6}. For many years, efforts at observing the Einstein shift therefore concentrated on comparing the frequencies of spectral lines from the Sun with those of the same lines produced in the laboratory. The great differences in conditions of line production (density and temperature of the gas concerned) unfortunately introduced other, larger, shifts which cannot be calculated accurately. Thus the test of the result of the compelling derivation given above had to wait until the exceedingly sharp gamma-ray lines produced through the Mössbauer effect finally enabled Pound and Rebka in 1960 to verify the

* Of course, looking from the bottom at light produced at the top, one would observe a blue shift.

theoretical predictions on the Earth, actually using a tower of about the height quoted above!

2.4 Though the gravitational red-shift is small, at least in all readily accessible situations, its mere existence has considerable consequences. It is therefore necessary to remind oneself that the theoretical derivation of the effect is not only logically compelling, but also requires only those parts of the theories concerned (see (i) to (vii) above) as have the most solid direct or slightly indirect experimental backing, and that moreover the effect itself has been tested with considerable precision.

The first major consequence arises from (v). A spectral line may be something sounding a little sophisticated, but in fact it is *the* means of measuring time. Also, whether one is talking of a super-accurate caesium clock, a quartz-controlled clock, an ordinary watch relying on its balance spring, or a nuclear clock, such as radio-carbon dating or estimation based on radioactivity of rock for geological time-scales, one is inevitably basing oneself on a source of time affected by the gravitational red-shift. Combining this with (v), it follows inescapably that *time is slower at the foot of a tower than at its top.*

Thus these considerations drive the disintegration of any universal time concept further beyond what is already familiar from special relativity. Time must never be thought of as pre-existing in any sense; it is a *manufactured* quantity. In special relativity one learns that each inertial observer manufactures his time, which is as perfect for him as that of any other inertial observer is for that second observer, but the two times are in no sense identical. However, whereas the discrepancy in time-keeping between inertial observers relates to their relative speed and vanishes if they are at relative rest, in gravitational theory we have a time discrepancy between observers at relative rest to each other but who are 'higher' or 'lower' than one another.

Although Einstein was no doubt a somewhat more complex personality than generally imagined, he was essentially a man of basic goodness and general kindliness. He was endowed with a robust sense of humour that survived the afflictions of life into old age. He was completely lacking in pomposity and in that sense of self-importance that so often corrupts lesser men. He was entirely free from the trappings of convention, not only in his thought but in his way of life.

Once when a too fulsome speech was being made at a dinner in America in his honour, he whispered to his neighbour, referring to himself, 'But he doesn't wear any socks!'

(G. J. Whitrow, *Einstein: The Man and His Achievement*)

2.5 Nothing in this discussion has had any relation to a relative acceleration of neighbouring particles. Thus no intrinsically observable property of the gravitational field is involved. Accordingly the entire Einstein shift can be abolished by free fall. Put the tower in a box falling freely down a shaft and remove the rotating wheel of mirrors. The compensation for the red-shift now occurs because during the time the light takes to travel 'up' (h/c), the box accelerates by gh/c. Thus the top has, at the moment a package of light arrives, this velocity relative to the motion of the bottom when the package of light started out, resulting in a fractional blue shift of gh/c^2, cancelling the gravitational red-shift. Thus in the freely falling box there is no gravitational effect, as is of course right for this condition of weightlessness.

As soon as an extended volume is considered, the observable of the gravitational field manifests itself. Relative accelerations will appear and the total abolition of the field can no longer be achieved through falling freely.

Indeed, if we consider an observer falling freely and vertically in one spot on the Earth, his motion cannot counterbalance the red shift observed a little distance away between the top and the foot of a tower. Thus there is an intrinsic connection between Einstein shift and the observable of the gravitational field. However, the mathematical treatment of Sections 2.6 and 2.7 is required to demonstrate that this connection requires a non-Euclidean geometry.

2.6★ To make equation (1) relativistic, we have to recall that the four-dimensional velocity vector

$$v^i = \frac{dx^i}{ds} \quad (7)$$

involves the differentiation of coordinate changes dx^i not with respect to coordinate time (dx^0), but with respect to the moving particle's proper time (ds), and is therefore of unit length, since

$$v^i v_i = g_{ij} v^i v^j = g_{ij} \frac{dx^i}{ds} \frac{dx^j}{ds} = 1 \quad (8)$$

where g_{ij} is the metric tensor, which for an inertial observer using Cartesian coordinates is

$$\begin{matrix} +1 & 0 & 0 & 0 \\ 0 & -1 & 0 & 0 \\ 0 & 0 & -1 & 0 \\ 0 & 0 & 0 & -1 \end{matrix} \quad (9)$$

The four-dimensional acceleration vector is defined by

$$f^i = \frac{dv^i}{ds} \quad (10)$$

and satisfies, by equation (8)

$$f^i v_i = 0 \quad (11)$$

With this condition, acceleration can never lead to a particle transgressing the speed of light, but equally it is clear that f cannot, as in equation (1), depend only on the displacement, for then it could not satisfy equation (11). Thus we try

$$\delta f^i = b^i{}_{jk} \delta x^j v^k \quad (12)$$

To ensure satisfaction of equations (8) and (11), we must have

$$(v^i + \delta f^i \Delta s)(v_i + \delta f_i \Delta s) = 1 \quad (13)$$

for any small Δs. Thus b has to have a structure such that

$$0 = \delta f_i v^i = b_{ijk} \delta x^j v^i v^k \quad (14)$$

Einstein's distaste for the 'merely personal' was not just a peculiarity. Like other extraordinary men, he felt that the personal, moment-to-moment existence, dominated by ever-changing wishes, hopes, and primitive feelings, is a chain that one should try to cast off in order to free one for the contemplation of the world 'that stands before us like a great, eternal riddle'. In the simplified but lucid image of the world that can thereby be gained, he once said, a person could hope to 'place the center of gravity of his emotional life, in order to attain the peace and security that one cannot find within the narrow confines of swirling personal experience.'
(Gerald Holton, *The Scientific Imagination: Case Studies*)

for all v^i. Thus b_{ijk} has to be anti-symmetric in its first and last suffixes. But the same argument as in Section 1.5. (as well as the need to reduce to equation (2) for slow motions) implies that b_{ijk} is symmetric in its first two suffixes. It is readily established that these two symmetry conditions are incompatible for any non-zero b. Thus we are forced to abandon equation (12) and try the next simplest, viz.:

$$\delta f^i = c^i{}_{jkl}\delta x^j v^k v^l \qquad (15)$$

'To punish me for my contempt for authority, Fate made me an authority myself.'

(A.E.)

Applying the same arguments again we find that c_{ijkl} must be symmetric in its first two suffixes, but anti-symmetric against an interchange of the second and third ones. Moreover, by its definition in equation (15), c_{ijkl} should be taken to be symmetric between its last two suffixes. These various symmetry properties reduce the number of freely choosable components of c_{ijkl} from the terrifying 256 of a general four-dimensional tensor of rank 4 to a mere 21, which should be compared with the 6 freely choosable components of a_{ij} in the non-relativistic analysis of Section 1.5. But now we are describing a much richer system, not only of slow particles, but of particles moving at any speed, and light itself as well.

2.7★ In special relativity, one works with the Euclidean (Minkowski) metric

$$ds^2 = g_{ij}dx^i dx^j = (dx^0)^2 - (dx^1)^2 - (dx^2)^2 - (dx^3)^2 \qquad (16)$$

(in Cartesian-type coordinates). Of course coordinate transformations of a vast variety can be made. To describe the gravitational red-shift we need (considering height z as the only relevant spatial dimension)

$$ds^2 = f^2(z)dt^2 - g^2(z)dz^2 \qquad (17)$$

Since, along a light ray, $ds = 0$, the t coordinate becomes some function of z, plus an arbitrary constant. Thus the difference in t values along successive light rays is height independent, i.e. it does not vary along the ray. Each observer's clock measures his ds. He is fixed in height, so that his $dz = 0$. Thus the gravitational red-shift implies that $f(z)$ is an increasing function of z.

Considering the Earth as spherically symmetrical we add the other dimensions and complete equation (17) to

$$ds^2 = f^2(r)dt^2 - g^2(r)dr^2 - r^2(d\theta^2 + \sin^2\theta d\phi^2) \qquad (18)$$

where the radial coordinate r has been calibrated to make the surface of a sphere r = constant have the area $4\pi r^2$. Of course ϕ is the longitude and θ the co-latitude, i.e. the latitude counted from 0° at the North Pole to 90° at the Equator to 180° at the South Pole.

We know that $f(r)$ cannot be a constant, but we also know that at large distances from the Earth the gravitational red-shift cannot go on increasing without limit (since the potential V tends to a limit) and so $f(r)$ tends to a constant as $r \to \infty$.

It turns out that such an $f(r)$ makes it impossible to fit equation (18) into a Euclidean 4-space, i.e. whatever $g(r)$, there is no way, with an $f(r)$ not constant but tending to a limit at infinity, in which equation (18) can be transformed into equation (8) with equation (9). There is still a metric tensor, as in equation (8), but in general there is no way in which the coordinates can be transformed so that the metric tensor assumes the form of (9).

2.8 The results of the analysis outlined in the two preceding sections can be summarized in a simple but profound statement: *Relativistic gravitation is incompatible with a Euclidean geometry*. It can still be true that a Euclidean geometry of space–time is applicable over a limited region. If so, this region is called *flat*; elsewhere space–time is said to be *curved*. However, any general theory of gravitation must be based on a non-Euclidean geometry.

'There were two kinds of physicists in Berlin: on the one hand was Einstein, and on the other all the rest.'

(Rudolf Ladenburg)

The simplest non-Euclidean geometry is Riemannian, as exemplified by the geometry of the surface of a sphere. It is of course well known from geography that this surface cannot be unrolled into a plane. There are no straight lines; the nearest analogue is a great circle, i.e. a circle passing through a given point on the surface, that results from the intersection of the sphere with a plane through its centre. A vector is now a direction in the surface of the sphere, and we say that it has suffered parallel displacement from P to Q if it makes the same angle with the great circle through P and Q at both points.

The essential way in which the curvature of the surface of the sphere can be discovered *intrinsically*, i.e. without going outside the surface, is through parallel displacement of a vector round a closed curve (see Diagram 4.3). Take P to be the North Pole and Q, R to be two points on the

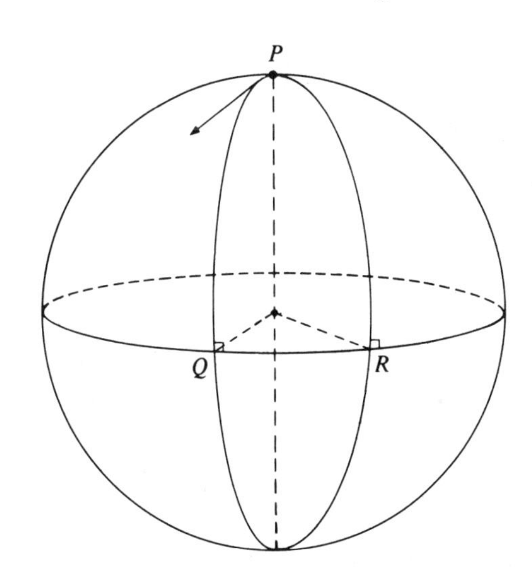

Diagram 4.3
An intrinsic method of discovering the curvature of the surface of a sphere.

Equator. Then the meridians PQ, PR, and the piece QR of the Equator are all parts of great circles. Now take the vector at P pointing along PQ. It makes a zero angle with PQ at P. Parallel displacement to Q will mean that it still makes a zero angle with the meridian PQ at Q, i.e. that it is there south pointing and at right angles to the Equator. Continuing by parallel displacement to R, it will be south-pointing there so that it will be along the meridian PR produced. Returning it now by parallel displacement to P, it will be along PR and thus be at an angle to its original direction equal to the angle between the two meridians at P. Thus the curvature of the surface can be detected without leaving it, and therefore this so-called Gaussian curvature is intrinsic to the surface. Its value is defined as the ratio of the angle through which the vector has been turned on suffering parallel displacement round a closed curve and the area round which it has been taken. This, as the reader can easily verify from the above example, turns out to be the reciprocal of the square of the radius of the sphere. For a more general case, where the curvature varies from point to point, the loop round which the vector should be taken must be chosen to be small. (Note that a developable surface, like that of a cylinder or a cone, is flat by this definition, as it can be unrolled into a plane.)

Einstein devoted ten years of his life to this problem [that of gravitation] when no one else was interested in it. . . . To ponder on a problem for ten years without any encouragement from the outside requires strength of character. This strength of character, perhaps more than his great intuition and imagination, led to Einstein's scientific achievements.
(L. Infeld, *Quest: The Evolution of a Scientist*)

Going from two to four dimensions does not change anything intrinsically, but adds complexity. Our analogon to a straight line is now a *geodesic* (a curve of extremal length). Since the area round which the vector is taken involves two directions, the vector itself has one, and the change in its direction is a fourth one, the curvature is now expressed by a 4-suffix curvature (or Riemann–Christoffel) tensor

$$R_{ijkl} \tag{19}$$

This tensor when multiplied by a vector and an element of area (a 2-suffix entity) yields the change in the vector on being taken round the area. Another, and very useful, application is to the vector δf^i joining a geodesic with tangent vector $v^k = \mathrm{d}x^k/\mathrm{d}s$ to a neighbouring geodesic a displacement δx^j away:

$$\delta f^i = R^i{}_{jkl} v^k v^l \delta x^j \tag{20}$$

This equation of geodesic deviation is identical with what we derived as equation (15). Moreover the curvature tensor has all the symmetry properties previously demanded of the tensor C_{ijkl} (plus, as it turns out, one more, reducing the number of free components to 20, plus a differential property to be discussed in Section 3).

Thus *we identify the paths of freely falling particles with geodesics, and the observable of the gravitational field with the curvature tensor*. Hence the gravitational field as defined in this article is represented completely by the curvature of space–time. Given a field, we can therefore calculate the paths of particles and of light rays.

It should be mentioned that the curvature tensor can be constructed from the second derivatives of the metric tensor g_{ij} (involving also the tensor itself and its first derivatives). There is therefore some analogy between the metric tensor and the Newtonian potential, since in both cases the observable can be constructed by twice differentiating from this 'potential', but the analogy should not be driven too far.

3. THE SOURCES OF THE GRAVITATIONAL FIELD

3.1. In Newtonian theory the gravitational field emanates from its sources, which are the masses present. If we consider a limited volume, allowance must be made, through appropriate boundary conditions, for the field coming in from beyond the border of our volume. More commonly one considers an infinite volume and imposes boundary conditions at infinity. Though these usually are a vanishing of the field, in fact this is only appropriate if the volume on which attention is concentrated (e.g. the solar system) has a mean density of matter far exceeding the average of the universe. After all, the long-range character of gravitation implies that the consideration of gravitational situations can only rarely be wholly separated from cosmology.

As was pointed out in Section 1.4, the triple concept of mass—as inertia, as responsive to gravitation, as causing gravitation—is well defined in the Newtonian scheme of things. Relativistically, the situation is not quite so simple. After all, the mass of a moving body can be taken to be either its rest mass or its total mass which includes the mass of its kinetic energy. Which is relevant for gravitation? For inertia there is no doubt: it is a well tested part of special relativity that inertia is given by the total mass. Thus for a body when hot, the heat energy (i.e. the fast internal motions of its constituent atoms) makes its inertia greater than for the same body when cold. Since hot bodies and cold bodies fall equally fast it follows that the passive gravitational mass is given by the total mass. Though the link between action and reaction is more complex in relativity, yet it follows that the gravitation-producing properties of matter must also be measured by total mass.

The contrast is strongest for light, which has zero rest mass, but a total mass given by its energy. It would be illogical to expect light not to exert a gravitational force, as well as being itself subject to gravitational influences.

The special theory was limited in one very important respect, and in the attempt to remove this limitation Einstein created the *general* theory of relativity, perhaps the most original scientific conception ever formed by the mind of a single man.
(D. W. Sciama, in G. J. Whitrow, *Einstein: The Man and His Achievement*)

3.2 The importance for gravitation of the mass conservation law was stressed in Section 1.4. This, and the law of conservation of momentum, need to be considered when examining methods of incorporating sources into our relativistic theory of gravitation. There is also the question of the positiveness of mass, which was raised earlier. Ideally it should find clear expression in a theory. Finally, the complexity of the relativistic gravitational field (Section 2.6) may or may not be mirrored in the complexity of the description of the sources. Though the treatment is necessarily rather mathematical (see below), the outcome is a description of the sources in Einstein's theory of gravitation that involves not only mass, but momentum and stress as well, that very beautifully includes the laws of conservation of mass and of momentum as tautologies, but that fails to give a clue why mass is always positive. Perhaps this can only come from a deeper, perhaps quantized, theory of the sources.

3.3* We start by looking for an analogy with equation (6) where a particular linear combination of the observables of the Newtonian gravitational field was put proportional to the source density, i.e. to the density of matter. When we look at the relativistic observable, there are just two basic reasonably simple linear combinations of R_{ijkl}, viz.:

'What I'm really interested in is whether God could have made the world in a different way; that is, whether the necessity of logical simplicity leaves any freedom at all.'

(A.E. to Ernst Straus)

$$R_{ij} = R_{ijk}{}^{k} \tag{21}$$

(a symmetrical 2-suffix tensor having 10 freely choosable components), and

$$R = R^{i}{}_{i} \tag{22}$$

(a scalar).

The simplest approach would be to make R proportional to the matter density. However, this will not work. First, it imposes too few constraints on the observables, so the sources would not determine the field. Secondly, there is no way in which total mass density can be made a scalar, since it changes with the motion of the observer. Thus only rest mass could function as source, which is unacceptable, as has been pointed out above.

Thus we are led to the more complex equation (22) as a possible basis for a link between source and field. In fact it turns out that a combination of equations (21) and (22) viz.:

$$G_{ij} = R_{ij} - \tfrac{1}{2} g_{ij} R \tag{23}$$

is the most promising. Like R_{ij}, this is a symmetrical 2-suffix tensor with 10 choosable components, involving only a linear combination of the observables and the common background tensor g_{ij}. But the essential property of this so-called Einstein tensor G_{ij} arises from the differential relation satisfied by the curvature tensor and hinted at towards the end of Part 2 (Section 2.8): G_{ij} has a vanishing divergence, i.e. it satisfies four conservation laws which are like the conservation of mass-energy (one scalar) and of momentum (a 3-vector). Thus we are led to describe the source by an entity having the same structure as G_{ij} and satisfying empirically the corresponding conservation laws. This is the energy–momentum tensor, which in the simplest case (dust) is

$$T^{ij} = \Sigma \rho v^{i} v^{j} \tag{24}$$

where ρ is the rest-mass density, v^{i} the velocity vector of the particles, and the sum means that one averages over volumes containing dust particles. This readily generalizes to a fluid with pressures (arising through the summation over differently moving particles) and equally readily a normal limiting process allows light to be included, its vanishing rest mass being compensated for by the (infinite) velocity vector. For a fluid the vanishing of the divergence of T^{ij} implies the Eulerian (or Navier–Stokes) equations of hydrodynamics together with the equation of continuity. Thus we are led to

$$G_{ij} = -8\pi k T_{ij}, \tag{25}$$

which are *Einstein's field equations*. The constant k involves the constant of gravitation and the speed of light and, in suitable units, equals one. For slowly moving matter of modest volume and density equation (25) effectively reduces to Poisson's equation.

Thus the formulation of Einstein's theory of gravitation can be regarded as complete by adding equation (25) to equation (20). The background and the interpretation of the symbols have been indicated sufficiently to make it clear that this is the least complex theory possible that is based on Galileo's principle and is relativistic.

3.4 The preceding development of Einstein's gravitational field equations differs radically from Einstein's own; it corresponds to the approach adopted by Fock. The essential basis of it is the existence of observable mutual gravitational accelerations between separated objects.

It is curious that Einstein, who in other areas of physics (notably special relativity) criticized anything that transcended actual experience, should in the case of gravitation have insisted instead on the physical equivalence of accelerated frames. Such equivalence does not in fact hold; accelerated clocks can behave very differently from unaccelerated ones, and may in fact be destroyed by the acceleration. But Einstein proceeded from the Principle of Equivalence to his general relativistic theory of coordinate transformations, and ignored the fact that gravitation *cannot* be completely transformed away by a general acceleration of an extended region (however small).

The magnitude of Einstein's achievement in creating a theory of gravitation is of course not lessened by such considerations. Indeed, one perhaps does more honour to him by presenting a different approach to the problem than by regurgitating his own derivations.

The equivalence principle demonstrated

THE PROBLEM

Eric M. Rogers

While I was living in Princeton, my wife and I would from time to time take a small puzzle involving physics to our neighbour Professor Einstein—often as a birthday present.

The last of these, presented on his seventy-sixth birthday, was, I believe, original. It was derived from an old-fashioned toy for small children: a ball on a string is tied to a cup in which the child has to catch the ball. But our modification was for Einstein a problem which he enjoyed, and solved at once.

A metal ball attached to a smooth thread is enclosed in a transparent globe. There is a central, transparent, cup in which the ball could rest; but initially the ball hangs by the thread outside the cup (as shown in the diagram). The thread runs from the ball up to the rim of the cup and down through a central pipe. Below the globe the thread is tied to a long, weak, spiral spring protected by a transparent tube which ends in a long pole—a broom-handle.

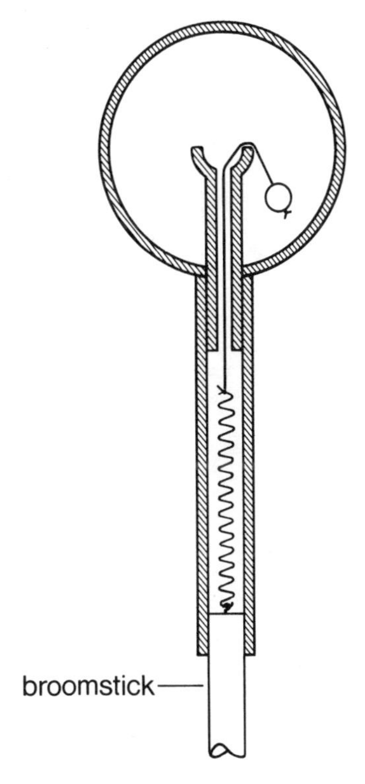

THE PROBLEM

Starting with the ball hanging down, get it into the cup by a 'sure-fire' method.

The boundary conditions and information

1. The globe and the transparent tube should not be opened.
2. The ball is made of solid brass.
3. The spring is already stretched, in a state of tension, even when the ball is in the cup; but it is not strong enough to pull the heavy ball up into the cup.
4. The broomstick is long.
5. There is a method which will succeed every time—in contrast with occasional success by random shaking.

And 6, for readers here, a relevant reminder: it was made as a present for Einstein—who solved it in a real experiment, with delight.

THE SOLUTION

I. Bernard Cohen

. . . At last I was taking my leave. Suddenly [Einstein] turned and called 'Wait. Wait. I must show you my birthday present'.

Back in the study I saw Einstein take from the corner of the room what looked like a curtain rod five feet tall, at the top of which was a plastic sphere about four inches in diameter. Coming up from the rod into the sphere was a small plastic tube about two inches long, terminating in the centre of the sphere. Out of this tube there came a string with a little ball at the end. 'You see,' said Einstein, 'this is designed as a model to illustrate the equivalence principle. The little ball is attached to a string, which goes into the little tube in the centre and is attached to a spring. The spring pulls on the ball, but it cannot pull the ball up into the little tube because the spring is not strong enough to overcome the gravitational force which pulls down on the ball.'

A big grin spread across his face and his eyes twinkled with delight as he said, 'And now the equivalence principle'. Grasping the gadget in the middle of the long brass curtain rod, he thrust it upwards until the sphere touched the ceiling. 'Now I will let it drop,' he said, 'and according to the equivalence principle there will be no gravitational force. So the spring will now be strong enough to bring the little ball into the plastic tube.' With that he suddenly let the gadget fall freely and vertically, guiding it with his hand, until the bottom reached the floor. The plastic sphere at the top was now at eye level. Sure enough, the ball rested in the tube.

With the demonstration of the birthday present our meeting was at an end. . . .

(From 'An interview with Einstein', *Scientific American*, 1955, **193**, July, 69–73.)

5 Einstein and the development of quantum physics

Martin J. Klein

> The great discoverers can readily be classed under two types of mentality: those who dig deep and those who range wide. Those who possess the gift of combining depth with breadth are rare indeed. Albert Einstein was one of them'.
>
> (François le Lionnais, 'From Plurality to Unity', in *Science and Synthesis*)

Towards the end of his life Albert Einstein wrote to his oldest friend that fifty long years of 'conscious brooding' over the question, 'What are light quanta?', had brought him no closer to its answer. As usual Einstein was not exaggerating: the problem of understanding discreteness as well as continuity in the natural world occupied him throughout his career. That Einstein spent so much time and energy wrestling with the quantum theory may well surprise many, and even many physicists. His creation of the special and general theories of relativity and his long series of attempts to construct a still more general theory, a unified field theory, have overshadowed his other achievements. Yet anyone who knows Einstein's work is likely to agree with Max Born, one of the major figures in the development of quantum mechanics, who wrote: 'In my opinion he would be one of the greatest theoretical physicists of all times even if he had not written a single line on relativity.' That opinion is based mainly on the papers in which Einstein reported the remarkable results of his 'conscious brooding' over the problems of quanta.

Einstein was the one who, in 1905, first proposed the idea of light quanta. It was simply heretical at that time to suggest that light sometimes behaved as though it consisted of localized particles of energy, and years went by before this suggestion won any acceptance. As Einstein probed further, and worked out the consequences of Max Planck's radiation law, he saw that a new theory of light was needed, one in which the dual nature of light—wave and particle—would be accounted for. By 1908 he was already convinced that these problems were 'so incredibly important and difficult' that every physicist should devote his efforts to trying to solve them. Einstein was also the first to realize that a quantum theory of matter was needed, as well as a new theory of radiation. His early effort in this direction—a quantun theory of the specific heats of solids—led to new and unexpected connec- tions among the thermal, optical, and elastic properties of solids, helping tc convince other physicists that the quantum theory must be taken seriously Einstein's papers in this field over a period of twenty years influenced anc inspired Niels Bohr, Louis de Broglie, and Erwin Schrödinger, amon others, in their own contributions to the great synthesis that created a quantum physics in the 1920s.

'I have greatly admired the papers published by Mr Einstein on questions dealing with modern theoretical physics. Moreover, I believe that the mathematical physicists all agree that these works are of the highest order. . . . If one considers that Mr Einstein is still very young, one has every right to justify the greatest expectations from him, and to see in him one of the leading theoreticians of the future. . . .'
(Marie Curie)

In this article I will sketch Einstein's role in this development, describin the works just mentioned, emphasizing the questions Einstein was tryin to answer, and the deep concern with the foundations of physics underlyin all his efforts. But the story does not end there. When the new quantun physics was developed, Einstein greeted it sceptically even though he ha done as much as anyone to bring it into being. He recognized its grea successes, but he never accepted it as the new fundamental theory it claime to be. Einstein wrote relatively little on this subject during the second hal of his career, concentrating on his search for a unified field theory. Hi critical comments during this period cannot, however, be ignored; the were important to his opponents, especially to Bohr, in helping to clarif just what the new quantum physics did mean. They are also important i understanding Einstein's own goals as a physicist for, as Born remarked 'Einstein's conception of the physical world cannot be divided into water- tight compartments'.

II

In June 1905, *Annalen der Physik* published an article by Einstein entitle 'On a Heuristic Viewpoint Concerning the Production and Transformatio of Light'. Physicists usually refer to this as 'Einstein's paper on the photo- electric effect', but that description does not do it justice. Einstein himsel characterized it at the time as 'very revolutionary', and he was right. Thi is the paper in which he proposed that light can, and in some situation must, be treated as a collection of independent particles of energy—ligh quanta—that behave like the particles of a gas. Einstein was well aware tha a great weight of evidence had been amassed in the course of the previou century showing light to be a wave phenomenon. He knew, in particular that Heinrich Hertz's experiments, carried out less than twenty years earlier had confirmed Maxwell's theoretical conclusion that light waves wer electromagnetic in character. Despite all this evidence Einstein argued tha the wave theory of light had its limits, and that many phenomena involvin the emission and absorption of light 'seemed to be more intelligible' if hi idea of quanta were adopted. The photoelectric effect was one of severa such phenomena which he analysed to show the power of his new hypo- thesis. But even granting the success of that hypothesis, what prompte Einstein to make this extraordinary suggestion?

Einstein devoted the greater part of his paper to answering just thi juestion, presenting the arguments that led him to his new 'heuristic viewpoint' of quanta. These arguments, at once simple and daring, embody ome of the essential features of his whole approach to physics. His deepest oncern, expressed in the opening sentences of his paper, was the very oundation of his science. Let us look briefly at the background for this oncern.

When Einstein was a student at the Polytechnic in Zürich just before 900, working eagerly in the laboratory but skipping many of the lectures o read the works of the great physicists on his own, he absorbed the spirit hat had guided the development of physics through three centuries. I refer o 'the mechanical world view', the conviction that all natural phenomena re to be explained in terms of a single underlying theory—mechanics. The uccesses of this approach were evident to the young Einstein. 'What made he greatest impression upon the student,' he wrote many years later, 'was . . the achievements of mechanics in areas which apparently had nothing o do with mechanics: the mechanical theory of light . . . and above all the inetic theory of gases. . . . These results supported at the same time nechanics as the foundation of physics and of the atomic hypothesis. . . . It vas also of profound interest that the statistical theory of classical mechanics vas able to deduce the basic laws of thermodynamics, something which vas in essence already accomplished by Boltzmann.' The vision of a single undamental theory as the basis for all the diverse aspects of the world aptured Einstein's imagination, as it had captured the imagination of heorists long before him.

. . . he never hesitated to change his opinion when he found that he had made a mistake and to say so. Indeed, there was an occasion when somebody accused him of saying something different from what he had said a few weeks previously, and Einstein replied, 'Of what concern is it to the dear Lord what I said three weeks ago?' It was just a way of saying that it did not matter. It was wrong, and now he knew better.
(Otto Frisch, in G. J. Whitrow, *Einstein: The Man and His Achievement*)

By 1900, however, it was no longer possible to accept the goal of xplaining all phenomena in mechanical terms, and Einstein recognized this oo in his early years. He read Ernst Mach, whose criticism of the mechanical rogramme, carried out with 'incorruptible scepticism and independence', hook Einstein's 'dogmatic faith'. He also studied Maxwell's theory of lectromagnetism, finding it 'the most fascinating subject at the time that was a student'. This theory made a shift in basic concepts, a shift that instein called nothing less than 'revolutionary', from the idea of forces cting at a distance to that of fields acting locally. Although Maxwell and is immediate successors thought of the electromagnetic field as acting hrough a mechanical medium whose structure could eventually be deternined, all attempts to determine that structure proved fruitless. Electronagnetism was not successfully explained in mechanical terms and, as instein put it: 'One got used to operating with these fields as independent ubstances without finding it necessary to give one's self an account of heir mechanical nature; thus mechanics as the basis of physics was being bandoned, almost unnoticeably, because its adaptability to the facts resented itself finally as hopeless.'

Einstein was very conscious of this disturbing dualism in the foundations f physics, with two kinds of basic theories of quite different character—nechanics, and the electromagnetic field theory. It was this dichotomy he pointed to at the beginning of his 1905 paper, 'On a Heuristic Viewpoint': 'There is a profound formal difference between the theoretical ideas which physicists have formed concerning gases and other ponderable bodies and the Maxwell theory of electromagnetic processes in so-called empty space.' He referred to the contrast between the discrete mechanics of matter which is atomic in structure and in which a finite number of mechanical quantities specify the state of a system, and the continuous field theory of electromagnetism in which a set of continuous functions are needed to specify the state of the field. This dualism between particle and field, between mechanics and electromagnetism, was the starting point of his considerations. It was a disturbing dualism because it could lead to serious problems when the two disparate fundamental theories had to be brought to bear together. Einstein immediately gave an example of one of these problems, so serious that his friend Paul Ehrenfest later gave it the dramatic name, 'the ultraviolet catastrophe'. Einstein's example involved the blackbody radiation recently studied in detail by Max Planck using quite another approach. Let us examine Einstein's treatment of this situation.

During one of the lectures, Paul Ehrenfest passed on a note to Einstein, saying 'Don't laugh! There is a special section in purgatory for professors of quantum theory, where they will be obliged to listen to lectures on classical physics for ten hours every day.' To which Einstein replied, 'I laugh only at their naiveté. Who knows who would have the laugh in a few years?'
(J. Mehra, *The Solvay Conferences on Physics*)

He considered a volume, enclosed by reflecting walls, that contained a gas and also a number of harmonically bound electrons. These electrons, acting as charged harmonic oscillators, would emit and absorb electromagnetic radiation and, when the system came to thermodynamic equilibrium, this would be identical with the blackbody radiation. The oscillating electrons would also exchange energy with the freely moving molecules of the gas through collisions. These oscillating electrons served, in effect, as the link between the material system—the gas, described by mechanics—and the electromagnetic system—the radiation, described by Maxwell's theory. Both theories could be used to determine the average energy u of an oscillator of frequency ν when the system is in equilibrium at absolute temperature T. The statistical mechanics of the gas required an oscillator in equilibrium with the gas molecules to have an average energy proportional to T,

$$u = kT \tag{1}$$

where k is a universal constant, the gas constant per molecule (or Boltzmann's constant as it is now called). The electromagnetic theory required the average energy of the oscillator to be proportional to the energy density of the surrounding radiation, if absorption and emission were to be equal on the average. If $\rho(\nu, T)\mathrm{d}\nu$ is the energy of the radiation, per unit volume, having frequencies in the interval ν to $\nu + \mathrm{d}\nu$, then the average energy u of the oscillator must be given by

$$u = (c^3/8\pi\nu^2)\rho(\nu, T) \tag{2}$$

where c is the velocity of electromagnetic waves, or light.

Since equations (1) and (2) give two alternative expressions for the same quantity u, these expressions can be equated, giving the result

$$\rho(\nu, T) = (8\pi\nu^2/c^3)kT \tag{3}$$

This equation ought to have fixed the distribution of energy in the spectrum of blackbody radiation by determining the function $\rho(\nu, T)$. The result obtained, however, was not only in conflict with experiment, but it was intrinsically unacceptable. For if one tried to calculate the total energy of the radiation in a unit volume by integrating $\rho(\nu, T)$ over all frequencies, the result obtained from equation (3) was proportional to ${}_0\int^\infty \nu^2 d\nu$ which is infinite. The result of combining the mechanical and electromagnetic equations was really no result at all. Einstein saw this as a clear sign that physics could not rest on its present divided foundations, and that in some way or other the foundations must be unified.

A physical theory, in Einstein's conception, springs from the free creative activity of a man who sets up axioms to start with and need only justify them by their results, which are sometimes rather distant, and by a conviction of internal coherence when the proposed theory unites very wide areas of physics.
(André Lichnerowicz, 'From Plurality to Unity', in *Science and Synthesis*)

Since he saw no way of accomplishing that step at the time, what could be done? Einstein proceeded to analyse the implications of the radiation spectrum $\rho(\nu, T)$ as it was then known. As long as the frequency of the radiation considered was not too low (or the temperature too high), the spectrum could be described by the distribution law suggested by Wilhelm Wien in 1896:

$$\rho(\nu, T) = \alpha\nu^3 \exp[-\beta\nu/T] \tag{4}$$

where α and β are constants. To see the consequences of this distribution, Einstein treated the radiation as a thermodynamic system at equilibrium, a system having definite values of entropy as well as energy. He showed that if one considers the radiation of frequency ν, and keeps the energy E of this radiation fixed while slowly changing the volume of the container from V_0 to V, the entropy of this radiation changes from S_0 to S according to the equation

$$S - S_0 = (E/\beta\nu)\log(V/V_0) \tag{5}$$

This result was strikingly similar to the entropy change of an ideal gas of N particles whose volume is changed from V_0 to V at constant energy (or temperature),

$$(S - S_0)_{\text{gas}} = Nk\log(V/V_0) \tag{6}$$

where k is the same universal constant that appeared in equation (1). Was this a mere coincidence, or did it suggest something essential about the nature of radiation? The answer to that question depended on the significance of that logarithmic form for entropy. To explore this, Einstein turned to Ludwig Boltzmann's statistical interpretation of the entropy, according to which the entropy difference $S - S_0$ between two states of a macroscopic system is proportional to the relative probability W of the occurrence of those two states

$$S - S_0 = k\log W \tag{7}$$

'All these fifty years of conscious brooding have brought me no nearer to the answer to the question "What are light quanta?" Nowadays every Tom, Dick, and Harry thinks he knows it, but he is mistaken.'

(A.E. to Besso, 12 December 1951)

with the same constant k appearing. Now, regardless of the laws of motio that describe the motions of the gas particles and regardless of the nature o these particles, so long as they move independently of one another an show no preference for one part of the available volume compared t another, the probability of finding the N particles in a subvolume V of th total volume V_0 is clearly

$$W=(V/V_0)^N \quad (8$$

In other words, the logarithmic dependence of the entropy of a gas on it volume comes only from the independence of the gas particles.

Einstein's next step was to turn the argument around and apply it to th radiation: since the entropy of the radiation has exactly the same form a that of the gas, one can legitimately infer that the probability of finding a the radiation (of frequency ν) in the subvolume V must be given by th equation

$$W_{\text{rad}}=(V/V_0)^{N'} \quad (9$$

where the exponent N' is obtained by comparing equations (5) and (6),

$$N'=(E/k\beta\nu) \quad (10$$

Einstein drew what was, for him, the inescapable conclusion:

> Monochromatic radiation of low density (within the region of validit of the Wien distribution law) behaves with respect to thermal phen mena as if it consisted of independent energy quanta of magnitude $k\beta\nu$

This was the chain of reasoning that led Einstein to suggest treatin radiation as if it were composed of a collection of independent particles energy. He took the suggestion very seriously himself, applying it im mediately to several phenomena, one of which was the photoelectric effec The experimental material on the emission of electrons from a met surface when the surface is irradiated by ultraviolet light was very limite in 1905, but it was known that the energies of the electrons emitted we independent of the intensity of the incident light. This was quite uninte ligible if the light were considered to be a wave, since the intensity of wave is always a measure of the energy carried by it. If one accept Einstein's proposal, however, the process of photoelectric emission could thought of as a combination of independent events, the simplest of whi is the absorption of one quantum of energy by an electron in the met surface, and its conversion into the kinetic energy of the electron which thereby set free. The maximum energy of such a photoelectron would th be determined by the energy of one light quantum, which is $k\beta\nu$ Einstein's hypothesis. The maximum kinetic energy of the electron cou not be equal to $k\beta\nu$ because it would take a certain amount of work, P, remove the electron from the metal in which it was bound, and so t equation for the maximum energy of the photoelectrons, $(\text{K.E.})_{\text{max}}$, wou be

$$(\text{K.E.})_{\max} = k\beta\nu - P \qquad (11)$$

This argument immediately explains the independence of the electron nergy from the intensity of the incident light, since increasing that intensity ncreases the number of incident quanta without affecting the energy $k\beta\nu$ of he individual light quantum. The energy of the emitted photoelectron vould be less than the maximum predicted by equation (11) if the energy f a quantum were shared among several electrons, or if the electron merged from the interior of the metal rather than its surface. The maxi- num energy can be measured by determining the electric potential Y_{stop} eeded just to prevent any photoelectrons from reaching the collecting lectrode. If e is the charge on an electron, equation (11) can then be ewritten in the form

$$Y_{\text{stop}} = (k\beta/e)\nu - (P/e) \qquad (12)$$

he stopping potential should be a straight line when plotted against the requency of the incident (monochromatic) light. The slope of that line $k\beta/e$) should be the same for all emitting surfaces, and this universal slope epends only on universal constants determinable from experiments on ompletely different phenomena. Only the work P is characteristic of the articular metal surface used in the experiment.

Figure 22

Millikan's verification of Einstein's photoelectric equation

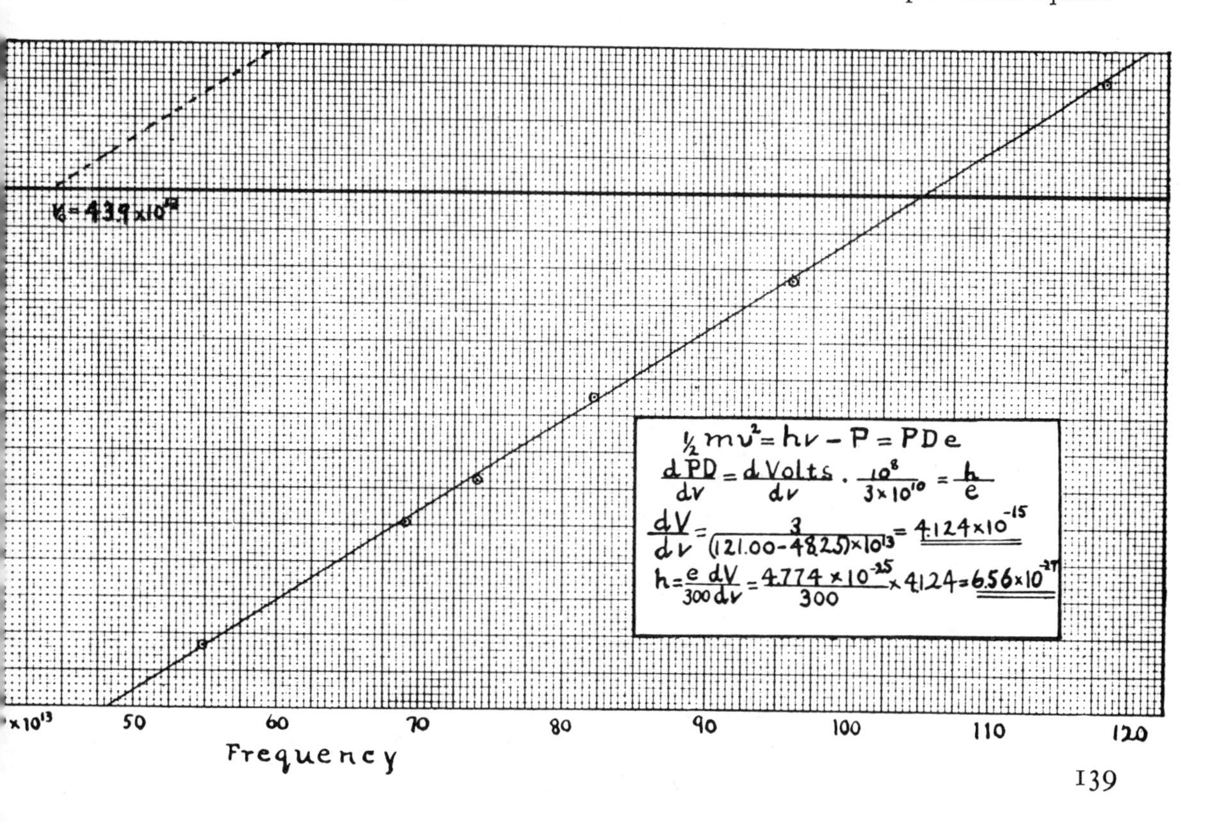

These predictions made by Einstein on the basis of his light quantun hypothesis were almost as remarkable as the hypothesis itself, since prac tically nothing was known in 1905 about the frequency dependence of th stopping potential for photoelectrons. It took a decade of difficult experi mentation before Einstein's equation (12) was fully confirmed, especiall by the work of Robert A. Millikan. Even in 1916, although Millika announced that Einstein had predicted 'exactly the observed results', h considered Einstein's idea of light quanta to be a 'bold, not to say reckless hypothesis', which had 'now been pretty generally abandoned'.

III

The year 1905 was Einstein's *annus mirabilis*. Because of his contributions, volume 17 of the *Annalen der Physik* of that year is now regarded as one of the most remarkable volumes of scientific literature ever published.
(G. J. Whitrow, *Einstein: The Man and His Achievement*)

When Einstein proposed the usefulness, the 'heuristic' value, of light quant on the basis of the arguments that have just been described, he had alread read Max Planck's papers on the theory of blackbody radiation. Planck ha been working on this problem since 1897 and in 1900 he announced a nev form for the radiation distribution, one which generalized Wien's lav (equation (4) above) and claimed validity for all frequencies an temperatures:

$$\rho(\nu, T) = \left(\frac{8\pi\nu^2}{c^3}\right)\frac{h\nu}{\exp(h\nu/kT)-1} \qquad (13$$

The constant h (Planck's constant), which appears in this radiation law related to the constants introduced earlier by the equation

$$h = \beta k \qquad (14$$

as one can easily see by looking at the limiting form of equation (13) fc large values of $(h\nu/kT)$, when it reduces to the Wien form. At the oth extreme, small values of $(h\nu/kT)$—low frequencies or high temperatures– Planck's result agrees with the inadequate result of mechanics and electr magnetic theory (equation (3) above) as Einstein pointed out in his 19c paper.

Planck's derivation of his distribution law was not easily disentangle however, and Einstein saw no direct connection between his own work an Planck's at that time. I say 'no direct connection' because Einstein had rea Planck's work and thought about it; it had stimulated him to find his ow way of dealing with radiation, a way quite different from Planck's. It w only in 1906 that Einstein realized that Planck, too, had introduced a ne discreteness into physics. In Planck's case it was not the energy of radiatic that was to be thought of as localized in particles or quanta, but rather t energy of those charged harmonic oscillators, the vibrating electrons th emitted and absorbed the radiation, that could only take on certain discre values rather than varying continuously. Planck had not been very cle about this point; he introduced the discreteness as a device to mal calculation possible, and did not insist on any physical significance of h 'elements of energy', as he called them, at the time he introduced them into physics in 1900.

Late in 1906, after Einstein had studied Planck's book on the theory of radiation and pursued his own ideas more deeply, he was ready to set forth some more startling consequences of his thinking. Planck's way of treating the charged oscillators in his theory was equivalent to saying that an oscillator of frequency ν could assume only the energies 0, $h\nu$, $2h\nu$, . . ., $nh\nu$, . . . and no others. The average energy u of such an oscillator in equilibrium at temperature T would no longer be given by equation (1), but instead by the equation

$$u = \frac{h\nu}{\exp(h\nu/kT) - 1} \qquad (15)$$

which reduces to the earlier result when $(h\nu/kT)$ is very small. This meant a modification in the kinetic molecular theory of heat, or statistical mechanics as we would now call it, a modification with major implications, as Einstein pointed out:

> While up to now molecular motions have been supposed to be subject to the same laws that hold for the motions of the bodies we perceive directly . . ., we must now assume that, for ions which can vibrate at a definite frequency and which make possible the exchange of energy between radiation and matter, the manifold of possible states must be narrower than it is for the bodies in our direct experience.

But this was not all, for Einstein went on to write:

> I now believe that we should not be satisfied with this result. For the following question forces itself upon us: If the elementary oscillators that are used in the theory of the energy exchange between radiation and matter cannot be interpreted in the sense of the present kinetic molecular theory, must we not also modify the theory for the other oscillators that are used in the molecular theory of heat? There is no doubt about the answer in my opinion. If Planck's theory of radiation strikes to the heart of the matter, then we must also expect to find contradictions between the present kinetic molecular theory and experiment in other areas of the theory of heat, contradictions that can be resolved in a similar fashion.

Einstein saw that what Planck had found was only the beginning, and that this unexpected discreteness of the energy would have to prevail in a variety of other situations. In other words, Einstein saw the need for a quantum theory which, when it was achieved, would clarify the properties of matter as well as those of radiation. He could not construct such a theory in general, but he could and did point to one of those 'contradictions between the present kinetic molecular theory and experiment' that already existed, and show how it could be resolved with the help of the new discreteness in energy. The contradiction concerned the specific heats of solids.

The essence of Einstein's profundity lay in his simplicity; and the essence of his science lay in his artistry—his phenomenal sense of beauty. 'This was sometime a paradox, but now the time gives it proof,' as Hamlet said in a different connection.

(Banesh Hoffmann: *Albert Einstein: Creator and Rebel*)

The calorimetric measurements made by Dulong and Petit early in the nineteenth century had shown that the heat capacities of the elements in the solid state had a common value, if each of these heat capacities were taken for a gram atomic weight (or mole) of the substance in question. This Dulong–Petit rule had provided a useful method for estimating atomic weights, and it found a simple explanation in the kinetic molecular theory. If the thermal motions of the atoms in a solid were taken to be simple harmonic vibrations about positions of equilibrium, there would be three independent motions per atom, or $3N_0$ oscillations for one mole of the substance. (N_0 is Avogadro's number, the number of atoms in a gram atomic weight.) Each vibration in a solid at temperature T would have an average energy of kT, as required by equation (1), and so the total thermal energy of one mole of the solid would have to be $3N_0kT$, or $3RT$, where R is the usual gas constant per mole. The rate of change of this thermal energy with temperature is the specific heat per mole, and it has the value $3R$, or approximately 6 calories per degree, the Dulong–Petit value. So far there is no contradiction. But this explanation of the Dulong–Petit rule proved too much, since the rule is only a rule and a number of elements were known to have specific heats much smaller than the Dulong–Petit value. These exceptions occur particularly among the lightest elements such as boron and carbon, and it was also known before 1900 that their specific heats vary rapidly with temperature, approaching the Dulong–Petit value well above room temperature.

During the first of my talks with Einstein an amusing incident occurred. I was very nervous and still very shy and after we had been talking for about twenty minutes the maid came in with a huge bowl of soup. I wondered what was happening and I thought that this was probably a signal for me to leave. But when the girl left the room Einstein said to me in a conspiratorial whisper: 'That's a trick. If I am bored talking to somebody, when the maid comes in I don't push the bowl of soup away and the girl takes whomever I am with away and I am free.
(L. L. Whyte, in G. J. Whitrow, *Einstein: The Man and His Achievement*)

There was also another problem, perhaps even more disturbing than these exceptions to the Dulong–Petit rule. By 1906 it was clear that atoms had an internal structure and that they 'contained', in some way, electrons. The frequencies at which ultraviolet light was absorbed in solids had been associated with electronic motions, just as the infrared absorption frequencies were associated with ionic vibrations. Why did these electronic motions contribute nothing at all to the specific heat of the solid, instead of the amount k per vibration that the classical theory seemed to require?

Einstein resolved all these difficulties with one stroke. For if he was right in thinking that all oscillations on the atomic scale had to have quantized energies ('if Planck's theory strikes to the heart of the matter'), then each oscillator has an average energy given by equation (15) instead of the classical value kT. The electronic oscillations at ultraviolet frequencies can be seen at once to make negligible contributions at any reasonable temperature, since at such high frequencies $(h\nu/kT)$ is a large number and the average energy given by equation (15) is, in effect, practically zero, as is its temperature derivative. As for the atomic vibrations, Einstein made the simplest possible assumption (recognizing explicitly that he might be oversimplifying): he took all these vibrations to be independent and of the same frequency ν. The energy U of one mole of the solid would then be given by the equation

$$U = \frac{3N_0 h\nu}{\exp(h\nu/kT) - 1}. \qquad (16)$$

The specific heat is obtained by differentiating U with respect to temperature. If this specific heat is plotted as a function of temperature, or rather of $(kT/h\nu)$, one obtains a curve that rises smoothly and monotonically from zero at the origin and approaches the value $3R$, the Dulong–Petit value, asymptotically when $(kT/h\nu)$ becomes large. Roughly speaking, the specific heat is negligibly small when $(kT/h\nu)$ is less than about 0.1, and it has about the full value of $3R$ when $(kT/h\nu)$ is appreciably greater than one. Since light atoms would be expected to have higher vibration frequencies than heavier ones, other things being equal, this result already explained qualitatively why the light elements have anomalously low specific heats at room temperature.

This theory of specific heats suggested an important and previously unsuspected connection between the optical and thermal properties of solids. Einstein identified the vibration frequency of the atoms with the frequency of optical absorption, at least for those crystals in which such absorption occurred. The data available to him were consistent with this relationship and in several cases he was able to make reasonably accurate predictions of the absorption frequency from the measured specific heat and his equation for its temperature dependence.

Even more important than this relationship between optical and thermal properties was the general theorem implied by Einstein's theory: the specific heats of all solids must become vanishingly small at sufficiently low temperatures. The solids that had been labelled as exceptions because they did not obey the Dulong–Petit rule were not to be thought of as exceptional at all; they merely exhibited the universal decrease of specific heats with decreasing temperature at relatively high temperatures, because of their light atoms and correspondingly high vibrational frequencies. Carbon in the form of the diamond crystal, for example, did not acquire the full Dulong–Petit value of its specific heat unless it was heated above 1000 °C, and its specific heat was only about a tenth of that value when it was cooled to only −50 °C. Einstein used the data on diamond, whose specific heat had been measured as a function of temperature, for a test of his theoretical equation. He could not, however, test it on other materials, particularly those that did obey the Dulong–Petit rule at room temperature, because no data for the behaviour of specific heats at low temperatures were available.

Such experiments were made a few years later by Walther Nernst and his collaborators in Berlin, not in order to test Einstein's theory of specific heats but to confirm Nernst's own ideas on the thermodynamic properties of matter near the absolute zero of temperature. Nernst found in 1910 and 1911 that all the many specific heats he measured did fall off at low enough temperatures, and learned that this had been predicted by Einstein's quantum theory of specific heats. Nernst was properly impressed by this

It has been said that common sense is the prerogative of the good, and the bad are destroyed by their lack of it. We may wonder if something similar does not apply to truth—that truth is the prerogative of the simple, and only those who are in a certain sense without guile are able to recognize it. In the case of someone like Einstein we cannot but feel that there is indeed an inner and necessary connection between the extraordinary theoretical simplicity of his work and the personal simplicity of the man himself. We feel that only someone himself so simple could have conceived such ideas.

(Henry Le Roy Finch, in *Conversations with Einstein*)

and became a staunch advocate of the importance of the new quantum theory of Planck and Einstein, even if he did refer to it as 'a very odd rule (for calculation), one might even say a grotesque one'.

The whole subject was discussed at the first of the famous Solvay Conferences on Physics, initiated and funded by the Belgian industrial chemist, Ernest Solvay. This first conference, held in 1911, had as its topic 'Radiation Theory and Quanta'. Lorentz, Planck, Nernst, and Einstein were among those who presented papers; the title of Einstein's paper was 'The Present State of the Problem of Specific Heats'.

IV

The history of physics offers many classic cases where the non-scientific attitude of 'disciples' is quite unmistakable, and the study of such cases might very well give the physicist a 'feel' for recognizing similar patterns occurring in our days. It was reflecting upon one such case—the difference of attitude between Newton and his successors—that made Einstein remark: 'Newton himself was better aware of the weaknesses inherent in his intellectual edifice than the generation of learned scientists which followed him. This fact has always aroused my deep admiration.'
(Stanley L. Jaki, *The Relevance of Physics*)

Einstein later summed up his feelings about the state of physics during this period in these words: 'It was as if the ground had been pulled out from under one's feet, with no firm foundation on which to build to be seen anywhere'. He devoted much of his effort to a continued probing of the consequences of Planck's distribution law for blackbody radiation, searching for what it implied about the structure of radiation and about the status of the electromagnetic field theory. In 1909 he reported some results of this probing at the annual meeting of German scientists, held that year at Salzburg. It was his first address to a major scientific gathering, and the first occasion for him to meet many of the physicists whose works he had studied.

In his address Einstein emphasized how much Planck had departed from classical ideas on radiation in his theory of the distribution law for blackbody radiation. Planck's answer, the law expressed in equation (13), was well confirmed by experiments over the whole accessible spectrum, but one might still have some doubts. 'Would it not be conceivable,' Einstein asked, 'that Planck's radiation formula was indeed correct, but that it could be derived by some method that was not based on such an apparently monstrous assumption as Planck had used? Would it not be possible to replace the hypothesis of light quanta by some other hypothesis by means of which one could do equal justice to the familiar phenomena? If it is necessary to modify the principles of the theory could one not at least retain the equations for the propagation of radiation and interpret only the elementary events of emission and absorption in a way different from that used previously?'

To all these questions Einstein's answer was 'No'. It was not possible to have Planck's satisfactory distribution law without the new and disturbing discreteness in nature. Einstein justified this assertion by extending his earlier application of Boltzmann's relation between entropy and probability (equation (7)). Given that the radiation was a thermodynamic system whose equilibrium state was described by Planck's law, one could calculate the fluctuations in its energy. If one considers that part of the blackbody radiation in a volume V, whose frequencies lie in a small interval between ν and $\nu + d\nu$, the mean square fluctuation in its energy $(\Delta E)^2$ is found to have the form

$$(\Delta E)^2 = V d\nu [h\nu\rho + (c^2/8\pi\nu^2)\rho^2], \qquad (17)$$

where ρ is given by Planck's law (equation (13)). Einstein was able to identify these two terms individually. The first is just the fluctuation to be expected in a collection of independent energy quanta, each of which has energy $h\nu$. The second is the fluctuation that would result from interfering waves.

Einstein commented that it was as though there were two independent causes producing the fluctuations, with their separate contributions being simply additive. In the high frequency, low temperature region, where Planck's law goes over to Wien's, the first or particle term predominates. In the low frequency, high temperature region, where the classical distribution is found, the second or wave term predominates. Einstein concluded that the particle-like behaviour in the high frequency region is a necessary consequence of Planck's distribution law. One cannot hope to avoid it by a new derivation of the distribution from alternative assumptions; the particle-like behaviour follows from the law itself. While Planck had introduced quantization as a *sufficient* condition for deriving his distribution, Einstein argued that it was a *necessary* implication of that distribution.

The fluctuation result with its two independent terms, which Einstein confirmed by other arguments of quite another sort, suggested something further. Einstein's earlier heuristic proposal of light quanta never purported to be more than that; he had never claimed that he was offering it as a new theory to replace Maxwell's theory of the electromagnetic field. But now there was at least a hint as to the proper direction in which progress might be made, since the wave and particle aspects of radiation appeared together in a single equation. 'It is my opinion,' Einstein announced, 'that the next phase of the development of theoretical physics will bring us a theory of light that can be interpreted as a kind of fusion of the wave and emission [particle] theories.' The problem was to take the next step since, as he remarked, 'the fluctuation properties . . . present small foothold for setting up a theory'. After all, if one had known nothing of interference or diffraction phenomena and had had only the second (wave) term in the fluctuations to go on, 'Who would have enough imagination to construct the wave theory of light on this foundation?'

Difficult as the task was, Einstein certainly tried. During the years from 1908 to 1911 he wrestled with the problem, trying to construct some sort of nonlinear equation that would allow him to introduce both the radiation constant h and the electronic charge e into the theory. He expected the discreteness of charge and the discreteness of energy to enter the theory together since the combination (e^2/hc) is dimensionless. Although he published nothing but a few passing remarks on his work we know from his correspondence during those years how intensively he worked on the radiation problem. This is especially true of his correspondence with H. A.

. . . in spite of so many touches which show his friendliness there is every sign that he was extraordinarily self-sufficient. Only a man as self-sufficient as he, could have worked out his first epoch-making discoveries in obscurity. But despite his friendships he was essentially a lonely figure. It was perhaps a penalty he had to pay for an endowment of genius of a magnitude which appears but rarely in the whole of recorded history.

(Christopher Sykes, in G. J. Whitrow, *Einstein: The Man and His Achievement*)

Photo Couprie, Bruxelles

GOLDSCHMIDT PLANCK RUBENS LINDEMANN HASENOHRL

NERNST BRILLOUIN SOMMERFELD M DE BROGLIE HOSTELET

SOLVAY KNUDSEN HERZEN JEANS RUTHERFORD

LORENTZ WARBURG WIEN EINSTEIN LANGEVIN

PERRIN Madame CURIE POINCARÉ KAMERLINGH ONNES

Figure 23 The First Solvay Conference on Physics, Brussels, 1911

Lorentz, whose electron theory was then much in Einstein's mind. In May 1911 Einstein wrote to his closest friend, Michele Besso, that he was no longer trying to construct quanta, 'because I now know that my brain is incapable of accomplishing such a thing'. It was at about this time that Einstein turned his full attention to the problem of gravitation, with historic consequences.

V

By the time Einstein took up the problem of radiation again in 1916, there had been major changes in the quantum theory. Niels Bohr's papers had shown that quantum concepts offered the possibility of understanding the structure of the atoms and the characteristics of the spectra they emit. Although Einstein did not work on these problems he was clearly influenced by Bohr's ideas, as Bohr had been by his. Einstein's new work was, in the first instance, a fresh derivation of the Planck distribution law. Einstein referred to it in print as 'astonishingly simple and general', and in a letter to Besso as perhaps '*the* derivation' of this important law. This new derivation avoided an inconsistency that marred Planck's own treatment, namely, the use of the electrodynamic result expressed in equation (2) in a situation where the assumptions underlying this equation were violated. Einstein had been aware of this difficulty since 1906, and now he had found a way of avoiding it.

The new derivation was based on statistical assumptions about the processes of emission and absorption of radiation, assumptions chosen so as to follow the pattern of the classical theory without adopting it in detail. It also employed the basic assumption of Bohr's theory, that atomic systems have a discrete set of possible stationary states. The proof then used the condition that the absorption and emission of radiation suffice to keep a gas of atoms in thermodynamic equilibrium. (This paper introduced the concept of stimulated emission into quantum physics and so is often referred to as having provided the basis for the laser.)

Einstein's new approach to the radiation problem also included arguments for the directional character of the radiation emitted by an atom. He showed that in each individual emission process in which a quantum of frequency ν is emitted, that quantum must carry away momentum $h\nu/c$ in a definite direction; spherical waves were ruled out. Einstein considered his theoretical proof that all radiation must be sharply directional to be the most significant aspect of this paper. There was no real experimental support for this result at the time, but it came a few years later in the form of the Compton effect, the increase in wavelength of X-rays scattered by effectively free electrons. In 1923 Arthur Compton and Peter Debye showed independently that the Compton effect could be explained if the scattering were treated as a collision, obeying the conservation laws, between a free electron at rest and a light quantum of energy $h\nu$ and momentum $h\nu/c$ in the direction of the incident beam. This successful treatment of the Compton effect made the light quantum acceptable to many physicists who had previously refused to take it very seriously.

During the 1920s the problems of applying the quantum theory to atomic structure and atomic spectra were at the centre of interest in physics. Einstein took no part in this development which was occupying so many of his colleagues, from Niels Bohr, Arnold Sommerfeld, and Max Born to their younger colleagues such as H. A. Kramers, Werner Heisenberg, and Wolfgang Pauli. Although his major concern in those years was the generalization of the theory of relativity, Einstein continued to think about the problems of quanta.

Einstein never liked his photon as tenderly as his beloved relativity. The photon was a natural child, a bastard born out of wedlock; Einstein remained a strong believer in differential equations in a continuous medium. Discontinuities and quanta seemed to him unnatural.
(Léon Brillouin, *Relativity Reexamined*)

In 1924 a new occasion for doing so arose when he received a paper in English from a young Indian physicist, S. N. Bose, setting forth a theory in which radiation was treated as a gas of light quanta. This approach had been tried before, but if the gas of quanta were treated by the usual statistical methods one ended up with Wien's distribution law rather than Planck's. By changing the statistical procedure for counting the states of the gas Bose had been able to obtain the proper Planck distribution. Einstein was much taken with this paper. He translated it into German and saw that it was published, and then applied Bose's idea to a gas of material particles. This Bose–Einstein gas, as it came to be called, showed a variety of novel and interesting properties.

While he was working out the behaviour of this gas Einstein received a copy of a doctoral dissertation written in Paris. The author, Louis de Broglie, inspired by Einstein's earlier studies of the wave–particle duality for radiation, had become convinced that this duality must hold for matter as well. His thesis developed the idea that every material particle has a wave associated with it, the frequency ν and wavelength λ of the wave being related to energy E and momentum p of the particle through the equations

$$E=h\nu \qquad p=h/\lambda \tag{18}$$

Since de Broglie had no experimental evidence for his matter waves, his work did not impress most physicists. Einstein, however, was quite taken with it, and realized that de Broglie had 'lifted a corner of the great veil'. He found that de Broglie's ideas fitted in very well with his current work on the new theory of the gas. Both were concerned with the parallels between the gas of quanta and the gas of material particles. The fluctuations in density of the Bose–Einstein gas, which Einstein calculated early in 1925 showed exactly the same two-term structure as the fluctuations in blackbody radiation. Einstein saw this as important evidence supporting de Broglie's matter waves, and went on to suggest a number of experimental possibilities for detecting the de Broglie waves.

VI

In that same year, 1925, Heisenberg proposed a new approach to the

Figure 24

Einstein and Niels Bohr deep in thought (taken by Ehrenfest in about 1927)

quantum theory, an approach quickly developed by him in collaboration with Born and Pascual Jordan into a quantum mechanics based on matrix algebra. Einstein was interested and impressed, but he was not convinced. 'The most interesting theoretical work produced recently is the Heisenberg–Born–Jordan theory of quantum states,' he wrote to Besso. 'It's a real witches' calculus, with infinite determinants (matrices) taking the place of Cartesian coordinates. Most ingenious, and adequately protected by its great complexity against being proved wrong.' The following year he expressed his negative opinion to Born: 'An inner voice tells me that it is still not the true Jacob,' a judgement that Born took as 'a hard blow'.

When Erwin Schrödinger introduced an alternative to the algebraic quantum mechanics with his wave equation, Einstein reacted much more favourably. 'I am convinced that you have made a decisive advance with your formulation of the quantum condition,' he wrote to Schrödinger, 'just as I am equally convinced that the Heisenberg–Born route is off the track.' This reaction of Einstein's is not too surprising since Schrödinger's work followed the direction pointed by de Broglie, and he had been much influenced by what he referred to as Einstein's 'short but infinitely far-seeing remarks' on the implications of de Broglie's thesis.

As it turned out, the two methods that seemed so different were mathematically equivalent, and both became part of the synthesis that constituted the new quantum mechanics. One of the key features of this synthesis was Born's statistical interpretation of Schrödinger's wave function. This meant that the new theory was intrinsically statistical and renounced as meaningless any attempt to go beyond the probabilities to obtain a deterministic theory. Bohr expressed the generally accepted opinion when he described quantum mechanics as a 'rational generalization of classical physics', a generalization that resulted from the 'singularly fruitful cooperation of a whole generation of physicists'.

There was one great dissenter from this general agreement—Albert Einstein. He never accepted the finality of the quantum mechanical renunciation of causality, or its claim to be the new fundamental theory. From the Solvay Conference of 1927, where the quantum mechanical synthesis had its first major discussion, to the end of his life, Einstein never stopped raising questions about this new approach to physics. At first he tried to propose conceptual experiments that would prove the logical inconsistency of quantum mechanics, but these attempts were all turned aside successfully by Bohr and his collaborators. In 1935 Einstein began to emphasize another basic limitation in quantum mechanics, as he saw it. He argued that its description of physical reality was essentially incomplete, that there were elements of physical reality that had no counterparts in the theory. Bohr's response to this was to reject Einstein's criterion of physical reality as ambiguous, and to claim that only through his own principle of complementarity could one arrive at an experimentally meaningful criterion of completeness.

Einstein recognized the power of quantum mechanics, calling it 'the most successful physical theory of our time', but he would not admit it as the basis for theoretical physics. He refused to give up the idea that there was such a thing as 'the real state of a physical system, something that objectively exists independently of observation and measurement, and which can, in principle, be described in physical terms'. Einstein was convinced that when a theory giving a complete physical description was developed, the position of quantum mechanics in the framework of this future physics would be analogous to that of statistical mechanics in the framework of classical physics. It would be the theory to use when only incomplete information was available or when only an incomplete description was wanted.

Einstein's colleagues could only regret that he had chosen to follow a path separate from the rest. As Born wrote: 'Many of us regard this as a tragedy—for him, as he gropes his way in loneliness, and for us, who miss our leader and standard-bearer.' To Einstein himself the choice was inevitable. He was prepared for the 'accusation' brought against him sometimes 'in the friendliest of fashions', but sometimes not: he was accused of 'rigid adherence to classical theory'. But, he wrote, it was not so easy to declare guilt or innocence of this charge 'because it is by no means immediately clear what is meant by "classical theory"'. Newtonian mechanics was a classical theory, but it had not been an acceptable claimant as the fundamental theory underlying physics since the introduction of field theory. Field theories were never completed—neither Maxwell's theory of electromagnetism nor his own theory of gravitation—since they were never extended to include the sources of the field in a non-singular way. Einstein did plead guilty to adherence to the programme of field theory; for it was his hope that a complete field theory would provide the basis for all of physics, giving that complete description he missed in the quantum mechanics he had helped so much to develop. He saw his whole career as striving to create a new unified foundation for physics. That was what he meant when he ended his scientific autobiography by writing that he had tried to show 'how the efforts of a life hang together and why they have led to expectations of a definite form'.

References

Bernstein, Jeremy, *Einstein* (New York: Fontana, 1973).
Einstein, A., *Ideas and Opinions* (New York: Dell, 1954).
—— *Out of My Later Years* (New York: Dell, 1950).
Einstein, A. and Besso, M., *Correspondance 1903–1955* (Paris: Hermann, 1972).
Frank, Philipp, *Einstein. His Life and Times* (New York: Knopf, 1947).
Hermann, Armin, *The Genesis of Quantum Theory (1899–1913)* (Cambridge, Mass.: MIT Press, 1971).
Hoffmann, Banesh and Dukas, Helen, *Albert Einstein: Creator and Rebel* (New York: Viking Press, 1972).
Jammer, Max, *The Conceptual Development of Quantum Mechanics* (New York: McGraw-Hill, 1966).
Schilpp, P. A. (Editor), *Albert Einstein: Philosopher-Scientist* (Evanston, Illinois: Library of the Living Philosophers, 1949). See especially Einstein's 'Autobiographical Notes', pp. 1–95, and also the essays by Louis de Broglie, Wolfgang Pauli, Max Born, and Niels Bohr.
Seelig, Carl, *Albert Einstein. A Documentary Biography* (London: Staples Press, 1956).

EDITOR'S NOTE: OTHER REFERENCES

An English translation of Einstein's 1905 paper on light quanta was published by Arons, A. B., and Peppard, M. B., *Am. J. Phys.*, **33**, 367–74 (1965).
The following articles are also recommended:
Klein, M. J., 'Einstein's first paper on quanta' in *The Natural Philosopher*, Volume 2 (New York: Blaisdell, 1963).
—— 'Einstein, specific heats and the early quantum theory', *Science*, **148**, 173–80 (1965).
—— 'Einstein and the wave-particle duality' in *The Natural Philosopher*, Volume 3 (New York: Blaisdell, 1964).

Figure 25 Cartoon by Low, 1929

6 'What, precisely, is "thinking"?' Einstein's answer

Gerald Holton

The task assigned to me is nothing less than writing about Albert Einstein's way of thinking in science—and to do so in a manner accessible to students. It is appealing and appropriate to put Einstein's own thoughts and words at the focus of a publication meant to honour his memory and achievement on the centennial of his birth. Yet, at first glance, it seems to be an impossible undertaking. His work was carried out at the very frontiers of physics and of human ability. And his mind was not open to easy study from the outside, even by those who worked with him—as was discovered by the physicist Banesh Hoffmann who, with Leopold Infeld, was Einstein's assistant in 1937. Hoffmann has given a delightful account of what it was like when he and Infeld, having come to an impassable obstacle in their work, would go and seek out Einstein to try to get help. At such a point, Hoffmann related, 'We would all pause and then Einstein would stand up quietly and say, in his quaint English, "I will a little think." So saying, he would pace up and down and walk around in circles, all the time twirling a lock of his long grey hair around his forefinger. At these moments of high drama, Infeld and I would remain completely still, not daring to move or make a sound, lest we interrupt his train of thought.' Many minutes would pass this way, and then, all of a sudden, 'Einstein would visibly relax and a smile would light up his face. . . . Then he would tell us the solution to the problem, and almost always the solution worked. . . . The solution sometimes was so simple we could have kicked ourselves for not having been able to think of it by ourselves. But that magic was performed invisibly in the recesses of Einstein's mind, by a process that we could not fathom. From this point of view the whole thing was completely frustrating.'

Einstein once wrote that the job of a lighthouse keeper would be ideal for a scientist, because he would be guaranteed the necessary spare time for thinking and working. I tried to convince him that there were perhaps only two or three men in the whole world who could carry out scientific work in such conditions, for man also needs company for his work. He did not seem, however, to understand my viewpoint. In actual fact he is the only scientist who could have lived quite contentedly as a lighthouse keeper.

(Carl Seelig, *Albert Einstein: A Documentary Biography*)

But if not accessible from the outside, Einstein's mind was accessible from the inside, since, like many of the best scientists, he was interested in the way the scientific imagination works and he wrote about it frankly. As far as possible, we shall follow the description in his own words of how he wrestled with theories of fundamental importance. Needless to say, we shall not be under any illusion that thereby one can imitate or even fully 'explain' his detailed thought processes, nor will we forget that other scientists have other styles. There are many sources to draw on, for Einstein discussed his view of the nature of scientific discovery, in a generally consistent way, on many occasions (notably in the essays collected in *Ideas and Opinions*, see references to further readings), and in letters written to his friends. He was also intrigued enough by this problem to discuss it with a number of researchers into the psychology of scientific ideas, including Wertheimer, Hadamard, and Piaget, and with many philosophers of science. Indeed, from his earliest student days, Einstein was deeply interested in the theory of knowledge (epistemology). He wrote, 'The reciprocal relationship of epistemology and science is of noteworthy kind. They are dependent upon each other. Epistemology without contact with science becomes an empty scheme. Science without epistemology is—in so far as it is thinkable at all—primitive and muddled' (see Schilpp, pp. 683–4).

'I am convinced that the philosophers have had a harmful effect upon the progress of scientific thinking in removing certain fundamental concepts from the domain of empiricism, where they are under our control, to the intangible heights of the *a priori*. This is particularly true of our concepts of space and time, which physicists have been obliged by the facts to bring down from the Olympus of the *a priori* in order to adjust them and put them in a serviceable condition'.

(A.E.)

An admirably suitable place to enter his thoughts is a set of pages near the beginning of an essay he wrote in 1946 as the opening article for the book *Albert Einstein: Philosopher-Scientist* (editor, P. A. Schilpp). It is the only serious autobiographical essay he ever wrote, and he called it jokingly his own 'obituary'. Obtain and read this essay if you possibly can; it gives a fascinating picture of Einstein's contributions as he viewed them, looking back at the age of sixty-seven. The essay is chiefly an account of his intellectual development, rather than an autobiography in the usual sense. We shall use this remarkable document to learn from his own words while avoiding the use of technical, philosophical terminology, as he himself avoided it.

THE COURAGE TO THINK

It is certainly curious to start one's autobiography, not with where and when one was born, the names of one's parents, and similar personal details, but to focus instead on the question: 'What, precisely, is "thinking"?' Einstein explains (see Schilpp, p. 33) why he has to start his 'obituary' in this way: 'For the essential in the being of a man of my type lies precisely in *what* he thinks and *how* he thinks, not in what he does or suffers.'

With this viewpoint, thinking is not a joy or a chore added to the daily existence. It is the essence of man's very being, and the tool by which the transient sorrows, the primitive forms of feeling, and the other 'merely personal' parts of existence can be mastered. For it is through such thought that man can lift himself up to a level where he can think about 'great, eternal riddles'. It is a 'liberation' which can yield inner freedom and security. When the mind grasps the 'extra-personal' part of the world—that part which is not tied to shifting desires and moods—it gains knowledge which all men and women can share regardless of individual conditions, customs, and other differences.

This, of course, is precisely why the laws of nature, towards which these thoughts are directed, are so powerful: their applicability can be demonstrated by anyone, anywhere, at any time. The laws of nature are utterly shareable. Insofar as his conclusions are right, the laws discovered by a scientist are equally valid, or *invariant* with respect to the personal conditions of different thinkers. Einstein's interest in this matter seems to be not unrelated to his work in the physics of relativity: the essence of relativity theory is precisely that it provides a tool for expressing the laws of nature in such a manner that they are invariant with respect to differently *moving* observers.

As his *Autobiographical Notes* show, Einstein was also aware that life cannot be all thought, that even the enjoyment of thought can be carried to a point where it may be 'at the cost of other sides' of one's personality. But the danger which the more ordinary person and young people generally face, is not that they will abandon their very necessary personal ties, but that the society surrounding them will not say often enough what Einstein here suggests to his wide audience: that the purpose of thinking is more than merely solving problems and puzzles. It is instead, and most importantly, the necessary tool for permitting one's strong side to come through, so that 'gradually the major interest disengages itself . . . from the momentary and merely personal'. Here Einstein is saying, 'have the courage to take your own thoughts seriously'.

I got the impression of a man with an unqualified devotion to truth in scientific matters, as ready to discard his own views as those of another, if they failed to measure up to the demands of reason or experience. This, of course, is the popular idea of the typical scientist. The peculiarity of Einstein was that he conformed to it.

(H. Dingle, in G. J. Whitrow, *Einstein: The Man and His Achievement*)

THINKING WITH IMAGES

Having touched on the *why* of thinking, the *Autobiography* takes up the *how* of thinking. In a similar essay elsewhere (*Physics and Reality*)* Einstein explains that he does not limit himself to science in this question: he was convinced that 'The whole of science is nothing more than a refinement of everyday thinking'. Both start with the same raw material. Adhering to one of several contesting traditions in psychology and philosophy current from Aristotle to the present (see Arnheim, Chapters 6 and 12), Einstein holds that the repeated encounter with images such as 'memory pictures' in different contexts, leads to the formation of 'concepts'. (Thus, a small child might form the concept 'glass' when he experiences that a variety of differently shaped, transparent objects break on being dropped.)

A concept must of course eventually be put into verbal form if the aim is to communicate the idea to others; but for private thought it is not necessary to wait for this stage. Indeed, for some people, including Einstein and probably also such physicists as Faraday and Rutherford, the most important part of thinking may occur not with the use of words: 'I have

* This essay, and the others cited later in this article, may be found in *Ideas and Opinions*, a collection of Einstein's writings, addresses, etc. Detailed page references may be found on page 164.

no doubts but that our thinking goes on for the most part without use of signs (words) and beyond that to a considerable degree unconsciously'. Such persons tend to think in terms of images to which words may or may not be assignable. Einstein tells of his pleasure in discovering, as a boy, his skill in contemplating relationships among geometrical 'objects'—triangles and other abstract objects of the imagination. He explains more fully in a letter to Jacques Hadamard:

> The psychical entities which seem to serve as elements in thought are certain signs and more or less clear images which can be 'voluntarily' reproduced and combined. . . . But taken from a psychological viewpoint, this combinatory play seems to be the essential feature in productive thought—before there is any connection with logical construction in words or other kinds of signs which can be communicated to others. The above-mentioned elements are, in my case, of visual and some muscular type. Conventional words or other signs have to be sought for laboriously only in a secondary stage, when the mentioned associative play is sufficiently established and can be reproduced at will.

Einstein found, as Spinoza did, that the human qualities of modesty, equanimity, universality, equality and kindness were actually promoted by a sense of the vast impersonality of truth and of a natural law and harmony far beyond human hopes, fears and wishes.

(Henry Le Roy Finch, in *Conversations with Einstein*)

GEDANKENEXPERIMENTS

Einstein's ability to visualize may have helped him in the brilliant use he made of 'thought experiments' (*Gedankenexperiments*). His first came to him at the age of sixteen, when he tried to imagine that he was pursuing a beam of light and wondered what the observable values of the electric and magnetic field vectors would be in the electromagnetic wave making up the light beam (see Schilpp, p. 53). He wrote later that in this problem 'the germ of the special relativity theory was contained'. Among other examples, Einstein related one which he said had led him to the general theory of relativity. In 1907 he was trying to modify the special theory of relativity, which applied to systems moving with constant velocity and therefore did not accommodate in a natural way the behaviour of accelerating objects, and hence was not applicable to gravitation.

> At that point there came to me the happiest thought of my life, in the following form: Just as is the case with the electric field produced by electromagnetic induction, the gravitational field has similarly only a relative existence. *For if one considers an observer in free fall, e.g. from the roof of a house, there exists for him during his fall no gravitational field*—at least in his immediate vicinity. For if the observer releases any objects they will remain relative to him in a state of rest, or in a state of uniform motion, independent of their particular chemical and physical nature. (In this consideration one must naturally neglect air resistance.) The observer therefore is justified to consider his state as one of 'rest'.
>
> The extraordinarily curious, empirical law that all bodies in the same gravitational field fall with the same acceleration received through this consideration at once a deep physical meaning. For if there is even a single

thing which falls differently in a gravitational field than do the others, the observer would discern by means of it that he is in a gravitational field, and that he is falling into it. But if such a thing does not exist—as experience has confirmed with great precision—the observer lacks any objective ground to consider himself as falling in a gravitational field. Rather, he has the right to consider his state as that of rest, and his surroundings (with respect to gravitation) as fieldfree.

The fact of experience concerning the independence of acceleration in free fall with respect to the material is therefore a mighty argument that the postulate of relativity is to be extended to coordinate systems that move non-uniformly relative to one another. . . .

(Holton, 'Notes towards the Psychobiographical Study of Scientific Genius')

In the dining-room there was always a little table set for him with milk, bread, cheese, cakes and fruit. 'What more can a man want than these things, plus a violin, a bed, a table and a chair?' he cried, delighted that he could stay here unobserved and in complete freedom.

(Carl Seelig, *Albert Einstein: A Documentary Biography*)

THE FREE PLAY WITH CONCEPTS

Having stressed the role of images and memory pictures, including Gedankenexperiments, in thinking and having defined 'concepts' as the crystallized products, the unvarying elements found to be common to many series of such memory pictures, Einstein makes a startling assertion: 'All our thinking is of this nature of a free play with concepts.' This has to be unravelled carefully, for it deals with the eternal antithesis between the two indispensable elements in all human thought, the empirical and the rational. Even if one grants that 'free play' is still play within definite, if only temporarily chosen, rules—similar to tentatively trying out a word to see if it fits into a crossword puzzle—by no means all philosophers would agree with Einstein's position. Some would argue that the external world imposes itself strongly on us and gives us little leeway for play, let alone for choosing the rules of the game. In Einstein's earlier years, most of his contemporaries argued that the boundaries of any such 'play' were fixed by final categories of thought such as those proposed by Immanuel Kant: concepts such as Newtonian absolute space or absolute time were absolute givens. Only a few disagreed, including Ernst Mach, who called absolute space 'a conceptual monstrosity, purely a thought-thing which cannot be pointed to in experience.'

Thus Einstein was struggling anew with the old question: what precisely is the relation between our knowledge and the sensory raw material—'the only source of our knowledge?' (*Remarks on Bertrand Russell's Theory of Knowledge*). If we could be sure that there is one unchanging, external, 'objective' world that is connected to our brains and our sensations in a reliable, causal way (a view held by many philosophers of antiquity and some today), then pure thought can lead to truths about physical science. But since we cannot be certain of this, how can we avoid falling constantly into error or phantasy? David Hume had shown that 'habit may lead us to belief and expectation but not to the knowledge, and still less to the understanding, of lawful relations' (*Remarks on Bertrand Russell's Theory of Know-*

ledge). Einstein concluded that 'In error are those theorists who believe that theory comes inductively from experience' (*Physics and Reality*).

In fact, he was sceptical about both of the major opposing philosophies. He wrote that there is an 'aristocratic illusion [of subjectivism or idealism] concerning the unlimited penetrating power of thought', just as there is a 'plebeian illusion of naive realism, according to which things "are" as they are perceived by us through our senses' (*Remarks on Bertrand Russell's Theory of Knowledge*). Like Ernst Mach (see Holton, *Thematic Origins*, p. 222), Einstein held that there is no 'real world' to which one can repair—the whole concept of the 'real world' is justified only insofar as it refers to the mental connections that weave the multitude of sense impressions into some connected net (*Physics and Reality*). Sense impressions are 'conditioned by an "objective" and by a "subjective" factor' (Schilpp, p. 673). Reality is a relation between what is in you and outside you: 'The real is not given to us, but put to us (by way of a riddle)' (Schilpp, p. 680).

You imagine that I look back on my life's work with calm satisfaction. But from nearby it looks quite different. There is not a single concept of which I am convinced that it will stand firm, and I feel uncertain whether I am in general on the right track.
(A.E. to Maurice Solovine, 28 March 1949)

Since the world that the scientist has to deal with is more complex than is allowed for in any of the current philosophies, Einstein thought that the way to escape illusion was to avoid being a captive of any one school of philosophy. He would take from any system the portions he found useful. Such a scientist, he realized 'therefore must appear to the systematic epistemologist as a type of unscrupulous opportunist: he appears as a *realist* insofar as he seeks to describe the world independent of the acts of perception; as *idealist* insofar as he looks upon the concepts and theories as the free inventions of the human spirit (not logically derivable from what is empirically given); as *positivist* insofar as he considers his concepts and theories justified *only* to the extent to which they furnish a logical representation of relations among sensory experiences. He may even appear as *Platonist* or *Pythagorean* insofar as he considers the viewpoint of logical simplicity as an indispensable and effective tool of his research' (Schilpp, p. 684).

But what justifies this 'free play with concepts'? There is only one justification: that it can result, perhaps after much labour, in a thought structure which gives us the testable realization of having achieved meaningful order over a large range of sense experiences, which would otherwise seem separate and unconnected. In the important essay *Physics and Reality*, which covers much the same ground and is strongly recommended, Einstein makes the same point with this fine image: 'By means of such concepts and mental relations between them, we are able to orient ourselves in the labyrinth of sense impressions.'

This important process is described by Einstein in a very condensed way in *Autobiographical Notes*. 'Imagine,' he says, 'on one side the totality of sense experiences', such as the observation that the needle on a meter is shown to deflect. On the other side, he puts the 'totality of concepts and propositions which are laid down in books', which comprises the distilled products of past progress such as the concepts of force or momentum, propositions or axioms that make use of such concepts (for example, the law of conservation of momentum), and, more generally, any concepts of ordinary thinking (for example 'black' and 'raven'). Investigating the relations that exist among the concepts and propositions is 'the business of logical thinking', which is carried out along the 'firmly laid-down rules' of logic. The rules of logic, like the concepts themselves, are of course not God-given but are the 'creation of man'; however, once they are agreed upon and are part of a widely-held convention—the rules of syllogism, for example—they tell us with (seemingly) inescapable finality that if all ravens are black and a particular bird is a raven, then the bird is black. They likewise allow us to deduce from the law of conservation of momentum that in a closed system containing only a neutron and a proton, the momentum gained by one is accompanied by the loss experienced by the other. Without the use of logic to draw conclusions, no disciplined thinking, and hence no science, could exist.

He always spoke quite openly of the aesthetic appeal, of the beauty and harmony, of certain conceptions of classical physics. It was this feeling, closely allied to his considerable musical talent, that guided him in his scientific thinking.
(E. H. Hutten, in G. J. Whitrow, *Einstein: The Man and His Achievement*)

But all such conclusions, Einstein warns, are empty of useful 'meaning' or 'content' until there is some definition by which the particular image e.g. ('raven' or 'neutron') is correlated with actual instances of the image which have consequences in the world of experience rather than in the world of words and logical rules. Necessary though the correlation or connection between concepts and sense experience is, Einstein warns that it is 'not itself of a logical nature'. It is an act in which, Einstein holds, intuition is one guide, even if not an infallible one. Without it one could not be led to the assertion that a particular bird, despite some differences in its exact size or degree of blackness from all other birds, does belong to the particular species raven; or that the start of a particular track, visible in the cloud chamber, is the place where a neutron has struck a proton.

One might wish that Einstein had used a notion more firm than the dangerous-sounding one of 'intuition'. But he saw no other way. He rejected the use of the word 'abstraction' to characterize the transition from the observation of individual black birds to the concept of 'raven', or the correlation of any sense experience with the corresponding concept. He rejected it precisely because, he said, 'I do not consider it justifiable to veil the logical independence of the concept from the sense experiences' (whereas the use of the term abstraction or induction might make it seem as if there were a logical dependence). He put it in terms of a marvellous analogy: the relation between sense experience and concept 'is not analogous to that of soup to beef, but rather of check number to overcoat' (*Physics and Reality*).

The danger is evidently that delusion or fantasy can and does make similar use of the same elements of thinking and since there are no hard, utterly reliable connections between the concepts, propositions, and experience, one cannot know with absolute certainty whether one has escaped the trap of false conclusion. That is why it was thought for so long that observations proved that the sun went around the earth; that time had a universal meaning, the same for all moving observers; and that Euclidean geometry was the only one that has a place in the physical world. But this is just where Einstein's view is most helpful: only those who think they *can* freely play with concepts can pull themselves out of such error. His message is even more liberal: the concepts themselves, in our thoughts and verbal expressions are, 'when viewed logically, the free creation of thought which cannot inductively be gained from sense experience' (*Remarks on Bertrand Russell's Theory of Knowledge*). We must be continually aware that it is not necessity but habit which leads us to identify certain concepts (for example, 'bread') with corresponding sense experience (feel, smell, taste, satisfaction); for, since this works well enough most of the time, 'we do not become conscious of the gulf—logically unbridgeable—which separates the world of sense experience from the world of concepts and propositions'. Einstein is perhaps so insistent on the point because he had to discover it the hard way: as a young man, he had to overcome the accepted meanings of such concepts as space, time, simultaneity, energy, etc., and to propose redefinitions which reshaped all our physics and hence our very concept of reality itself.

Einstein's was a firm belief 'that the history of scientific development has shown that of all thinkable theoretical structures a single one has at each stage of advance proved superior to all the others.' Such a system can always be recognized by its high degree of simplicity, and this is, as Einstein remarked, what physics looks for: 'the simplest possible system of thought which will bind together the observed facts.'
(Stanley L. Jaki, *The Relevance of Physics*)

Once a conceptual structure has tentatively been erected, how can one check whether it is scientifically 'true'? It depends on how nearly the aim of making the system deal with a large amount (ideally, cover the totality) of diverse sense experience has been achieved, and how economical or parsimonious the introduction of separate basic concepts or axioms into a system has been. Einstein doubted a physical theory, and would say that it failed to 'go to the heart of the matter', if it had to be jerry-built with the aid of *ad hoc* hypotheses, each specially introduced to produce greater agreement between theory and experience (experiment). He was also rarely convinced by theories that dealt with only a small part of the range of physical phenomena, applied only here or there under special circumstances. A really good theory, one that has high scientific 'truth' value, is correct not merely by virtue of not harbouring any logical contradictions; it also allows a close check on the correspondence between the predictions of the theory and a large range of possible experimental experiences. He summarized all this in the following way: 'One comes nearer to the most superior scientific goal, to embrace a maximum of experimental content through logical deduction from a minimum of hypotheses. . . . One must allow the theoretician his imagination, for there is no other possible way for reaching the goal. In any case, it is not an aimless imagination but a search for the logically simplest possibilities and their consequences' (*The Problem of Space, Ether and Field in Physics*, in the translation by Seelig).

This search may take 'years of groping in the dark'; hence the ability to hold on to a problem for a long time, and not to be destroyed by repeated failure, is necessary for any serious researcher. As Einstein once said, 'Now I know why there are so many people who love chopping wood. In that activity one immediately sees the results.' But for him, the goal of 'embracing a maximum of experimental content . . . with a minimum of hypotheses' meant nothing less than the simplification and unification of our world picture by producing fusions in hitherto separate fundamental concepts such as space and time, mass and energy, gravitation and inertial mass, electric and magnetic fields, and inertial and accelerating systems.

KEEPING ALIVE THE SENSE OF WONDER

Before further unravelling Einstein's views on how to think scientifically about the deep problems, let us consider, at least briefly, an engaging passage in the *Autobiographical Notes* in which Einstein speaks of the importance of the sense of marvel, of deep curiosity, of 'wonder', such as the two experiences he describes: when, at the age of 4 or 5, he was shown a magnetic compass by his father, and when, at the age of 12, a book on Euclidean geometry came into his hands. A person's thought-world develops in part just by mastering certain new experiences which were so inexplicable, in terms of the previous stage of development, that a sense of wonder or enchantment was aroused. As we learn more, through both science and other approaches, we progressively find that the world around us, as it becomes more rational, also becomes more 'disenchanted'. But Einstein repeatedly insisted in other writings that there is a limit to this progressive disenchantment, and even the best scientist must not be so insensitive or falsely proud as to forget it. For, as Einstein said in a famous paragraph (in *Physik und Realität*, 1936, properly translated): 'It is a fact that the totality of sense experiences is so constituted as to permit putting them in order by means of thinking—a fact which can only leave us astonished, but which we shall never comprehend. One can say: the eternally incomprehensible thing about the world is its comprehensibility.'

He went on: 'In speaking here of "comprehensibility", the expression is used in its most modest sense. It implies: the production of some sort of order among sense impressions, this order being produced by the creation of general concepts, by relations among these concepts, and by relations of some kind between the concepts and sense experience. It is in this sense that the world of our sense experiences is comprehensible. The fact that it is comprehensible is a wonder.'

That wonder, that sense of awe, can only grow stronger, Einstein implied, the more successfully our scientific thoughts find order to exist among the separate phenomena of nature. This success aroused in him a 'deep conviction of the rationality of the universe'; to this conviction he gave the name 'cosmic religious feeling', and he saw it as the 'strongest and noblest motive for scientific research' (*Religion and Science*).

After the publication of such sentiments, Einstein received a plaintive letter from one of his oldest and best friends, Maurice Solovine. They had come to know each other during Einstein's early years in Bern when Einstein was twenty-three years old, and they became close friends. Solovine was then a young philosophy student at the University of Bern, which he had come to from Rumania, and, together with Conrad Habicht,

As different as they apparently were as men, both Newton and Einstein shared a feeling in the fitness of their own intuitions. As Einstein once put it, 'To him who is a discoverer in this field the products of his imagination appear so necessary and natural that he regards them, and would like to have them regarded by others, not as creations of thought but as given realities.'

(Jeremy Bernstein, *Einstein*)

who was also a student at the University, they banded together to meet regularly for a weekly reading and discussion of works in science and philosophy. With high irony they called themselves the 'Olympia Academy'. Their 'dinners' were no banquets: they all lived on the edge of poverty, and Solovine tells us that their ideal of a special treat was two hard-boiled eggs each. But the talk was that much better, as they discussed works by Ernst Mach, J. S. Mill, David Hume, Plato, Henri Poincaré, Karl Pearson, Spinoza, Hermann Helmholtz, Ampère—and also those of Sophocles, Racine, and Dickens. Many of Einstein's epistemological ideas might be traced back to those discussions.

'What position does the world-picture of the theoretical physicist occupy among all those that are possible? He demands the greatest rigour and accuracy in his representation, such as can be gained only by using the language of mathematics. But for this very reason the physicist has to be more modest than others in his choice of material, and must confine himself to the simplest events of the empirical world, since all the more complex events cannot be traced by the human mind with that refined exactness and logical sequence which the physicist demands.'

(A.E.)

Now, half a century later, Maurice Solovine was worried. He asked Einstein how there could be a puzzle about the understandability of our world. For us it is simply an undeniable necessity, which lies in our very nature. No doubt Solovine was bothered that Einstein's remarks seemed to allow into science, that most rational activity of mankind, a function for the human mind which is not 'rational' in the sense of being coldly logical. But Einstein rejected as a 'malady' (*Remarks on Bertrand Russell's Theory of Knowledge*) the kind of accusation which implied that he was becoming 'metaphysical'. Instead, he saw it not as a weakness, but as a strength, that one could use *all* one's faculties and skills to do science. Certainly, he did not propose to abandon rationality, nor to guess in situations where one must puzzle things out in a careful, logical way. But he saw that there is, and has to be, a role for those other elements of thinking which, properly used, can help scientific thought. Specifically, this can happen at two points in Einstein's scheme. One is the courageous use of an intuitive feeling for nature *when there is simply no other guide available at all*—as when one has tentatively to propose an axiom that by definition is unproved (as Einstein did at the start of the first paper on relativity, where he simply proposed the principle of relativity and the principle of constancy of light velocity); or when one decides which sense experiences to select in order to make an operational definition of a concept. The other point is the sense of wonder at being able to discern the grand design of the world, a feeling that motivates and sustains many a scientist in his quest.

Einstein's reply to Solovine (30 March 1952) addresses this second point.

You find it remarkable that the comprehensibility of the world (insofar as we are justified to speak of such a comprehensibility) seems to me a wonder or eternal secret. Now, *a priori*, one should, after all, expect a chaotic world that is in no way graspable through thinking. One could (even *should*) expect that the world turns out to be lawful only insofar as we make an ordering intervention. It would be a kind of ordering like putting into alphabetic order the words of a language. On the other hand, the kind of order which, for example, was created through [the discovery of] Newton's theory of gravitation is of a quite different character. Even if the axioms of the theory are put forward by human agents, the success of such an enterprise does suppose a high degree of order in the objective world, which one had no justification whatever to expect *a priori*. Here lies the sense of 'wonder' which increases ever more with the development of our knowledge.

And here lies the weak point for the positivists and the professional atheists, who are feeling happy through the consciousness of having successfully made the world not only god-free, but even 'wonder-free'. The nice thing is that we must be content with the acknowledgment of the 'wonder', without there being a legitimate way beyond it. I feel I must add this explicitly, so you wouldn't think that I—weakened by age—have become a victim of the clergy.

EPILOGUE

You are now in a good position to look, if you wish, at other essays by Einstein on the subject (e.g. in *Ideas and Opinions*), other analyses of Einstein's epistemology (as in the Schilpp volume), or at an analysis of one of Einstein's other letters to Maurice Solovine (see page 271) in which Einstein goes over some of these questions, but this time with the aid of a diagram—as befits a person who prefers to think visually. In all these writings, Einstein asks his reader to take the business of making progress in science into one's own hands; to insist on thinking one's own thoughts even if they are not blessed by the crowd's acceptance; to challenge the presumed inevitability or orthodoxy of ideas which do not meet the test of an original mind; and to live and think in all three portions of our rich world—the level of everyday experience, the level of scientific reasoning, and the level of deeply felt wonder.

Of course, to do this with such grand results as he attained requires what Einstein called simply his 'scientific instinct' (*Physics and Reality*). Yet one cannot help but be touched by this message, the message to think and act with courage, independence, and imagination. For it was not for his own edification that he wrote on these matters, again and again. Rather, it was meant to help us, too, on the other side of the century.

Principal References

Einstein, A., *Ideas and Opinions*, new translations and revisions by Sonja Bargmann (New York: Dell, 1954). The essays specifically cited in this article are as follows (page numbers refer to the Dell edition):

Physics and Reality (pp. 283–315).

Remarks on Bertrand Russell's Theory of Knowledge (pp. 29–35).

The Problem of Space, Ether and Field in Physics (pp. 270–83).

Religion and Science (pp. 46–9).

Schilpp, Paul (Editor), *Albert Einstein: Philosopher-Scientist* (Evanston, Illinois: The Library of Living Philosophers, 1949). This includes (pp. 3–94) Einstein's 'Autobiographical Notes' and also (pp. 665–88) a set of supplementary comments by him from which some quotations in the present article are taken. The autobiographical essay is scheduled to be published in 1979 as a separate book.

Further Reading

Arnheim, Rudolf, *Visual Thinking* (Berkeley: University of California Press, 1971) chapters 6 and 12.

Einstein, A., and Infeld, L., *The Evolution of Physics* (New York: Simon and Schuster, 1938).

In addition to the essays from *Ideas amd Opinions* listed above, the following are also specially recommended:

'A Mathematician's Mind' (pp. 35–6);
'Principles of Theoretical Physics' (pp. 216–19);
'Principles of Research' (pp. 219–22);
'Geometry and Experience' (pp. 227–40)*;
'On the Method of Theoretical Physics' (pp. 263–70)*;
'The Fundaments of Theoretical Physics' (pp. 315–26);

Hadamard, Jacques, *The Psychology of Invention in the Mathematical Field* (New York: Dover, 1945).

Hoffmann, Banesh, *Albert Einstein, Creator and Rebel*, with the collaboration of Helen Dukas (New York: The Viking Press, 1972).

Holton, G., *The Scientific Imagination: Case Studies* (New York and Cambridge: Cambridge University Press, 1978).

—— *Thematic Origins of Scientific Thought: Kepler to Einstein* (Cambridge: Harvard University Press, 1973).

—— 'Notes towards the Psychobiographical Study of Scientific Genius', in *The Interaction between Science and Philosophy*, edited by Y. Elkana (New York: Humanities Press, 1975), pp. 370–1.

Mach, E., *Knowledge and Error: Sketches on the Psychology of Enquiry*, with an Introduction by Erwin N. Hiebert (Boston and Dordrecht: D. Reidel, 1976). (First published 1905.)

Poincaré, Henri, *Science and Hypothesis* (New York: Dover, 1952). (First published 1902.)

Schilpp, Paul (Editor), *Albert Einstein: Philosopher-Scientist* (Evanston, Illinois: The Library of Living Philosophers, 1949). In addition to Einstein's own contributions, this book contains an extensive set of 'Descriptive and Critical Essays on the Work of Albert Einstein'; among the contributors are Niels Bohr, Max Born, Percy Bridgman, Louis de Broglie, Max von Laue, Wolfgang Pauli, and Arnold Sommerfeld.

* 'On the Method of Theoretical Physics' and excerpts from 'Geometry and Experience' are contained in Part IV of this book.

Figure 26 Einstein at Pasadena in 1931

7 Einstein, science and culture

Boris Kuznetsov

1. KNOWLEDGE AND THE FOUR-DIMENSIONAL UNIVERSE

In all ages science has had a determining influence on the development of culture. Nevertheless, it is difficult to find historical antecedents for the influence exerted on the culture of the twentieth century by Einstein's ideas, and, as far as one can foresee, of future centuries. An analysis of this influence reveals the internal structure of non-classical physics: the relation, connection, and effect of its separate branches become components of the thought of contemporary physics itself. The theory of relativity is characterized by its influence not merely on separate cultural fields—economics, education, literature, art, etc.—but also on the common transformation of all these fields.

> It was my good fortune to work with Einstein. You would imagine that this would be a wonderful opportunity to see how his mind worked and so you would learn how to become a great scientist yourself. Unfortunately, no such revelation was forthcoming. Genius simply cannot be reduced to a set of simple rules for anyone to follow.
>
> (Banesh Hoffmann, in G. J. Whitrow, *Einstein: The Man and His Achievement*)

Which of Einstein's ideas is the most important for culture as a whole, for the structure of culture, for the reinforcement of the role of science on the transformation of the spiritual and material aspects of mankind? It is the idea of a four-dimensional universe, a picture of the universe that eliminates instantaneous distant action and eliminates the concepts of absolute time existing independently of space, and of absolute simultaneity.

The spatial order of the universe has become the spatial-temporal history of the universe. And even science itself has become to some degree a *history* of science by eliminating the fiction of definite categories that are independent of time, by bringing to light the endless approximations to absolute truth. The causal connection between Einstein's ideas and the style of scientific thought—science as a phenomenon of culture—began with the correspondence between a four-dimensional nature which acts in time and a system of knowledge that recognizes its dependence on time. This kind of connection has continued and science has remained not simply isomorphic to culture, but a real, causal influence on it.

The style of thought in physics, that is, the features of contemporary ideas in physics considered as an element of contemporary culture (such as the inclusion of time in human understanding), leads to an inevitable expansion of the ideas, to a transformation of the scientific conception in a world outlook. At the beginning of this century, when the theory of relativity had just appeared, Nernst said that this theory was not just a theory of physics, but a philosophy. This remark was made long before the development of the theory of atomic structure. Now, in a time of abundant experimental confirmation and practical application of the theory of relativity, it is hardly possible to underestimate its proper physical character. But in some sense, Nernst's remark captures the characteristic features of Einstein's theory—a conceptual shift of unprecedented magnitude given to a sweeping general principle and the ramifications of this principle. In his 1946 autobiography Einstein speaks of two criteria for the selection of a physical theory, two criteria for scientific truth. These are *external justification*, the agreement between observations and experiments, and *internal perfection*, the possibility of inferring a given theory by scientific means from the most general principles without additional propositions. The genesis of the theory of relativity was the synthesis of these criteria: an experiment did not agree with the old theory even with the help of special *ad hoc* propositions, and therefore forced a change in the most general principles. Lorentz's theory predicted the contraction of a moving rod from a specially derived electrodynamic hypothesis. Einstein appealed to the most general relations of space and time, and the length contraction acquired an internal perfection. The halo of a paradox (or, if you prefer, its odium) was carried over to experiments which demonstrated the constancy of the speed of light in moving systems and fitted naturally into this new view of space and time, of cosmos and microcosm, of matter and movement. Einstein called such a transfer 'a flight of wonder' since it showed the subordination of the most paradoxical observation to a cosmic principle, and along with this an inevitable modification of the principle. Such a principle, which underlies our belief in the internal perfection of science, ceased to be fixed. The fundamental laws of life proved to be connected with external justification, dependent on experiment, dynamic, changeable, and not at all *a priori*.

. . . although he [A.E.] rejected the churches he had a Spinoza-like belief in a cosmic religious force. He regarded this as an eternal spiritual being that communicates small details of itself to our weak and inadequate minds. As he once declared, 'This deep intuitive conviction of the existence of a higher power of thought which manifests itself in the inscrutable universe represents the content of my definition of God.' In other words, he had no more use for the shallow materialism that is the most widely accepted philosophy of scientists and others today than for the authoritarian views of the churches that once were so powerful.
(G. J. Whitrow, *Einstein: The Man and His Achievement*)

A contemporary physicist is like a certain American lawyer who explained that his knowledge of the general principle of the law was based on his knowledge of the laws individually, and said, 'And what shall I do if a law that I know is repealed?' The laws of physics have not been repealed, but they have been generalized and modified, and this feature of contemporary physics makes it non-classical, not only in content, but also in style. Indeed it can no longer *be* classical. It will always see as an ideal before it, not a finished picture of the universe, but the most rapid possible advance in the representation of the universe, the endless approach of the picture of the universe to its objective original.

In this respect the theory of relativity, like Einstein's no less paradoxical nsights about light quanta, was the beginning not only of non-classical hysics, but of all non-classical science. In the middle of this century the deas of relativity theory and quantum mechanics began to spread energetically to the study of the cosmos and microcosm, accompanied by a rowing synthesis of these ideas. In the middle of the century non-classical hysics entered even into the study of life, giving a great stimulus to nolecular biology. The practical application of non-classical physical concepts began in atomic energy, quantum electronics, and cybernetics. The nitial contours of new directions of technological progress, which were on-classical with respect to their scientific bases, were defined. Modern hysics was a little like Aristotle's in that it had a single causal explanation f the universe, but there was an important difference: the new physics not nly had an explanation of the universe, but was also transforming it. This vas a radical change in the role of physics in the development of culture, change in the influence of physics on the evolution of culture. In order to nderstand its transforming effect, its influence on production and on the tyle of human thought and culture, one must emphasize Einstein's basic remise for the definition of nonclassical science. The basic idea for the niversalization of both physical thought and of its influence on culture lies n a search for internal perfection, a transition to new, optimally general rinciples that find their external justification through experiment.

I believe with Schopenhauer that one of the strongest motives that leads men to art and science is escape from everyday life with its painful crudity and hopeless dreariness, from the fetters of one's own ever-shifting desires. A finely tempered nature longs to escape from personal life into the world of objective perception and thought; this desire may be compared with the townsman's irresistible longing to escape from his noisy, cramped surroundings into the high mountains, where the eye ranges freely through the still, pure air and fondly traces out the restful contours apparently built for eternity.
(A.E. at the celebration of Planck's 60th birthday)

The influence of knowledge on the culture of one period or another has requently served as the basis for naming the period. In other words, a eriod is defined by the role reason plays in it. I recall reading one phrase, ttered by the Russian biologist, Clement Timiriazev, in 1886 on the ccasion of a celebration for the French chemist Eugène Chevreul, who vas then 100 years old. The chemist had been born in the eighteenth entury, and his scientific activity had spanned almost the whole of the ineteenth century. Addressing Chevreul, Timiriazev said, 'Son of the Age f Reason, you have become a living embodiment of the Age of Science!' hese characterizations are correct for the two centuries concerned, but vhat is the correct characterization for the twentieth century?

The traditional distinction between reason and understanding proceeds rom the fact that understanding applies logical norms and fundamental aws of being, but reason changes them. From this point of view, nineteenth entury science was the apotheosis of understanding. It constructed a owerful and developed system of knowledge on the basis of firm logical orms and laws—this was the essence of classical science in every case. With he same degree of accuracy, bearing in mind the specific features of wentieth century science which affect the age and culture as a whole much nore intensively, one can say that the present century is the embodiment f reason as defined above, the embodiment of the human thought which hanges its canons. Of course, in the history of human thought there were arlier changes in its canons, but these were either sporadic or very slow, eing revealed only after the event. Laplace said that reason goes forward more easily if it extends into itself. Such an extension in the twentietl century has been an obvious and uninterrupted accompaniment of th forward movement of reason. When the canons of reason are modified w have, in some sense, a history of reason, because we include time as an axi of such a modification. If we compare a scientific conception to an inter section of logical lines, that is, to points in some n-dimensional 'space o knowledge', then contemporary science includes time in this 'space' as a additional $n+1$th dimension. This is somewhat analogous to the effect o the theory of relativity with respect to three-dimensional space.

> Just as the Roman Inquisition characterized and condemned the investigations of Copernicus and Galileo as 'philosophically false' because they did not fit into its conception of nature, many philosophers and physicists all over the world rejected Einstein's theory of relativity since they could not understand it from their mechanistic point of view. In both cases the reason for the condemnation was not a difference of opinion in the judgment of observations, but the fact that the new theory did not employ the analogies required by the traditional philosophy.
>
> (Philipp Frank, *Einstein: His Life and Times*)

2. THE TOPOLOGY OF CULTURE

The representations of space and time and their connections have alway been specific traits of a culture. The culture of antiquity was characterize by the idea of a *static* harmony of being. All aspects of the culture wer permeated with the ideas of the canon, an idea of complete perfection Herein lie the mythological sources of antiquity: the sculptured gods o Greek art were the embodiment of the ideal canons of beauty. The con ventions of poetry and drama were a part of it. And the acceptance of thi view by the Aristotelian philosophers determined the fate of that culture— its transformation into the dogmatic world outlook of the mediaeva scholastics.

Of course, the thought and art forms of antiquity were not completel static. We are talking only about certain 'invariant' traits that were commo to all aspects of the culture. Static harmony in this sense begins in th cosmology and physics of Aristotle, in which the movement of a body i determined by a static scheme of a universe with an immovable centre bounded and having a 'natural place'. Greek culture as a whole did no deny movement, did not exclude time from its picture of the universe The conventions of art did not close the way to the dynamism of the *Iliad* the tragedies of Sophocles, or the sculpture of Phidias. For the most part the principles of art, like those of logic and of the scheme of a fixe heavenly harmony, did not close the way to the Aristotelian conception o movement. Nevertheless the culture of antiquity can be called a culture o three-dimensional, purely spatial, concepts.

The predominant direction of the culture of the Middle Ages, th direction which is most specific and which is preserved in the transitio from one field to another, was to make the static tradition more categorica and to deprive it of the polyphonic accompaniment found in the environ ment of antiquity. For St Augustine time was finite, limited by the creatio of the universe and the end of the world, and in this sense absolute. Tim flows from the fall of man towards the redemption, and then its flov ceases. This limited conception of time was specific to the theology of th Middle Ages. It could be compared to a one-dimensional 'space', extendin from a 'top' in the realm of heaven to a 'bottom' in the realm of hell. Th cultural history of the Middle Ages shows how a similar one-dimensiona ierarchy was interpreted in religious terms, in literature, and in art, and ow closely it was tied to Aristotelian cosmology and physics.

The culture of the Modern Age repudiated the cosmological, moral, and esthetic assumptions so important for the culture of the Middle Ages. 'ime ceased to be something predetermined and separate from the physical vorld. Space became a universe of sensually conceivable things, which ould be used to provide a measure of physical time. However, the union f space and time, the transition to a spatial-temporal representation was ıcomplete; Newton's *Principia* retained the picture of a purely space-ependent distribution of interactions and a non-spatial absolute time vhose flow is independent of spatial events.

But specific to the Modern Age, sharply differentiating it from the Aiddle Ages and most important for the culture of this period, was the nion of space and time in the infinitely small; in other words, the differ-ntial representation of movement from point to point and from moment ɔ moment. Through this arose a space–time conceptualization of motion vhich became the basis of a causal picture of the universe and of deter-ıinism, characteristic of the new age—as expressed in Laplace's famous emarks to the effect that the whole future of the universe could in principle e calculated from a knowledge of the forces of nature and the positions and elocities of all objects at a given instant.

While Einstein is the only modern scientist one can begin to compare, from the point of view of achievements, with Newton, it is difficult to find very much that they had in common as men. Everyone who had real contacts with Einstein came away with an overwhelming sense of the nobility of the man. The phrase that recurs again and again is his 'humanity'—or, as trite as it may sound, the simple, lovable quality of his character. Nowhere in all of his professional life is there the remotest sense of the often bitter competitiveness, the struggle over claims to scientific invention, that cloud and sometimes destroy the lives of scientists.

(Jeremy Bernstein, *Einstein*)

The development of this point of view allows one to envisage a con-ection between the transformation of nature and the transformation of uman society itself, of culture in the widest sense. Engels said that in the ighteenth century science, when practised, led to the industrial revolution, nd the development of its ideas was a source of the political revolution. 'here are complicated but nonetheless clear connections between spatial-emporal determinism on one hand, and natural religion, natural ethics, nd social ideals, on the other. The ideal of cosmic and social harmony has ost its static character. Henceforth it is tied to time. Cosmic harmony is not spatial scheme of 'natural places', but a spatial-temporal picture of novements. Social harmony is carried into time—with Rousseau to the ast, with Voltaire to the future. The spatial-temporal determinism even enetrates literature. There is a very significant remark by P. Muratov on ıe novel, *Les Liaisons Dangereuses*, by Choderlos de Laclos: 'the chief haracters of this novel are indeed full of Newtonian confidence in the un-mbiguousness of the results of their actions and remarks'. But, of course, if ve apply the words 'spatial-temporal determinism' to the complicated henomena of culture, we mean by 'space' not ordinary space, but the eometric form of a very complex structure. Before turning to the problem f these complex structures, some preliminary remarks are in order.

In 1872, the mathematician, Felix Klein, in his *Comparative Considerations ı Recent Geometric Researches*, which became famous as 'The Erlanger rogramme', examined the hierarchy of geometries, with particular eference to the concept of an invariant which had been introduced twenty years earlier. In elementary geometry the invariant of a transformation use the distance between points, defined by some appropriate formula tha embodies the properties of the space. Certain features (e.g. the dimensionality of a geometric figure) remain invariant in the face of topologica transformations; such transformations express in the language of geometry the properties of very complex physical objects and processes. In Einstein's theory similar transformations have obtained a physical meaning. Einstein explained fundamental properties of the universe by moving from three-dimensional to four-dimensional space.

It is possible to imagine spaces for which the dimensionality increase without limit, and (as in Einstein's theory) the growth of dimensionality i able to express a radical change in the picture of the world. Does thi method go beyond the scope of physics? Can it represent the effect o physics on culture, its influence on the spiritual and material life of human society? An affirmative answer to this question, a demonstration of such a possibility, can only be constructive. It consists of the application of the concept of dimensionality to the phenomena of culture. But it is clear *a priori* that such an application changes the meaning of dimensionality; i ceases to be purely a collection of categories, and, essentially, the word requires quotation marks in the new context, while still preserving some analogy, some isomorphism, with the strict mathematical and physica concept.

... he had a character more like that of an artist than of a scientist as we usually think of them. For instance, the highest praise for a good theory or a good piece of work was not that it was correct nor that it was exact but that it was beautiful.
(H. A. Einstein, in G. J. Whitrow, *Einstein: The Man and His Achievement*)

We are not speaking about a superficial analogy between the dimensionality of the universe and the 'dimensionality' of culture, but about a true reflection of a picture of the universe in the structure and development of contemporary culture, which has been formed on the framework of non-classical science. Dimensionality, in its original topological meaning includes the transformation from a zero-dimensional space of isolated points to a one-dimensional space, a line which includes and consists of these points. In this original, purely topological meaning the concept of dimensionality reduces to a statement about the structuring of the universe the exclusion of absolutely isolated points from the picture of the universe the representation of points as elements of a set. And here it is possible to draw a unifying line, to see the isomorphism of this with a statement about the structuring of society. This statement would deny the possibility of the kind of individual autonomy propounded by the nineteenth century nihilist Max Stirner in his book, *The Ego and His Own* (1844); it would represent culture as a structure in which the individual is included in an ever-increasing number of intersecting material and ideological connections Skipping ahead a little, one must say that such an analogy becomes a causal statement if one considers the concept, or the image, of the universe to be a determining and moving force of culture, communicating to culture a developing structure, a growing complexity, like a reflection of the world which becomes more complex and which, in its turn, reflects the true infinite structure of the cosmos and microcosm. Here we can begin to see the connection between the concepts of the dimensionality of the universe, the dimensionality of knowledge, and the dimensionality of culture, which is the theme of this article.

3. THE IRREVERSIBILITY OF COSMIC EVOLUTION AND THE IRREVERSIBILITY OF CULTURE

The theory of relativity on one hand, and the representation of discrete fields on the other, are the beginning of a series of physical conceptions leading to a new understanding of the irreversibility of time, of its 'arrow', and to a broad understanding which includes both cosmos and microcosm and is represented by a very general geometric scheme. This understanding takes us from the structure of the cosmos to the structure of science itself as a phenomenon of culture, and thence to the structure of culture, to the irreversibility of its development.

The classical conception of the irreversibility of time did not have such a broad outlook. It saw the physical basis of the irreversibility of time, the irremovable difference between earlier and later, in the rise of entropy. This kind of evidence of irreversibility ignores the microcosm and encounters great difficulties when applied to the cosmos, to the 'universe as a whole'. The classical explanation was connected neither with the problem of a subjective, inner perception of irreversible time, nor with the problem of the irreversibility of understanding and the irreversibility of culture.

The theory of relativity still does not provide, in an unambiguous form, a scheme for the causal connection of the irreversibility of the cosmic evolution with microscopic processes, and the connection of the irreversible evolution of the cosmos with the irreversible evolution of its representations. But it offers the prospect of such a scheme. It can be hoped that Einstein's complaint (stated in his 1946 autobiography), that the theory of relativity does not derive its relationships from the atomic structure of matter, may be overcome in the further development of physics, and that a detailed description of the irreversible development of the universe will emerge. Although the study of such matters has not led to a theory of elementary particles and astrophysics, the irreversible development has a certain confirmation when speaking not about the cosmos, but about understanding it, about science as a phenomenon of culture and as a component part of the history of an understanding reason. Understanding moves along a line which resembles the track of a Brownian motion; science goes away from the path of truth, sometimes returning, but there are some irreversible transitions towards a more adequate representation of the universe. The history of the theory of relativity exhibits what is true for the history of science as a whole: science is the irreversible approach of reason to objective truth.

Does the irreversibility of knowledge guarantee the irreversibility of culture? The irreversibility of knowledge is related to the growth of the Einstein criterion for external justification and internal perfection of the

Einstein's influence was not confined to the technicalities of modern physics. For, just as it is inconceivable that there will be any general reversion to pre-Copernican, pre-Newtonian or pre-Darwinian assumptions concerning the general nature of the universe and man's place in it, there will likewise be no return to the world-view of Einstein's predecessors.

(G. J. Whitrow, *Einstein: The Man and His Achievement*)

representation of the universe. Just such a growth gives knowledge an irreversible character. It leads to what we have called the growth of the dimensionality of the picture of the universe: each separate scientific assertion proves to be an intersection of a growing number of more general logical series (internal perfection) and a generalization of a growing number of empirical data (external justification). The picture of the world becomes more united in its foundations and more differentiated in the details of its elements. What is the cultural effect of this development?

The example provided by Einstein's work as an answer to this question is clearly not a special example. Relativity theory is not only a sum of a whole series of answers to the question of the nature of space, time, motion, and matter. It is a summary of the development of science as a phenomenon of culture, of the evolution of its relation to other cultural elements, the evolution of the human value of science. Therefore, the analysis of the cultural effect of the theory of relativity reveals the connection between the irreversible growth on many levels of the dimensionality of the structure of science, on one hand, and the characteristics of culture, on the other. The theory of relativity asserts the physical reality of a local situation—the motion of a given body with a definite speed, introducing to the problem a frame of reference without which movement loses meaning. This general formulation of the theory of relativity is the clearest and fullest expression of knowledge striving for infinite multi-dimensionality. But we remember the traditional definition of truth, goodness, and beauty as a triadic embodiment of infinity. The development of culture also involves the development of these ingredients. I am not giving here a new definition of culture, because every definition whether clear or not embodies a recognition of the infinite nature of culture. The linking of man to a universe which is infinite in its complexity, the genesis and evolution of multi-dimensional man, this is what connects knowledge and the components of culture with goodness and beauty.

'Out of just a little string and matchboxes and so on, he could make the most beautiful things. As a matter of fact, he always liked to improvise things of that sort, just as he would also like to improvise in his work in a way: for instance, when he had to give a talk he never knew ahead of time exactly what he was going to say. It would depend on the impression he got from the audience in which way he would express himself and into how much detail he would go. And so this improvisation was a very important part of his character and of his way of working.'
(H. A. Einstein, in G. J. Whitrow, *Einstein: The Man and His Achievement*)

4. THE ECONOMIC AND SOCIAL EFFECTS OF NONCLASSICAL SCIENCE

When science is considered as one among many different components of culture, then, in addition to the methods and content of science, its value also becomes the object of analysis. The value of knowledge is its effect and influence on knowledge itself, on technology, economics, social relations, education, literature, art, teaching, and customs. But there is a causal connection among the components of culture, and the impulses for their development originate at the industrial, technological, and economic levels, the evolution of which serves as the immediate means of cultural progress.

If only the simplest physical and chemical techniques are used in factory and shop laboratories, the consequence tends to be a constancy of the technological and economic parameters, and hence a constant level of productive work. Constructive technological research can effect an undamped rate of development of the productivity of work, a non-zero time derivative. The cycles of physics provide a formula for such a search. Designers and technicians strive towards an optimum real embodiment of the ideal physical relations. If these ideal canons change, which would be a scientific result in the proper sense, then the productivity of labour accelerates, giving a non-zero *second* time derivative. In this manner the index of the effect of science will be some fundamental economic index:

$$\Omega=f(P, P', P'')$$

where P' and P'' are the first and second time derivatives of the productivity of work P. The maximum value of Ω corresponds to the optimum structuring of production, that is, to science used to the fullest measure.

And what role is played here by discoveries which change the most fundamental principles of science, which serve as a whole canon for science in the search for new physical cycles and relationships? In particular, what influences can one ascribe to Einstein's theory of relativity?

We can think of an economic structure, by analogy with an Einsteinian physical world, as being an n-dimensional space, in which n is the number of branches to be considered, and the coordinates of each point q of this space, $q_1, q_2, \ldots, q_n$, are the projections on each branch. The results of scientific investigations change the structure of production, that is, provoke a transition from one point of the space of economic structure to another. Fundamental discoveries change the metric of a given space. It is not necessary here to give an account of this concept of the economic effect of fundamental science. But in our consideration of the characteristics of non-classical physics as a phenomenon of contemporary culture, we are interested in another side of the unexpected utilization of the mathematical apparatus of the general theory of relativity in econometrics. It appears that all contemporary culture, and particularly philosophical, sociological, natural and technological scientific thought, is characterized by a tendency that finds its expression in the mathematical language of non-Euclidean hyperspaces.

Like contemporary scientific thought (which began with the theory of relativity), human thought in general is now much freer, operating as it does with billions of light years and a billion parts of a second. The atomic–cosmic age is not only a scientific–technical characterization of our time; it is also a logical–psychological characterization, a characterization of what can be called intellectual culture. At the same time the metagalactic and subnuclear worlds do not appear to be zones of a uniform hierarchy, in which structure and laws are repeated, on a larger or a smaller scale. The verses of the Russian poet, Valeria Brusov, which picture the electron as the condensed repetition of the earth ('perhaps these electrons are worlds, where the five continents . . .') are not at all typical of contemporary thought. The atomic age is accustomed to the paradoxes that may be

When judging a scientific theory, his own or another's, he asked himself whether he would have made the universe in that way had he been God. This criterion may at first seem closer to mysticism than to what is usually thought of as science, yet it reveals Einstein's faith in an ultimate simplicity and beauty in the universe. Only a man with a profound religious and artistic conviction that beauty was there, waiting to be discovered, could have constructed theories whose most striking attribute, quite overtopping their spectacular successes, was their beauty.

(Banesh Hoffmann, *Albert Einstein: Creator and Rebel*)

expected at certain degrees of the infinitely large and the infinitely small, the paradoxes of physics. At present a hundred thousand or, perhaps, a million people work in areas where the relativistic and quantum paradoxes of physics serve as the basis of production, and a still greater number in other areas draw upon scientific–technical information from atomic engineering, electronics, the construction of space instruments, of accelerators, of astrophysical observatories. There has developed a mass culture, not limited to professional physicists, based on a real paradox applied in practice, which gives it an external justification, which causes the paradox to be removed and to acquire internal perfection. Because of this, the teaching of physics in our time moves in the same direction as the development of production based on nonclassical physics. But about this later. Now we must discuss something else—the influence of nonclassical science on man's work, on its object, content, structure, and producer.

I remember a beautiful remark of his when he criticized a well-known American physicist. Einstein said he 'couldn't really understand how anybody could know so much and understand so little'! Einstein always emphasized that you could know too many facts and get lost among them. Nevertheless, there existed no field of physics about which he could not immediately speak without hesitation. It did not matter whether it was a fashionable part of physics or some almost forgotten part, his listeners felt that he had the whole of physics spread out before his eyes. And yet I am quite sure that Einstein never realized what an exceptional man of genius he was.

(E. H. Hutten, in G. J. Whitrow, *Einstein: The Man and His Achievement*)

Man's work, as we have already said, consists of a suitable organization of the forces of nature. First of all there is the selection of the sources of energy, the sources of raw material, and all other natural resources of production. Contemporary science has led to a significant regrouping of these resources, a situation which demands new economic and ecological criteria specific to this age. Science has changed the character of work by intellectualizing it. Science has changed the structure of work—the relations among the branches of production. And finally, science has changed the producer of work—man himself. Here we have arrived at a very fundamental criterion of culture, perhaps the most fundamental. This criterion is connected with the concept of humanism. From its appearance in the fifteenth century, when the *City of the World* was contrasted with *The City of God* by Saint Augustine, this concept has changed significantly. But its basic meaning has remained: man, his interests, and his capacities are in principle unlimited; the development of these capacities, the goal of culture, has characterized this concept up to the present. All the components of culture are humanistic; the interests of man define the value of culture as a whole. But our time is characterized not so much by the value of the achievements of culture from the point of view of man's interests, as by some relatively stable complex of intellectual potentials and moral, aesthetic norms, and by an interaction of culture with man, which constitutes the *dynamic humanism* of contemporary culture.

5. SCIENCE AND THE PROBLEM OF INDIVIDUALITY

Let us return to the topological framework of an n-dimensional space, where n grows irreversibly, and this growth forms an $n+1$th coordinate axis which is irreversible time. Such a framework allows a clearer representation of the relation of a person in his uniqueness, his individuality, unreduced to universals. 'Man' is that individuality which, in the fifteenth century, became the banner of humanism. There exists an unquestionable connection between the concept of the individual in physics, which is the particle, and the individual in humanistic culture, which is the human personality. The declaration of the autonomous physical individual—an atom—was already for Epicurus and Lucretius a weapon that defended the autonomy of man. Lucretius wrote about the spontaneous deviation of atoms from prescribed macroscopic laws which had been proposed by Epicurus so that man would not be wholly subordinated to necessity and so that 'it would not be necessary only to endure and suffer and bow before being defeated'. In his dissertation, *The difference between the Natural Philosophy of Democritus and the Natural Philosophy of Epicurus*, Marx says that an atom would not have become a basic concept of the philosophical picture of the world if spontaneous deviation had not been attributed to it. Lenin compared human caprice to the movement of electrons. From the time of the ancient atomists to the nonclassical physics of the twentieth century, the idea of autonomous individuality has had natural philosophical and physical equivalents. But along with this there was also a contradictory idea: individual fate was ignored in the presence of all-powerful elemental social laws, just as the fate of molecules was ignored in macroscopic thermodynamics.

For Einstein the relation between the idea of autonomous individuality and the world of physics acquired a new form. For him human individuality became deeper and more interesting as it became to a greater degree connected to the 'extra-personal'. In this respect the first paragraphs of his 1946 autobiography, *Autobiographical Notes*, have maintained their meaning for a lifetime, and, in particular, the paragraph dedicated to the objective extra-personal world, the contemplation of which brings deliverance to man. I quote these lines, which characterize the most important cultural effect of science:

> Out yonder there was this huge world, which exists independently of us human beings and which stands before us like a great, eternal riddle, at least partially accessible to our inspection and thinking. The contemplation of this world beckoned like a liberation, and I soon noticed that many a man whom I had learned to esteem and admire had found inner freedom and security in devoted occupation with it. The mental grasp of this extra-personal world within the frame of the given possibilities swam as highest aim half consciously and half unconsciously before my mind's eye.

For a contemporary physicist the 'inner freedom' of which Einstein writes is associated not with spontaneous bias, but rather with degrees of freedom that are connected with the dimensionality of the space in which a particle moves. Inner freedom is represented by a set of bonds with the extra-personal world, with the infinitely complex, infinitely measured universe, the knowledge of which liberates the individual consciousness. Culture is the genesis of the *multi-dimensional man*; it is the dimensionality, complexity, levels, and the number of degrees of 'internal freedom', which

While philosophers are prone to overestimate the power of ideas and to submit to the fascination of words, scientists often underestimate the value of theoretical work and become enamoured of gadgets. While philosophers may, on occasion, indulge in phantasies, scientists sometimes refuse to employ their imagination. Newton's famous dictum '*hypotheses non fingo*' is misunderstood. It is all very well to stick to the facts provided one doesn't get stuck in them. This hard-headed attitude was rebuked by Einstein: 'Everything that they learned up to the age of eighteen is believed to be experience. Whatever they hear about later is theory and speculation'.

(E. H. Hutten, *The Language of Modern Physics*)

grow infinitely and irreversibly. Indeed, 'Here is Hegel and book wisdom and the philosophical thought of all.'

Here is the foundation of the connection between the irreversibility of culture and the irreversibility of knowledge. Knowledge is a liberation, a growth and realization of internal freedom; this is the definition of a man, *homo cogitans*. The history of knowledge is the process of the humanization of man, the process of cultural development. Or more exactly, it is the substratum that makes this cultural development irreversible.

For Einstein, in his statements on the problems of culture, social life and ethics, the human individual is not an isolated 'single one', as for Stirner, but he is an intersection of interests, impressions, conceptions, forms, and emotions that are connected with the personal and supra-personal, and this is what serves as a measure of progress. This view is related to the content of Einstein's scientific ideas. The theory of relativity, the exclusion of absolutely isolated bodies from the conception of motion, the idea of particles as elements of a field, created the frame for a general philosophical consideration of the whole and the included elements. On the other hand, a protest against ignoring individual human fate was reflected in that original, intuitive trend from which the scientific ideas came. In this connection one should remember the famous phrase of Einstein: 'Dostoevsky gave me more than any thinker, more than Gauss!' A comparative analysis of Einstein's scientific, social, and ethical conceptions on one hand and the work of Dostoevsky on the other allows us to see here a deep isomorphism. The question which passes through Dostoevsky's novels is the question of the fate of a private man in the face of the blind laws of being, those dictates from the ecumenical harmony which do not compensate the torment of one crushed human being. This is the question of culture as a whole, of all its development given in a general sharp form: could, perhaps, a universal harmony be established that excludes individual tragedy? Just this question attracted Einstein's attention and the great physicist searched for an answer in the contemplation of the extra-personal world—a quest that acquired an ethical accompaniment. Here it is appropriate to remember Einstein's position on the question of the relation of science and ethics. In 1954 Einstein wrote to his friend, Maurice Solovine:

One of the clichés about Einstein's theory is that it shows that everything is relative. The statement that everything is relative is as meaningful as the statement that everything is bigger. As Russell pointed out, if everything were relative, there would be nothing for it to be relative to.

(James R. Newman, *Science and Sensibility*)

'That which we call science pursues one single goal: the establishment of that which exists in reality. The determination of that which *should* be is a task to a certain degree independent of the first.'

This independence characterizes science, which is a compilation of objective statements. The movement of science, its development, and science as a phenomenon of culture, depend on ethical self-consciousness. 'Just in this,' says Einstein, 'appears the moral side of our nature—that internal striving towards the attainment of truth, which under the name *amor intellectualis* was so often emphasized by Spinoza' ('Einstein, Science and God', *Forum*, 1930, **83**, 373–437).

Nonclassical science, with its special mobility in the most general and deep foundations of knowledge, is more closely tied than classical science to *amor intellectualis*, to moral selfconsciousness, to that component of culture which in the greatest measure unites human individuality with extra-personal being.

6. THE POETRY OF SCIENCE

That revolution in science and, through science, in the culture of man, which was Einstein's theory of relativity, changed the relation of truth to goodness and to beauty, and the relation of science to production, economics, ethics and aesthetic values. At the foundation of the new relation of science, as a phenomenon of culture, with the other components of culture lay a sweeping transformation of the world picture, the scope of which can be expressed as a topological transformation of the dimensions of knowledge or (if one prefers a more traditional and more philosophical Laplace-like formulation) as a deepening of reason into itself. In this regard fundamental science became closer not only to economics and to ethics, but also to art. In contemporary science strict mathematical proofs are used so intensively that intuition plays a larger role than in the science of the nineteenth century. This is again related to the synthesis of the Einsteinian criteria of external justification and internal perfection, which characterizes the theory of relativity and all contemporary science.

'The emotional state which leads to such achievements resembles that of the worshipper or the lover; the daily struggle does not arise from a purpose or a programme, but from an immediate need.'
(A.E. to Max Planck, 1918)

When into the consciousness of a contemporary thinker there flashes a new paradoxical scheme to explain a given experiment, this scheme does not yet have an internal perfection, such that it could be inferred from a more general principle covering complex sets of different extremely varied experiments. When in the consciousness of a thinker there arises a new *logical* deduction, it demands for its internal perfection ramified chains of new deductions. In both cases the realizations of the Einsteinian criteria are related to the transformation of a metric, the transformation of a topology, the transformation of a mathematical/logical apparatus. A particular experiment, a particular deduction, is accompanied by a preliminary intuitive attainment of infinity, which is hidden in the still unrealized experiments and deductions. In this intuitive achievement of the infinite is the 'illumination' which is characteristic of artistic creation. Intuitive illumination creates the poetry of science and makes it related to music, in which, in the words of Leibniz, the soul calculates, itself not yet knowing. The capacity for such illumination is called an inspiration. It is always a condition for both artistic and scientific creativity, but in nonclassical science it appears as an *obvious* condition. The reason is in the form of the logical/mathematical norms to which nonclassical science is subordinated.

The theory of relativity changes the mathematical basis of knowledge. But its further development, in particular the removal of the inadequacies about which Einstein wrote in his 1946 autobiography, the convergence with quantum theory, clearly demands a transformation of logical norms, a rejection of the law of the excluded middle, what is called a metalogical transformation. When a given logic undergoes a metalogical transformation, it cannot accomplish this transformation within the framework of the old logical norms; here it is necessary 'to hear the unwritten symphony', to intuitively imagine the results of the transformation. These moments of intuitive scientific illumination do not have an exclusively heuristic value. They exert a significant influence on the culture of our age. Poetry, as a component of culture, acquires a clear reconstructive function.

7. THE IMMORTALITY OF FORM AND THE INVARIANTS OF CULTURE

Professor Einstein was sitting next to an 18-year-old girl at an American dinner party. When the conversation flagged his next door neighbour asked: 'What are you actually by profession?' 'I devote myself to the study of physics', replied Einstein, whose hair was already white. 'You mean to say you study physics at your age?' said the girl, quite surprised. 'I finished mine a year ago.'
(Carl Seelig, *Albert Einstein: A Documentary Biography*)

The analysis of nonclassical physics as a cultural phenomenon, and of the influence of science on contemporary culture as a whole, is not separate from the analysis of the psychology of scientific creativity, of a statement of the specific peculiarities of a thinker who has arrived at nonclassical ideas. This relation, between science and the individuality of the scientist, which is closer than it was in the nineteenth century, follows from the characteristic properties of the style and methods of contemporary physics. Style, as a common cultural category which encompasses both science and art, is an invariant on which are imprinted the properties of the producer of knowledge and creativity: the properties of an artist, a thinker, a school, a direction, a given national medium, a given age. Together with the style of art, which includes the properties of the artistic skill that are preserved in the transition from topic to topic, and which allows one to recognize the artist, school, medium, and age, there exists a style of scientific creativity which is preserved when transferred from one cycle of problems to another, from one object of investigation to another.

Classical science shifted the emphasis from style to method, an invariant of creativity which depends on a given object of investigation and which is preserved when transferred from one investigator to another. The distinctive feature of Newton's creativity was to liberate science from personal idiosyncrasy, to impose an inductive/empirical and logical/mathematical rigour, and to exclude the subjective colouring of experimental results and of logical deductions. The theory of relativity did not return to the subjective self-expressiveness of the Renaissance. It has made the objectivization of knowledge still deeper. But at the same time the theory of relativity does not really reject 'observers'; it relates different systems of reference through the invariants of the transformations from system to system. Quantum mechanics continues this tendency, disclosing the active role of the experiment in the understanding of the objective structure of the microcosm. Classical science did not exclude man from nature, but nonclassical science has included his influence on nature in the process of understanding it more clearly, distinctly and in a more obvious and tangible manner. For this reason, the method and the style of scientific creativity are no longer opposed, but have drawn closer together. The producer of knowledge introduces to the description of an object everything that he receives from society, from history, from predictions, from the culture of the time. Herein lies a significant concrete definition for the concept of 'science as a phenomenon of culture', which we mentioned previously. By studying contemporary science, together with its contents—the objective description of the structure of the universe—one discovers the subjective properties of creativity, which are the result and condensed expression of contemporary culture.

What then is the style of Einstein's scientific creativity? And a second question related to this: why do we feel that Einstein has secured immortality, that his image will be preserved forever in the memory of man?

To use the language of science, immortality can be regarded as an 'invariant', analogous to a physical invariant that survives through transformations of the representation of the universe. In the development of the theory of relativity there was such a transformation of representations, based upon the constancy of the speed of light in vacuum and the pseudo-Euclidean character of space–time. It is very likely that already in our century there will be some more fundamental convergences of the theory of relativity with the theory of the microcosm. But at the same time there are preserved with these or other modifications the methodological substrata of the theory of relativity, namely a basic, non-classical variability of the geometric properties of space and time, involving an evolution of the most fundamental principles of science, a transition from metric to topological concepts in the search for external justification and internal perfection. The transition to a still more complex and multi-dimensional structure in understanding the universe as a mapping of its real structuralization remains invariant. *Amor intellectualis* is expressed in the whole appearance of Einstein, in his work (in particular, in the search for a unified field theory), in his ethical position, and in his personal charm. All this characterizes the individuality of Einstein and makes his image immortal.

The quality that dominated his personality was a very great and genuine modesty. When anybody contradicted him he thought it over and if he found he was wrong he was delighted, because he felt that he had escaped from an error and that now he knew better than before.
(Otto Frisch, in G. J. Whitrow, *Einstein: The Man and His Achievement*)

It is important to emphasize that the invariants of Einstein's creativity are inseparable from the invariants of culture, which, like those of physics, are not metric but topological; they preserve not some kind of numerical indicators, but the direction of a qualitative evolution.

8. THE CULTURAL EFFECT OF THE TEACHING OF PHYSICS

Two conclusions follow from the above: one about the meaning of Einstein's image in the teaching of contemporary physics and one about the meaning of the teaching of physics in contemporary culture. We begin with the first. The specialist in physics, now, in the last quarter of the twentieth century, needs to have a very high cultural level, a high ability to proceed from physics in the proper sense to the contemporary analogue of Aristotelian *physis*. The physics of our century, especially of its middle period and later, has extremely specific outputs to contiguous disciplines. The theory of relativity has led to the very energetic emergence of physical constants and concepts in mechanics. Later, with relativistic cosmology, physics went into astronomy and, in the framework of the quantum theory of the atom, into chemistry and biology. Physics has even entered into mathematics. Beginning with Einstein there quickly developed the physical understanding of mathematical axiomatics, a tendency about which we have already spoken. But this is not all. Contemporary physics has entered into production technology much faster and more immediately than classical physics did and in an ever growing number of branches and with an ever larger restructuring effect.

Another process is also occurring: the rapid development of new situations in physics, the appearance of experimental results which demand extremely radical transformations of the fundamental ideas, and the appearance of theoretical investigations which demand new experiments. There was a time when a physicist had in his initial training very consistent representations which, in the course of his life, developed and became more concrete; but such a time has passed. Now there is a need for a plasticity of thought incomparably greater than before; in particular, a need for the ability to look forward, to feel the movement of science, to determine the way that lies before us. Prognostic thought is needed. Data for such thought are established by the study of the movement of physics, its phylogenesis and ontogenesis, in those moments when the evolution of ideas can be observed in the framework of some scientific revolution, observed in the 'illumination' about which we spoke earlier. In the history of science in the twentieth century there is no more effective material for the training of plasticity of thought in physics than the ontogenesis of the theory of relativity, the development of this theory in Einstein's brain.

Our second conclusion is about the meaning of the teaching of physics for contemporary culture. This is part of a very large problem, the meaning of which is hard to overestimate. The task placed before humanity by non-classical physics, the task of the atomic age, is for the moral potential of this age to rise to the level of the intellectual potential which is embodied in the contemporary scientific-technical revolution. One way to solve this problem is for the population of the world to understand where physics is going and to receive from adolescence in the schoolroom what is necessary for this knowledge. This is the knowledge not only of physics in the general and traditional sense, but also of *physis*, that is, of the general scientific picture of the universe and the cultural and emotional accompaniment of science. The image, life, and creativity of Einstein is a distillation and epitome of all of this.

Bibliography

Einstein, A., *Sobranie nauchnyh trudov* (4 volumes), edited by I. Tamm, B. Kuznetsov, and I. Smorodinski (Moscow: 1965–7).

—— *The World as I See It* (New York, 1934).

—— *Out of My Later Years* (New York: Philosophical Library, 1950).
—— *Ideas and Opinions* (New York: Dell, 1956).
Kuznetsov, B., *Einstein and Dostoyevsky* (London, 1972).
—— *Philosophy of Optimism* (Moscow, 1977).
—— *Leben-Tod-Unsterblichkeit* (Berlin, 1977).

8 Einstein and world affairs

A. P. French

Truth, *independent of man*,
independent of consciousness,
independent of sense experience,
independent of morality—this was
Einstein's 'religion'.

(Henry Le Roy Finch, in
Conversations with Einstein)

1. INTRODUCTION

Although Einstein did not go out of his way to mix in the day-to-day rough-and-tumble of worldly affairs, he was no ivory-tower scientist. On the contrary, he maintained throughout his life a passionate concern with social justice and the preservation of world peace. In politics he was firmly committed, both emotionally and intellectually, to socialism and a controlled economy. He was also a convinced internationalist, and felt that the only solution to the world's problems lay in a substantial surrender of autonomy by individual countries and governments. He was a wholehearted pacifist until the threat to civilization posed by Hitler's Germany led him to advocate rearmament of the West in self-defence, and later to let his name be associated with the proposal to develop atomic energy for military purposes. But after the war he resumed his fundamental commitment to the cause of disarmament, and almost the last act of his life, only a few days before he died, was to subscribe to an anti-war statement prepared by Bertrand Russell and subsequently signed by a number of eminent scientists.

In the face of all the discouraging evidences of man's inhumanity to man, he was at most times a basically optimistic realist who believed that individuals of goodwill could, by speaking out, bring about changes for the better in human society, and he never shrank from lending his own immense prestige to causes that he thought worthy.

In what follows we offer a brief account of this side of his life, based mainly on his published utterances and on the book *Einstein on Peace* (edited by Nathan and Norden, 1960).

2. IN GERMANY: WORLD WAR I AND AFTER

It was the outbreak of World War I in August 1914, soon after Einstein had moved from Switzerland to take up his chair at Berlin, that first made him fully alive to the evils of militarism and chauvinism. Not long afterwards he became part author of a 'Manifesto to Europeans' that was circulated and discussed (but received little support) within Germany. It was Einstein's first overt political act. In this same connection he was a founding member of a group calling itself the New Fatherland League, dedicated to Internationalism and a just peace.

An interesting sidelight on the circumstances of the time is that, despite the war, Einstein was able in 1915 to make a visit to Switzerland, where he held a long conversation with the novelist Romain Rolland, who became a lifelong friend. Rolland recorded in his diary that Einstein spoke with great frankness about the war, and expressed his hopes for an Allied victory to destroy the power of Prussian militarism. Later, in 1918, Einstein welcomed the abdication of the Kaiser and the establishment of a Republic immediately after the Armistice was declared.

Einstein was not only a great scientist, he was a great man. He stood for peace in a world drifting towards war. He remained sane in a mad world, and liberal in a world of fanatics.
(Bertrand Russell, in G. J. Whitrow, *Einstein: The Man and His Achievement*)

In the domestic chaos in Germany following the end of the war, Einstein tried to exert his influence on behalf of political prisoners and other social causes. On the international scene he allied himself to pacifist causes and travelled widely. In particular, in 1922, while animosity between Germany and France still ran high, he made a visit to Paris and was well received by both scholars and politicians (though not by members of the government). During informal discussions he emphasized the importance of both cultural and political cooperation. Shortly afterwards a delegation of Frenchmen came to Berlin to participate in a pacifist rally. Einstein, addressing the meeting on the floor of the Reichstag, said:

> I should like to describe our present situation . . . as though we were fortunate enough to witness the happenings on this miserable planet from the vantage point of the moon.
>
> First, we might ask ourselves in what sense the problems of international affairs require today an approach quite different from that of the past—not just the recent past, but the past half-century. To me, the answer is quite simple: due to technological developments, the distances throughout the world have shrunk to one tenth of their former size. The production of commodities in the world has become a mosaic composed of pieces from all over the globe. It is essential and altogether natural that the increased economic interdependence of the world's territories, which participate in mankind's production, be complemented by an appropriate political organization.
>
> The famous man in the moon would not be able to comprehend why mankind, even after the frightful experience of the war, was still so reluctant to create such a new political organization. Why is man so reluctant? I think the reason is that, where history is concerned, people are afflicted with a very poor memory.

It is a strange situation. The common man, exposed to events as they happen, has relatively little trouble adjusting himself to great changes, while the learned man who has soaked up much knowledge and serves it up to others faces a more difficult problem. In this respect, language plays a particularly unfortunate role. For what is a nation but a group of individuals who are forever influencing one another by means of the written and spoken word. The members of a given language community may scarcely notice it when their own peculiar outlook on the world becomes biased and inflexible.

I believe the condition in which the world finds itself today makes it not only a matter of idealism but one of dire necessity to create unity and intellectual co-operation among nations. Those of us who are alive to these needs must stop thinking in terms of 'What should be done for our country?' Rather, we should ask: 'What must our community do to lay the groundwork for a larger world community?' For without that greater community no single country will long endure.

'In my opinion it is not right to bring politics into scientific matters, nor should individuals be held responsible for the government of the country to which they happen to belong.'
(A.E. to Lorentz, 1923)

As a result of such activities Einstein exposed himself to attack by anti-Semitic extremists—a foretaste of things to come.

In these post-war years the League of Nations was trying, without much success, to establish an effective role for itself. Einstein was invited to join its Committee on Intellectual Cooperation. With some misgivings he accepted, but resigned within a year because he felt that the League was functioning 'as a tool of those nations which, at this stage of history, happen to be dominant powers'. However, he remained in sympathy with the aims of the League, and in 1924 was persuaded to rejoin the Committee on Intellectual Cooperation. He was encouraged by this second experience of it, and was glad to find an increased readiness to readmit Germany to the worlds of both politics and culture. He was distressed, however, to find that individual artists and scientists could be more narrowly nationalistic than many men of affairs, and he terminated his connection with the Committee in 1930.

In the meantime Einstein was beginning to associate himself actively with pacifist organizations. His commitment to pacifism received special prominence when, on a visit to the United States at the end of 1930, he gave a speech on the subject in New York. In it, he called for deeds, not words, and urged true pacifists to refuse military service even in peacetime. During an interview later in the same visit, he is recorded as having said:

It may not be possible in one generation to eradicate the combative instinct. It is not even desirable to eradicate it entirely. Men should continue to fight, but they should fight for things worth while, not for imaginary geographical lines, racial prejudices and private greed draped in the colours of patriotism. Their arms should be weapons of the spirit, not shrapnel and tanks . . .

We must be prepared to make the same heroic sacrifices for the cause

> of peace that we make ungrudgingly for the cause of war. There is no task that is more important or closer to my heart.
>
> Nothing that I can do or say will change the structure of the universe. But maybe, by raising my voice, I can help the greatest of all causes—good will among men and peace on earth.

And a few months later, in an article in the *New York Times*, he wrote:

> Let me begin with a confession of political faith: that the state is made for man, not man for the state. This is true of science as well. These are age-old formulations, pronounced by those for whom man himself is the highest human value. I should hesitate to restate them if they were not always in danger of being forgotten, particularly in these days of standardization and stereotype. I believe the most important mission of the state is to protect the individual and make it possible for him to develop into a creative personality.
>
> The state should be our servant; we should not be slaves of the state. The state violates this principle when it compels us to do military service, particularly since the object and effect of such servitude is to kill people of other lands or infringe upon their freedom. We should, indeed, make only such sacrifices for the state as will serve the free development of men.

One consequence of Einstein's preoccupation with the problems of wars was an exchange of correspondence with Sigmund Freud (subsequently published as a pamphlet entitled *Why War?*). Einstein, seeing war as a product of what he called 'the dark places of human will and feeling', wondered whether this instinct could be understood well enough in psychological terms to be controlled or eliminated. Freud's reply was lengthy but pessimistic; he felt (like Einstein) that the only practical solution that could be contemplated in the near future was to set up effective supranational organizations.

3. TO AMERICA: THE YEARS 1933–1940

Although Einstein moved permanently to America in 1933, he retained a deep concern with what was by then happening in Europe and particularly in Germany. To a resurgence of militarism was added the Nazi persecution of the Jews. On both questions he was to speak and act in every way he considered effective. It was at this stage that he began to modify his view that the use of force in international affairs could never be justified. He hoped that a compromise solution could be reached through the use of small professional armies and an international police force, so that the individual's right to refuse military service could be preserved. But within a short space of time he had reached the view that, to safeguard such freedoms in the future, it was reasonable to show readiness to fight against Nazi domination of Europe. He declined, in fact, to support the cause of some young conscientious objectors in Belgium; he saw clearly, six years before the actual outbreak of war, the magnitude of the menace and the need to meet it. At about this time (actually en route to America in the autumn of 1933) he met Winston Churchill in England and found him similarly far-seeing.

In 1933 also, not long before moving to America, Einstein made a public announcement of his intention to resign from the Prussian Academy of Sciences and to renounce his Prussian citizenship (which he acquired in 1913). His decision drew a hostile letter from the Academy, upbraiding him for helping to spread slanderous rumours about Germany in other countries, when he could have put in a good word for the German people. In his reply Einstein wrote:

> You have also remarked that a 'good word' on my part for 'the German people' would have produced a great effect abroad. To this I must reply that such a testimony as you suggest would have been equivalent to a repudiation of all those notions of justice and liberty for which I have stood all my life. Such testimony would not be, as you put it, a good word for the German people; on the contrary, it would only have helped the cause of those who are seeking to undermine the ideas and principles which have won for the German people a place of honour in the civilized world. By giving such testimony in the present circumstances I should have been contributing, even if only indirectly, to moral corruption and the destruction of all existing cultural values.
>
> It was for this reason that I felt compelled to resign from the Academy, and your letter only shows me how right I was to do so.

Almost immediately afterwards he resigned also from the Bavarian Academy, saying:

> The primary duty of an Academy is to further and protect the scientific life of a country. And yet the learned societies of Germany have, to the best of my knowledge, stood by and said nothing while a not inconsiderable proportion of German scholars and students and also of academically trained professionals have been deprived of all chance of getting employment or earning a living in Germany. I do not wish to belong to any society which behaves in such a manner, even if it does so under external pressure.

Einstein's abandonment of total pacifism led to sharp differences with former allies in the peace movement. He spent much time and effort explaining that the potential destruction of our intellectual and cultural heritage was too high a price to pay for the avoidance of war, and furthermore that the best deterrent to Nazi aggression would be military strength in the democracies.

Einstein's move from Europe to America took place when the western world was struggling towards recovery from the economic crisis that began with the crash on Wall Street in 1929. At about this time he made public some of his views on economic problems, saying, disarmingly: 'If there is anything that can give a layman in the sphere of economics the courage to express an opinion on the nature of the alarming economic difficulties of the present day, it is the hopeless confusion of opinions among the experts.'

He went on to advocate, not a totally planned economy (he felt that Russia showed the deficiencies of a system in which the competitive element was suppressed) but rather a moderate amount of control aimed at limiting working hours to achieve full employment, and limiting prices in cases where monopolistic practices would lead to abuses. 'My personal opinion,' he wrote, 'is that those methods are in general preferable which respect existing traditions and habits so far as that is in any way compatible with the end in view.' This remark was typical of Einstein's approach to problems. He did not let himself be carried away by simplified appeals to general principles; each individual problem was submitted to objective scrutiny by his constantly analytical and critical mind. Also, in human affairs, he did not believe that any hard-won cultural heritage should be lightly discarded.

Despite his interest in economic and social problems, it was however the preservation of world peace that remained Einstein's chief concern outside physics. He felt that the main hope had to be reposed in the League of Nations, since some kind of international organization was essential and this was the only one extant. The absence of the United States from the League clearly weakened it seriously, and in 1934 he made a public statement urging Americans to exert their influence to persuade the United States to join. Again and again he emphasized that merely talking about peace was not enough. Nor, despite his great admiration for Mahatma Gandhi, did he believe that Gandhi's tactics of passive resistance would be of any avail against the Nazi tyrants. He resigned himself, in fact, to the overwhelming probability of war, and became deeply critical of pacifism for its own sake. In a letter dated 1937 to the American League against War and Fascism he wrote, 'It must be said that, of late, pacifists have harmed rather than helped the cause of democracy. This is especially obvious in England, where the pacifist influence has dangerously delayed the rearmament which has become necessary because of the military preparations in Fascist countries.'

The Nazi annexation of Austria in 1938 made the threat of war closer, and also brought the plight of Jews in Europe to a desperate level. Einstein tried to stimulate actions to help them, but found little desire on the part of Americans to become involved with troubles so far away.

In August 1939, however, four weeks before Hitler precipitated World War II through his invasion of Poland, Einstein signed the famous letter to Roosevelt proposing a research and development programme on nuclear chain reactions as the basis of bombs of unprecedented power. It is ironic

Albert Einstein
Old Grove Rd.
Nassau Point
Peconic, Long Island

August 2nd, 1939

F.D. Roosevelt,
President of the United States,
White House
Washington, D.C.

Sir:

Some recent work by E.Fermi and L. Szilard, which has been communicated to me in manuscript, leads me to expect that the element uranium may be turned into a new and important source of energy in the immediate future. Certain aspects of the situation which has arisen seem to call for watchfulness and, if necessary, quick action on the part of the Administration. I believe therefore that it is my duty to bring to your attention the following facts and recommendations:

In the course of the last four months it has been made probable - through the work of Joliot in France as well as Fermi and Szilard in America - that it may become possible to set up a nuclear chain reaction in a large mass of uranium,by which vast amounts of power and large quantities of new radium-like elements would be generated. Now it appears almost certain that this could be achieved in the immediate future.

This new phenomenon would also lead to the construction of bombs, and it is conceivable - though much less certain - that extremely powerful bombs of a new type may thus be constructed. A single bomb of this type, carried by boat and exploded in a port, might very well destroy the whole port together with some of the surrounding territory. However, such bombs might very well prove to be too heavy for transportation by air.

-2-

The United States has only very poor ores of uranium in moderate quantities. There is some good ore in Canada and the former Czechoslovakia, while the most important source of uranium is Belgian Congo.

In view of this situation you may think it desirable to have some permanent contact maintained between the Administration and the group of physicists working on chain reactions in America. One possible way of achieving this might be for you to entrust with this task a person who has your confidence and who could perhaps serve in an inofficial capacity. His task might comprise the following:

a) to approach Government Departments, keep them informed of the further development, and put forward recommendations for Government action, giving particular attention to the problem of securing a supply of uranium ore for the United States;

b) to speed up the experimental work,which is at present being carried on within the limits of the budgets of University laboratories, by providing funds, if such funds be required, through his contacts with private persons who are willing to make contributions for this cause, and perhaps also by obtaining the co-operation of industrial laboratories which have the necessary equipment.

I understand that Germany has actually stopped the sale of uranium from the Czechoslovakian mines which she has taken over. That she should have taken such early action might perhaps be understood on the ground that the son of the German Under-Secretary of State, von Weizsäcker, is attached to the Kaiser-Wilhelm-Institut in Berlin where some of the American work on uranium is now being repeated.

Yours very truly,
A. Einstein
(Albert Einstein)

Figure 27 The letter to President Roosevelt from Einstein, 2 August 1939 that only a few years earlier he, like Rutherford, had dismissed as absurd the notion of using nuclear energy for practical purposes.

Einstein's letter was finally put into Roosevelt's hands in October 1939, along with a more technical statement by Leo Szilard, who had been the driving force behind these developments and who had written the first draft of Einstein's own letter. As in many other instances, Einstein's main contribution was his great prestige, which may indeed have played a decisive role in Roosevelt's decision to set up an Advisory Committee on Uranium, from which the whole atomic bomb project developed.

After this initial step Einstein had nothing more to do with the development of the bomb, but in March 1945, when the feasibility of the bomb was assured, he again wrote to Roosevelt on behalf of Szilard and other scientists who by then had decided that the bomb need not and should not be used for military purposes. Whether Roosevelt would have been influenced by this it is impossible to say; he died less than three weeks after the letter was sent, and the decision passed into other hands.

It may be somewhat astonishing that a theoretically-oriented mind as that of Albert Einstein would be interested in technical matters. But he thoroughly enjoyed learning about clever inventions and solutions, as he had always loved to solve certain types of puzzles. Maybe both, inventions and puzzles, reminded him of the happy, carefree and successful days at the patent office in Bern, the days before the first world war and all that followed.
(H. A. Einstein, in G. J. Whitrow, *Einstein: The Man and His Achievement*)

4. AFTER 1945: THE ATOMIC AGE

The dreadful success in the development and use of atomic weapons led Einstein to renew his efforts to stimulate the creation of effective international cooperation to prevent war. In 1945 he was reported as saying: 'The release of atomic energy has not created a new problem. It has merely made more urgent the necessity of solving an existing one. As long as there are sovereign nations possessing great power, war is inevitable. This does not mean that one can know *when* war will come, but only that one is sure that it *will* come. That was true even before the atomic bomb was made. What has changed is the destructiveness of war.'

He proposed the formation of a world government, led by the United States, the Soviet Union and Great Britain, and to which these nations should commit all their military resources. He feared the release of the 'secret' of the atomic bomb to the multitude of countries in the newly formed United Nations Organization. He did not believe that Russians should be let into the secret either (of course we now know that they were already well informed through unofficial channels). He called upon the public to press for the kind of partial surrender of sovereignty that he envisaged.

In this whole situation he saw a special responsibility on the part of physicists and other scientists. He spelled this out in a speech in New York in December 1945:

> Physicists find themselves in a position not unlike that of Alfred Nobel. Alfred Nobel invented an explosive more powerful than any then known—an exceedingly effective means of destruction. To atone for this 'accomplishment' and to relieve his conscience, he instituted his awards for the promotion of peace. Today, the physicists who participated in producing the most formidable weapon of all time are harassed by a

similar feeling of responsibility, not to say guilt. As scientists, we must never cease to warn against the danger created by these weapons; we dare not slacken in our efforts to make the peoples of the world, and especially their governments, aware of the unspeakable disaster they are certain to provoke unless they change their attitude toward one another and recognize their responsibility in shaping a safe future. We helped create this new weapon in order to prevent the enemies of mankind from achieving it first; given the mentality of the Nazis, this could have brought about untold destruction as well as the enslavement of the peoples of the world. This weapon was delivered into the hands of the American and the British nations in their roles as trustees of all mankind, and as fighters for peace and liberty; but so far we have no guarantee of peace nor of any of the freedoms promised by the Atlantic Charter. The war is won, but the peace is not.

He wrote to the President of the USSR Academy of Sciences, asking for contributions to a book, *One World or None*, that was then in preparation and was published in 1946. The reply expressed sympathetic interest but no contributions were received from Russia. Einstein's own contribution to the book began by reviewing the unhappy developments following World War I, and the ineffectiveness of the League of Nations and the International Court of Justice at the Hague. He feared that the United Nations, based on moral authority alone, might prove equally ineffectual. He therefore proposed various specific steps, based upon a strong supranational organization. These steps included mutual inspection of military installations, exchange of technical information and personnel, and the absorption of individual armed forces into an international peace-keeping force.

The theme of world government never ceased to dominate Einstein's thinking on international problems. In 1946 he attached his signature to a document ('Appeal to the Peoples of the World') which had as its leading recommendation the proposal 'That the United Nations be transformed from a league of sovereign states into a government deriving its specific powers from the peoples of the world.'

Within the United States, Einstein agreed in May 1946 to serve as chairman of the Emergency Committee of Atomic Scientists. The main function of this committee was to raise funds to support a massive programme of public education and policy development in areas relating to atomic energy. The committee remained in existence until 1951, and during its lifetime took first place among Einstein's non-scientific activities. He wrote many articles and gave interviews and radio broadcasts, all on the theme of the need for cooperation in world government in face of the threat of atomic holocaust. He emphasized the obligation for America, having a temporary lead in atomic technology, to take the leading role also in making atomic war impossible. In an eloquent statement in June 1946, he is quoted as saying:

Einstein was one of the most unusual of all human beings. To me he appears as out of comparison the greatest intellect of this century, and almost certainly the greatest personification of moral experience. He was in many ways different from the rest of the species. Meeting him in old age was rather like being confronted by the Second Isaiah—even though he retained traces of a rollicking, disrespectful, common humanity and had given up wearing socks.

(C. P. Snow, in *Conversations with Einstein*)

... There has been too much emphasis on legalisms and procedure. It is easier to denature plutonium than it is to denature the evil spirit of man. ...

Before the raid on Hiroshima, leading physicists urged the War Department not to use the bomb against defenceless women and children. The war could have been won without it. The decision was made in consideration of possible future loss of American lives; but now we have to consider the possible loss, in future atomic bombings, of millions of lives. The American decision may have been a fatal error, for men accustom themselves to thinking that a weapon which was used once can be used again. ...

Science has brought forth this danger, but the real problem is in the minds and hearts of men. We will not change the hearts of other men by mechanical devices; rather we must change our own hearts and speak bravely. We must be generous in giving the rest of the world the knowledge we have of the forces of nature, after establishing safeguards against possible abuse. We must not merely be willing, but must be actively eager to submit ourselves to the binding authority necessary for world security. We must realize we cannot simultaneously plan for war and for peace.

When we are clear in heart and mind—only then shall we find courage to surmount the fear which haunts the world.

In reading the early papers of Einstein one has—perhaps erroneously—the sense of being close to the thinking processes of the man. They are full of such phrases, as 'In a memoir published four years ago I tried to answer the question whether the propagation of light is influenced by gravitation. I return to this theme, because my previous presentation of the subject does not satisfy me. ...' We have the continual sense that these papers have been written by a human being, and that we are witness to his 'personal struggle' with the puzzles and mysteries of the natural universe.

(Jeremy Bernstein, *Einstein*)

Near the end of 1946, the Emergency Committee of Atomic Scientists held a conference at the end of which the following statement was issued:

These facts are accepted by all scientists:

1. Atomic bombs can now be made cheaply and in large number. They will become more destructive.
2. There is no military defense against the atomic bomb and none is to be expected.
3. Other nations can rediscover our secret processes by themselves.
4. Preparedness against atomic war is futile, and if attempted will ruin the structure of our social order.
5. If war breaks out, atomic bombs will be used and they will surely destroy our civilization.
6. There is no solution to this problem except international control of atomic energy and, ultimately, the elimination of war.

The programme of the committee is to see that these truths become known to the public. The democratic determination of this nation's policy on atomic energy must ultimately rest on the understanding of its citizens.

This statement was widely reported, and provided a basis for subsequent efforts to raise public consciousness of the political problems generated by the atom. The going, however, was hard. In an interview in late 1947, Einstein was reported as saying:

Since the completion of the first atomic bomb nothing has been accomplished to make the world safer from the threat of war, while much has been done to increase the destructiveness of war. I am not able to speak from any first-hand knowledge about the development of the atomic bomb since I do not work in this field: but enough has been said by those who do to indicate that the bomb has been made more effective. Certainly, one can envisage the possibility of building a bomb of far greater size, capable of causing destruction over a larger area than heretofore.

His concern was prophetic. Less than three years later in 1950, after the Soviet Union had exploded their own first fission bomb, President Truman authorized the crash programme to develop a hydrogen bomb, and in November 1952 it became an achieved fact.

One of the developments that dismayed Einstein particularly was the re-emergence of Germany as a military power. After the end of the war, Einstein had been quick to declare his deep opposition to the restoration of military strength to Germany, and even to its industrial rehabilitation. He was convinced that, unless Germany was kept in check, it would once again become an aggressor nation, seeking revenge for its defeat in the war. When the remilitarization of the Federal Republic of Germany was being fostered from 1950 onward, mainly as a consequence of the Cold War and to make it a bulwark against the Russians, Einstein restated his opposition and his misgivings on many occasions. This opposition derived extra strength from his objective judgement that the resurgence of the Federal Republic of Germany as a military power would cripple the chance of bringing the United States and the Soviet Union together. However, he also had an intense personal feeling about Germany and things German, which persisted with him until the end of his life. When his old friend Max Born after many years in Scotland (as Professor in Edinburgh) returned to Germany in 1953 to live out his retirement, Einstein wrote to him in critical and almost harsh terms.

In the face of the mutual fear and suspicion between the United States and Russia, Einstein tried on various occasions to involve Russian scientists in the campaign for cooperation and world government, but it proved an impossible task. He criticized America for failing to give adequate recognition to Russia's need to feel secure, but he also rebuked the Russians for their resistance to the idea of a supranational regime. His comments on this latter question drew an open letter from four Russian scientists, attacking his point of view. At about the same time he took part in unsatisfactory discussions with some of his American scientific colleagues, including members of the Emergency Committee. His frustrations expressed themselves not in anger or depression, but in the following mock statement:

The friendship of my thesis professor, Rudolf Ladenburg, and Einstein was a long one. They had been together in Berlin before coming to Princeton. Ladenburg liked to tell of their first meeting, about 1908, when he called upon Einstein in the Swiss Patent Office. Einstein told him that he was the first physicist that he had seen in five years. During these years Einstein had done much of his important work. He pulled out one drawer of his desk and said that it was his theoretical physics office. His duty of reading patents took little time, and he worked upon physics whenever he was free.

(Yardley Beers, *Am. J. Phys.*)

Resolution

We American scientists, after three days of careful consideration, have come to the following conclusions:

We do not know

(a) What to believe;
(b) What to wish for;
(c) What to say; and
(d) What to do.

Appendix

On the basis of an open letter signed by Russian scientists, we may construct a parallel resolution for them:

After careful consideration, and after due consultation with our government, we do know

(a) What not to believe;
(b) What not to wish for;
(c) What not to say; and
(d) What not to do.

Of course this humorous summary did not betoken any weakening of his deeply serious concern with the world's problems, as he made clear in several statements published during 1948. The main themes remained the same: world government and the reduction of military organizations. But his views ran counter to the general trend of international relations in the climate of post-war suspicions, and in his last years he began to be discouraged. He found himself forced to admit that governments and the general public alike were largely deaf to the pleadings and arguments of concerned individuals, however eloquent and perceptive. Weapons, not words, were the basis of international debate. And with typical philosophic detachment he wrote (in 1952) to his friend Maurice Solovine: 'If all our efforts are in vain and man goes down in self-destruction, the universe will shed no tears.'

During the very last months of his life, Einstein was approached by Bertrand Russell, who proposed the issuance of a statement, signed by a small number of people of the highest scientific attainment, giving a solemn warning concerning the appalling consequences of war with nuclear weapons. Einstein associated himself enthusiastically with this plan, and wrote to Niels Bohr in Copenhagen to enlist his participation. Einstein died before the statement was issued in its final form in July 1955, but his commitment to it was clear and definite. The statement ended with an appeal, in the broadest possible terms, to the governments of the world: 'We urge [them] to realize, and to acknowledge publicly, that their purpose cannot be furthered by a world war, and we urge them, consequently, to find peaceful means for the settlement of all matters of dispute between them.' It was a fitting epitome of Einstein's lifelong advocacy of rationality and decency in world affairs.

CONCLUDING REMARKS

An article as short as this cannot do justice to Einstein's many-sided concern with the welfare of individuals and nations, with human rights, and with the preservation of peace and freedom. In a self-deprecating statement written little more than a year before he died, he said:

> In a long life I have devoted my faculties to reach a somewhat deeper insight into the structure of physical reality. Never have I made any systematic effort to ameliorate the fortunes of men, to fight injustice and oppression, or to improve the traditional forms of human relations. The only thing I did was this: At long intervals I have publicly expressed opinions on such conditions in society which I considered to be so bad and unfortunate that silence would have made me feel guilty of complicity . . .

In fact his contribution, though admittedly limited for the most part to written and spoken statements, was impressive and represented a substantial fraction of his activities. (His deep concern with Jewish affairs, scarcely mentioned in our discussion, is the subject of a separate article.)

Einstein was criticized in many quarters for holding so unswervingly to the theme of world government and the partial surrender of national autonomy. This was seen as an unrealistic and therefore unfruitful position to take. His own view, however, was that compromise on the basic principle would surely prevent its ultimate attainment. This was in keeping with his related views on disarmament. As Otto Nathan has remarked (in the preface to *Einstein on Peace*): 'He had never believed that disarmament by small stages was a practicable policy against war, a policy which would ever lead to total disarmament and peace; he was convinced that a nation could not arm and disarm at the same time.' One should not be surprised at such an attitude in Einstein, for it was just the same approach which, in scientific matters, led him to cut through all patched-up compromises and to arrive at new fundamental theories.

Einstein felt strongly that scientists, *qua* scientists, were not the ones to change the course of human affairs. In 1952 he wrote:

> Betterment of conditions the world over is not essentially dependent on scientific knowledge but on the fulfilment of human traditions and ideals. I believe, therefore, that men like Confucius, Buddha, Jesus, and Gandhi have done more for humanity with respect to the development of ethical behaviour than science could ever accomplish.

But, applying this to himself, one can claim that Einstein the man did indeed do a great deal as a spokesman for truth and goodness in our troubled century.

9 Einstein and Zionism

Gerald E. Tauber

Concern for the man himself and his fate must always form the chief interest of all technical endeavours. Never forget this in the midst of your diagrams and equations (Einstein, *Mein Weltbild*).

'Mankind has lost its finest son, whose mind reached out to the ends of the universe but whose heart overflowed with concern for the peace of the world and the well-being, not of humanity as an abstraction, but of ordinary men and women everywhere.'
(Dr Israel Goldstein)

t was that concern which singled Einstein out amongst the great scientists, man who spoke out openly for his beliefs and principles, who took his bligations to society seriously, and never forgot his people and its spirations.

Einstein spent his youth in a Jewish, but by no means religious, home. Ie attended the local Catholic elementary school, which was cheaper and iore convenient than the distant Jewish private school. Nevertheless, his ewish education was not neglected and he received private lessons, and hus at an early age became acquainted with the teachings of both Moses nd Jesus. Anti-Semitism was not foreign to Einstein and his contemporaries, nd as he wrote later (see Hoffmann, 'Einstein and Zionism') 'Physical ssaults and insults were frequent on the way to school, though not really nalicious. Even so, however, they were enough to confirm, even in a child f my age, a vivid feeling of not belonging.'

However, it was not until Einstein became professor at Prague in 1911 hat he came into contact with Jews—who lived and thought like Jews— nd began to understand the particular problems that beset them. There he lso came into contact with Zionists who formed 'a small circle of philoophical and Zionist enthusiasts which was loosely grouped around the niversity' (see Frank, *Einstein—His Life and Times*), but he was not intersted, at that time, in the problems of Jewry on a world basis.

The pursuit of knowledge for its own sake, an almost fanatical love of justice and the desire for personal independence—these are the features

> of the Jewish tradition which make me thank my stars that I belong to it (see Einstein, 'Jewish Ideals', *Mein Weltbild*).

In Germany, even more than in Prague, Einstein realized that anti-Semitism could not be fought by assimilation, but would have to be combated by more knowledge.

> Before we can effectively combat anti-Semitism, we must first of all educate ourselves out of it and out of the slave-mentality which it betokens. We must have more dignity, more independence, in our own ranks. Only when we have the courage to regard ourselves as a nation, only when we respect ourselves, can we win the respect of others; or rather, the respect of others will then come of itself . . . (see Einstein, 'Assimilation and Nationalism').

'As long as I have any choice in the matter, I shall live only in a country where civil liberty, tolerance, and equality of all citizens before the law prevail. Civil liberty implies freedom to express one's political convictions, in speech and writing; tolerance implies respect for the convictions of others whatever they may be.'
(A.E., on hearing that President Hindenburg had asked the National Socialists to form a German government in 1933)

Nor did he have much patience with the Central Association of German Citizens of the Jewish Persuasion which tried to pawn off Judaism as a mere religious persuasion:

> When I come across the phrase 'German Citizens of the Jewish Persuasion', I cannot avoid a melancholy smile. What does this high-falutin description really mean? What is this 'Jewish persuasion'? Is there, then, a kind of non-persuasion by virtue of which one ceases to be a Jew? There is not. What the description really means is that our *beaux esprits* are proclaiming two things: First, I wish to have nothing to do with my poor Jewish brethren, Secondly, I wish to be regarded not as a son of my people, but only as a member of a religious community. Is this honest? Can an 'aryan' respect such dissemblers? I am not a German citizen, nor is there anything about me that can be described as 'Jewish persuasion', but I am a Jew, and I am glad to belong to the Jewish people, although I do not regard it as 'chosen'. Let us just leave anti-Semitism to the non-Jews, and keep our own hearts warm for our kith and kin (see Einstein, 'Assimilation and Nationalism').

Perhaps it is then not surprising that Einstein was eventually attracted to Zionism. In 1897 Theodore Herzl, the Austrian journalist and author of the 'Judenstaat', had launched political Zionism at the Congress at Basel which resolved 'to secure for the Jewish people a home in Palestine guaranteed by public law'. In 1917 that dream seemed to become a reality, when the British government issued, through its Foreign Minister Lord Balfour, the so-called Balfour Declaration according to which 'His Majesty's Government views with favour the establishment in Palestine of a national home for the Jewish people, and will use their best endeavours to facilitate the achievement of this object . . .' After the cessation of hostilities Palestine became a British Mandate under the League of Nations and Great Britain was charged with the implementation of her pledge, an implementation which was to take thirty years and many bloody confrontations and wars n the meantime the Zionist movement, whose headquarters had moved, fter Herzl's death in 1904, from Vienna to Germany (first to Cologne, and inally to Berlin in 1911), tried to attract prominent Jews to its ranks. instein, naturally, was amongst possible candidates, although at that time e had not yet achieved world fame resulting from the experimental erification (by solar eclipse) of general relativity. At first Einstein, the pponent of nationalism, was lukewarm towards the idea of a national ome for Jews, but he eventually became convinced of the need for a ewish national home. In one of his many discussions with Kurt Blumenfeld, Zionist leader, he said, 'I am against nationalism but in favour of Zionism. he reason has become clear to me today. When a man has both arms and e is always saying I have a right arm, then he is a chauvinist. However, vhen the right arm is missing, then he must do something to make up for he missing limb. Therefore I am, as a human being, an opponent of ationalism. But as a Jew I am from today a supporter of the Jewish Zionist fforts' (see Blumenfeld, *Erlebte Judenfrage*)—Einstein had become a Zionist.

Once Einstein had become convinced of the correctness of his decision e became an outspoken supporter, as in all causes he espoused.

> I am a national Jew in the sense that I demand the preservation of the Jewish nationality as of every other. I look upon Jewish nationality as a fact, and I think that every Jew ought to come to definite conclusions on Jewish questions on the basis of this fact. I regard the growth of Jewish self-assertion as being in the interest of non-Jews as well as of Jews. That was the main motive of my joining the Zionist movement. For me Zionism is not merely a question of colonization. The Jewish nation is a living thing, and the sentiment of Jewish nationalism must be developed both in Palestine and everywhere else. To deny the Jews' nationality in the Diaspora is, indeed, deplorable. If one adopts the point of view of confining Jewish ethnic nationalism to Palestine, then to all intents and purposes one denies the existence of a Jewish people.

was this secondary, but more important facet of Zionism which Einstein nphasized.

> . . . But the main point is that Zionism must tend to enhance the dignity and self-respect of the Jews in the Diaspora. I have always been annoyed by the undignified assimilationist cravings and strivings which I have observed in so many of my friends (see Einstein, 'Assimilation and Nationalism', pp. 29, 30).

In 1921 Einstein was asked to join Chaim Weizmann, a biochemist and esident of the Zionist Organization, on a fund-raising tour to the USA 1 behalf of the Hebrew University to be established in Jerusalem. Einstein st demurred saying that he was no orator and that 'you will only be using y name'. Apparently his sense of duty intervened and he finally agreed to), although it meant missing the next Solvay Congress, the first since the

'I should demand the introduction of compulsory practical work. Every pupil must learn some *handicraft*. He should be able to choose for himself which it is to be, but I should allow no one to grow up without having gained some technique, either as a joiner, bookbinder, locksmith, or member of any other trade, and without having delivered some useful product of his trade.'

(A.E.)

Figure 28

Albert Einstein with Chaim Weizmann in 1921

end of the war. When it became known that Einstein was to travel t
America he was inundated with invitations and honorary degrees. Wha
was to be a simple fund-raising campaign—with Einstein as the 'show piec
—turned out to be a major lecture tour. Einstein did more than just lend h
presence; knowing first-hand the '*numerus clausus*' facing Jewish students i
Eastern and Central Europe, he could speak with authority on the need fo
a Jewish University for Jews and run by Jews: 'the greatest thing in Palestin
since the destruction of the Temple in Jerusalem'. He even predicted i
eventual central role,

> . . . but it is at any rate permissible to hope that in the course of time th
> Jerusalem University will grow into a centre of Jewish intellectual lif

which will be of value not for Jews alone (see Einstein, 'The Jews and Palestine').

Ie summed up his experiences in a letter to his friend Michele Besso.

> Two frightfully exhausting months now lie behind me, but I have the great satisfaction of having been very useful to the cause of Zionism and of having assured the foundation of the University. We found special generosity among the Jewish doctors of America (*ca* 6000) who provided the funds to create the medical school . . . I had to let myself be exhibited like a prize ox, to speak an innumerable number of times at small and large gatherings, and to give innumerable scientific lectures. It is a wonder I was able to hold out. But now it is over, and there remains the beautiful feeling of having done something truly good, and of having intervened courageously on behalf of the Jewish cause, ignoring the protests of Jews and non-Jews alike.

Einstein's travels were not over; in fact, they had just begun. The ·llowing year he and his wife (Elsa Einstein always accompanied her ısband to protect him from the many curiosity seekers) went on a trip ·hich was to take them to Japan, and Palestine. It was a memorable visit, ɔt only to the thousands who thronged the streets to have a glimpse of ·e distinguished visitor and who filled every place where Einstein appeared · spoke, but also to Einstein himself. At a reception at the Lemel school : said:

> I consider this the greatest day of my life. Hitherto I have always found something to regret in the Jewish soul, and that is the forgetfulness of its own people—forgetfulness of its being, almost. Today I have been made happy by the sight of the Jewish people learning to recognize themselves and to make themselves recognized as a force in the world. This is a great age, the age of liberation of the Jewish soul, and it has been accomplished through the Zionist movement, so that no one in the world will be able to destroy it (*Palestine Weekly*, 9 February 1923).

ıe highlight of the trip was a visit to Mount Scopus, the site of the future ebrew University, where Einstein was to give the inaugural address. From e 'lectern that had waited for him for two thousand years' Einstein spoke French, later repeated his address in German, but as he wrote in his diary, had to begin with a greeting in Hebrew, which I read off with great fficulty', and so the first official words spoken from the university had ·en in Hebrew.

Einstein, like everyone else, was deeply shocked by the continuous Arab :acks and, in particular, by the massacre of Yeshivah students in Hebron 1929:

> Shaken to its depths by the tragic catastrophe in Palestine, Jewry must now show that it is truly equal to the great task it has undertaken. It

> goes without saying that our devotion to the cause and our determinatio
> to continue the work of peaceful construction will not be weakened i
> the slightest by any such set-back (see Einstein, 'Jew and Arab').

However, he continued to plead for cooperation between Arab and Je
and in 'Letter to an Arab' even made some practical suggestions:

> A Privy Council is to be formed to which the Jews and Arabs shall eac
> send four representatives, who must be independent of all politic
> parties: each group to be composed as follows:
> A doctor, elected by the Medical Association.
> A lawyer, elected by the lawyers.
> A working men's representative, elected by the trade unions.
> An ecclesiastic, elected by the ecclesiastics.

The purpose of this 'Privy Council', he then continued, 'was to brin
about the gradual composition of differences, and secure a united repre
sentation of the common interests of the country before the mandator
power, clear of the dust of ephemeral politics.' Needless to say and, as
turned out, unfortunately, Einstein's advice was not followed up and th
country continued to be in turmoil.

When Hitler came to power in 1933 Einstein was in Pasadena on a visi
ironically sponsored by a fund to further German–American relations. H
refused to return to Germany, stating:

> As long as I have any choice, I will stay only in a country where politic
> liberty to express one's political opinion orally or in writing, and
> tolerant respect for any and every individual opinion, are the ru
> ('Manifesto—March 1933').

He severed all connections with German institutions and spoke out again
the oppression of the Nazis with a fervour reminiscent of the Jewis
prophets. Even after the war Einstein refused to have anything to do wi
German organizations:

> Because of the mass murder that the Germans inflicted on the Jewi
> people, it is evident that any self-respecting Jew could not possibly wi
> to be associated in any way with any official German institution (s
> Hoffmann, *Einstein and Zionism*).

Einstein was offered many positions including, of course, one at th
Hebrew University which he refused, since he believed that place should b
made for young comparatively unknown scholars who needed a place
refuge. He himself accepted a position at the Institute for Advanced Stuc
at Princeton, but not until he made sure that his young Jewish assista
Walther Mayer could join him. He was only the first of many friends ar
strangers Einstein was able to save from death by the Nazis. He w
deluged with invitations to address countless meetings and dinners f

charity or lend his name to numerous causes. All were refused, except those which helped the ever-growing stream of refugees or helped the Jews in their land.

In 1939, only months before the outbreak of World War II, the British Government published the infamous *White Paper* which curtailed immigration, and in fact closed the doors of Palestine to the Jewish refugees from Germany and other occupied countries. Einstein's already deep identification with his people was even more intensified when the full impact of the holocaust became known. He appeared before the Anglo-American Committee of Inquiry on Palestine and entered a strong plea for a Jewish Homeland. When the United Nations voted for partition in 1947 and for the State of Israel, established in May 1948, he heralded the event as the

Figure 29

Einstein playing the violin during a charity concert in a Berlin synagogue, 1930 'fulfilment of an ancient dream and to provide conditions in which the spiritual and cultural life of a Hebrew society could find free expression'.

When Chaim Weizmann, who became the first president of the State of Israel, died in 1952 Einstein was asked whether he would accept the presidency, if offered by the Knesset (Parliament)—his acceptance making this a mere formality. Einstein, deeply moved by the offer, declined, pointing out that he lacked 'both the natural aptitude and the experience to deal properly with people and to exercise official functions'. These reasons alone, he continued, not even mentioning his preoccupation with his work, would make him unsuited for that high office, even if advancing age was not making increasing inroads on his strength.

> I am the more distressed over these circumstances because my relationship with the Jewish people has become my strongest human bond, ever since I became fully aware of our precarious situation among the nations of the world (quoted from a letter to Abba Eban, 18 November 1952).

On the occasion of Israel's Seventh Anniversary of Independence in 1955, Einstein was asked to prepare a statement stressing Israel's cultural and scientific achievement which could be broadcast as part of the celebrations. 'I should very much like to assist the cause of Israel in the difficult and dangerous conditions prevailing today,' he answered, but felt that such a statement should touch upon the Arab-Israeli relations rather than Israel's cultural and scientific developments.

> I feel, therefore, that to make any impact on public opinion, such an address should attempt to appraise the political situation. In fact, I tend to believe that a somewhat critical analysis of the policies of the Western nations with regard to Israel and the Arab states might be most effective. I realize that it is easier for me to offer such remarks than for someone officially connected with Jewish organizations (from a letter to the Israeli Consul, 4 April 1955).

To make it meaningful, Einstein suggested that it be prepared in cooperation with responsible Israeli officials. As a result of his suggestion Ambassador Abba Eban and Consul Reuven Dafni met with Einstein on 11 April and again on 13 April. Two hours after the visit Einstein collapsed, and eventually was moved to the Princeton Hospital. He had his notes put at his bedside in the hope of writing the speech, and although an unfinished draft of one page exists (a facsimile is reproduced in *Einstein on Peace*), Einstein never completed that task—he died 18 April 1955 after spending a peaceful night.

Now, as we commemorate the centenary of Einstein's birth, we might well ask how he would look at Zionism and Israel of today. Many of his dreams and predictions have been fulfilled, but others are still unfulfilled. Foremost, perhaps, is our universal desire for peace:

One of these ideals is peace, based on understanding and self-restraint, and not on violence. If we are imbued with this ideal, our joy becomes somewhat mingled with sadness, because our relations with the Arabs are far from this ideal at the present time. It may well be that we would have reached this ideal, had we been permitted to work out, undisturbed by others, our relations with our neighbours, for we *want* peace and we realize that our future development depends on peace (from Einstein, 'The Jews of Israel').

Bibliography

Blumenfeld, Kurt, *Erlebte Judenfrage* (Stuttgart, 1962).

Einstein, A., 'Assimilation and Nationalism', 'Jew and Arab', and 'The Jews and Palestine', in *About Zionism: Speeches and Letters*, translated and edited by Sir L. Simon (New York: Macmillan, 1931).

—— 'Jewish Ideals', 'Letter to an Arab', and 'Manifesto—March 1933', in *Mein Weltbild* (Amsterdam: Querido Verlag, 1934).

—— 'The Jews of Israel', broadcast on radio for the UJA, 27 November 1949, in *Out of My Later Years* (New York: Philosophical Library, 1950).

Frank, P., *Einstein: His Life and Times* (New York: Knopf, 1947).

Hoffmann, B., 'Einstein and Zionism', in *General Relativity and Gravitation*, edited by G. Shaviv and J. Rosen (Jerusalem: John Wiley, The Keter Press, 1975). This book also contains the letter to M. Besso and excerpts from Einstein's Diary.

Nathan, O., and Norden, H., *Einstein on Peace* (New York: Simon and Schuster, 1960).

10 Einstein and the academic establishment

Martin J. Klein

'Conviction is a good mainspring, but a bad regulator.'
(A.E. to Willem De Sitter)

The French novelist Stendhal began his most brilliant novel with this entence: 'On May 15, 1796, General Bonaparte made his entrance into Milan at the head of that youthful army which had just crossed the bridge of Lodi, and taught the world that after so many centuries Caesar and Alexander had a successor.' In its military context, the quotation is rrelevant here, but it can be adapted: almost exactly a century later Milan aw the arrival of another young foreigner who would soon teach the world hat after so many centuries Galileo and Newton had a successor. It would, however, have taken superhuman insight to recognize the future intellectual onqueror in the boy of fifteen who had just crossed the Alps from Munich. For this boy, Albert Einstein, whose name was to become a symbol for profound scientific insight, had left Munich as what we would now call a 'high school dropout'.

He had been a slow child; he learned to speak at a much later age than the verage, and he had shown no special ability in elementary school—xcept a talent for day-dreaming. The education offered at his secondary chool in Munich, one of the highly praised classical gymnasia, did not ppeal to him. The rigid, mechanical methods of the school appealed to him ven less. He had already begun to develop his own intellectual pursuits, but the stimulus for them had not come from school. The mystery hidden n the compass given to him when he was five, the clarity and beauty of Euclidean geometry, discovered by devouring an old geometry text at the ge of twelve—it was these things that set him on his own road of independent study and thought. The drill at school merely served to keep him rom his own interests; after some months he was fed up with it and resolved o leave. His leaving was assisted by the way in which his teachers reacted o his attitude towards school. 'You will never amount to anything,

Einstein,' one of them said, and another actually suggested that Einstein leave school because his very presence in the classroom destroyed the respect of the students. This suggestion was gratefully accepted by Einstein, and he set off to join his family in Milan. The next months were spent gloriously loafing, hiking around northern Italy, enjoying the many contrasts with his homeland. With no diploma, and no prospects, he seemed a very model dropout. But it is sobering to think that no teacher had sensed his potentiality.

Einstein had decided to leave school, but he had not lost his love for science. Since his family's resources, or lack of them, would make it

Figure 30

Einstein (at the right of the front row) in the classroom at Aarau, with Dr Jost Winteler

necessary for him to become self-supporting, he decided to continue his scientific studies in an official fashion. He therefore presented himself for admission at the renowned Swiss Federal Institute of Technology in Zürich. Since he had no high school diploma he was given an entrance examination —and he failed. He had to attend a Swiss high school for a year in order to make up his deficiencies in almost everything except mathematics and physics, the subjects of his own private study. And then, when he was finally admitted to the Polytechnic Institute, did he settle down and assume what we would consider to be his rightful place at the head of the class? Not at all. Despite the fact that the courses were now almost all in mathematics and physics, Einstein missed most of the lectures. He did enjoy working in the laboratory, but he spent most of his time in his room studying the original works of the masters of nineteenth-century physics, and pondering what they set forth.

The lectures on advanced mathematics did not hold him because in those days he saw no need or use for higher mathematics as a tool for grasping the structure of nature. Besides, mathematics appeared to be split into so many branches, each of which could absorb all one's time and energy, that he feared he could never have the insight to decide on one of them, the fundamental one. He would then be in the position of Buridan's ass, who died of hunger because he could not decide which bundle of hay he should eat.

Physics presented no such problems to Einstein, even then. As he wrote many years later, 'True enough, physics was also divided into separate fields, each of which could devour a short working life without having satisfied the hunger for deeper knowledge. . . . But in physics I soon learned to scent out the paths that led to the depths, and to disregard everything else, all the many things that clutter up the mind, and divert it from the essential. The hitch in this was, of course, the fact that one had to cram all this stuff into one's mind for the examination, whether one liked it or not.'

As he put it, 'This coercion had such a deterring effect upon me that, after I had passed the final examination, I found the consideration of any scientific problems distasteful to me for an entire year.' And he went on to say, 'It is little short of a miracle that modern methods of instruction have not already completely strangled the holy curiosity of inquiry, because what this delicate little plant needs most, apart from initial stimulation, is freedom, without that it is surely destroyed. . . . I believe that one could even deprive a healthy beast of prey of its voraciousness, if one could force it with a whip to eat continuously whether it were hungry or not. . . .'

For almost two years after his graduation from the Polytechnic in 1900 Einstein seemed to be headed for no more success than his earlier history as a dropout might have suggested. He applied for an assistantship, but it went to someone else. During this period he managed to subsist on the odd jobs of the learned world: he substituted for a Swiss high school teacher who was doing his two months of military service, he helped the professor of astronomy with some calculations, he tutored at a boys' school. Finally, in the spring of 1902, Einstein's good friend Marcel Grossmann came to his rescue. Grossmann's father recommended Einstein to the director of the Swiss Patent Office at Bern, and after a searching examination he was appointed to a position as patent examiner. He held this position for over seven years and often referred to it in later years as 'a kind of salvation'. It freed him from financial worries, he found the work rather interesting, and sometimes it served as a stimulus to his scientific imagination. And besides, it occupied only eight hours of the day, so that there was plenty of time left free for pondering the riddles of the universe.

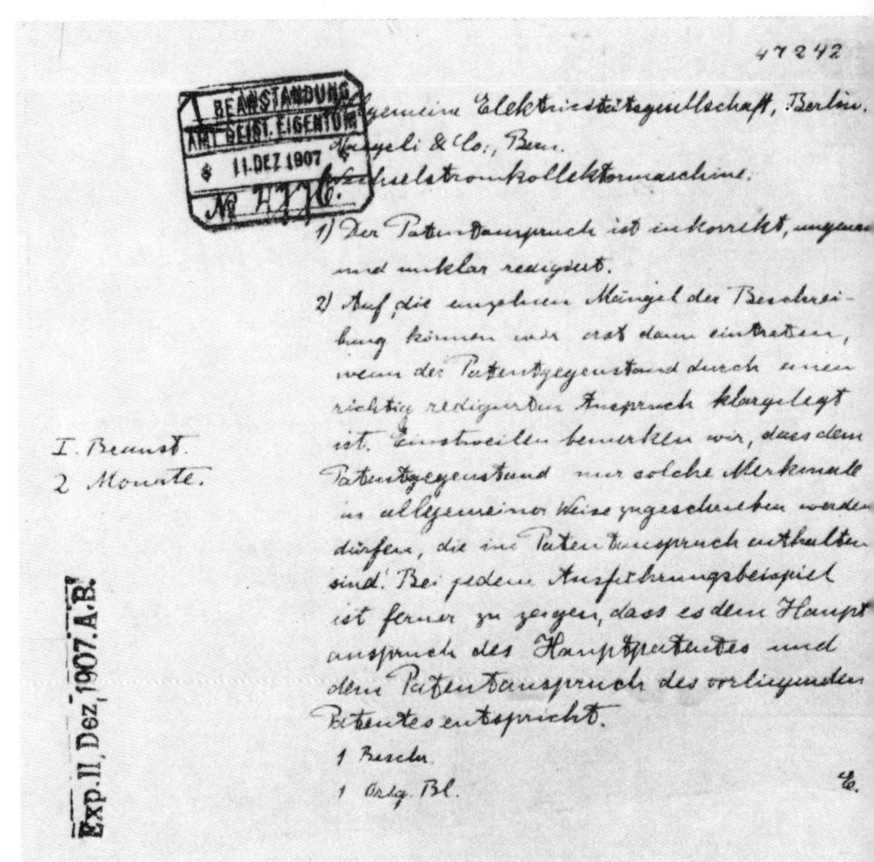
47242

BEANSTANDUNG
AMT f. GEIST. EIGENTUM
11.DEZ.1907

gemeine Elektricitätsgesellschaft, Berlin.
geli & Co., Bern.
chselstromkollektormaschine.

I. Beanst.
2 Monate.

1) Der Patentanspruch ist inkorrekt, ungenau und unklar redigiert.

2) Auf die einzelnen Mängel der Beschreibung können wir erst dann eintreten, wenn der Patentgegenstand durch einen richtig redigierten Anspruch klargelegt ist. Einstweilen bemerken wir, dass dem Patentgegenstand nur solche Merkmale in allgemeiner Weise zugeschrieben werden dürfen, die im Patentanspruch enthalten sind. Bei jedem Ausführungsbeispiel ist ferner zu zeigen, dass es dem Hauptanspruch des Hauptpatentes und dem Patentanspruch des vorliegenden Patentes entspricht.

1 Beschr.
1 Orig. Bl.

E.

Exp. II, Dez, 1907. A.B.

Figure 31

A patent evaluation written out by Albert Einstein, stamped 11 December 1907

In his spare time during those seven years at Bern, the young paten examiner wrought a series of scientific miracles: no weaker word is ade quate. He did nothing less than to lay out the main lines along whicl twentieth-century theoretical physics subsequently developed. What i more, Einstein did all this completely on his own, with no academic con nections whatsoever, and with essentially no contact with the elders of hi profession. Years later he remarked to Leopold Infeld that until he wa almost thirty he had never seen a real theoretical physicist. To which, o course, we should add the phrase (as Infeld almost did aloud, and a Einstein would never have done), 'except in the mirror!'

I suppose that some of us might be tempted to wonder what Einstei might have done during those seven years, if he had been able to worl 'under really favourable conditions', full time, at a major university nstead of being restricted to spare time activity while earning his living as minor civil servant. We should resist the temptation: our speculations vould be not only fruitless, but completely unfounded. For not only did Einstein not regret his lack of an academic post in these years, he actually onsidered it a real advantage. 'For an academic career puts a young man nto a kind of embarrassing position,' he wrote shortly before his death, 'by equiring him to produce scientific publications in impressive quantity—a eduction into superficiality which only strong characters are able to withtand.' Einstein was even a little reluctant about accepting a research rofessorship at Berlin, partly because Prussian rigidity and academic ourgeois life were not to his Bohemian taste. But he was also reluctant ecause he knew very well that such a research professor was expected to e a sort of prize hen, and he did not want to guarantee that he would lay ny more golden eggs. . . .

The way in which physics is taught is deeply influenced by our views on ow and why physics is done. Einstein, who was sceptical about the rofessionalization of research, was unswerving in his pursuit of fundanental understanding, he was a natural philosopher in the fullest sense of hat old term, and he had no great respect for those who treated science as game to be played for one's personal satisfaction, or for those who solved roblems to demonstrate and maintain their intellectual virtuosity. If hysics is viewed in Einstein's way, it follows that it should be taught as a rama of ideas and not as a battery of techniques. It follows too that there hould be an emphasis on the evolution of ideas, on the history of our ttempts to understand the physical world, so that our students acquire ome perspective and realize that, in Einstein's words, 'the present position f science can have no lasting significance'. Do we keep this liberal view of ur science, or is it lost in what we call necessary preparation for graduate vork and research?

One of Einstein's last public statements was made in answer to a request hat he comment on the situation of scientists in America. He wrote: Instead of trying to analyse the problem I should like to express my feeling n a short remark. If I were a young man again and had to decide how to nake a living, I would not try to become a scientist or scholar or teacher. would rather choose to be a plumber or a peddler, in the hope of finding hat modest degree of independence still available under present circumtances.'

We may wonder how literally he meant this to be taken, but we cannot elp feeling the force of the affront to our entire institutionalized life of the ntellect.

As we pride ourselves on the success of physics and physicists in today's vorld, let us not forget that it was just that success and the way in which it vas achieved that was repudiated by Einstein. And let us not forget to ask vhy: it may tell us something worth knowing about ourselves and our ociety.

11 Einstein and education

Arturo Loria

In my opinion anyone who wishes to understand Einstein's ideas on education must, before doing anything else, read pp. 315–18 of this book; they are taken from a talk he gave on 15 October 1936 and are quoted under the heading 'On Education' in his book *Out of My Later Years*. It would hardly be suitable to paraphrase or summarize what he says there, so I shall confine myself to quoting a few of the opening remarks. The rest of the quotations from Einstein in the present article come from writings not specifically devoted to education. To these quotations I shall add one or two comments suggested by, among other things, Einstein's autobiographical essay (1946).

> 'He often told me that one of the most important things in his life was music. Whenever he felt that he had come to the end of the road or into a difficult situation in his work he would take refuge in music and that would usually resolve all his difficulties.'
> (H. A. Einstein, in G. J. Whitrow, *Einstein: The Man and His Achievement*)

1

The address 'On Education' was delivered in Albany, New York, at a celebration to mark the tercentenary of higher education in the United States. It might have seemed obligatory for the speaker on that occasion to remind his audience of the most important names and dates in the history of American education. Einstein, however, cleverly shirked this obligation and went on to deal with the topics that most interested him, topics of a very general nature and not confined to particular times and places.

But there remains a doubt, a perplexity*:

> From what source shall I, as a partial layman in the realm of pedagogy, derive courage to expound opinions with no foundations except personal experience and personal conviction? If it were really a scientific matter, one would probably be tempted to silence by such considerations.

* All quotations in this article are either from *Ideas and Opinions* or from *Out of My Later Years*.

He goes on:

> However, with the affairs of active human beings it is different.

Thus, Einstein arrives at an affirmative reply which is perhaps onl seemingly tinged with modesty. The suggestive comparison that Einstei made (in 'On Education') between truth and a marble statue threatene with being buried under the sands of the desert implies, in fact, that eve before one sets to work to preserve the splendour of the statue, experienc and personal conviction have already led to a discovery of the truth abou education.

> To the sphere of religion belongs the faith that the regulations valid for the world of existence are rational, that it is comprehensible to reason. I cannot conceive of a genuine scientist without that profound faith. The situation may be expressed by an image: science without religion is lame, religion without science is blind.
>
> (A.E.)

After these introductory remarks, 'On Education' goes straight to th heart of the matter, as the reader will have seen.

We note that Einstein claims the right for a person to pronounce o Education, even if he has not devoted himself to the scientific study of it And this is not a remark thought up for the particular occasion; on th contrary, it shows his deep conviction, notoriously unshared by many, tha when the human condition is threatened each of us should make his ow personal contribution to the discussions and decisions on which its fate ma eventually depend. On this ground—and he often says so explicitly— Einstein takes a stand on political and social questions. An eloquent exampl is the article 'Why Socialism?' (1949), which begins like this:

> Is it advisable for one who is not an expert on economic and social issue to express views on the subject of socialism? I believe for a number o reasons that it is.

Later in this article he states:

> For these reasons, we should be on our guard not to overestimate scienc and scientific methods when it is a question of human problems; and w should not assume that experts are the only ones who have a right t express themselves on questions affecting the organization of society.

The same ideas are also put forward elsewhere, in different words, of course and sometimes with some variations, together with other ideas which the may presuppose or from which they may follow.

On freedom in research and teaching, addressing himself to one o Mussolini's ministers, Einstein said:

> . . . we both admire the outstanding accomplishments of the Europea intellect and see in them our highest values. Those achievements are base on the freedom of thought and of teaching, on the principle that th desire for truth must take precedence over all other desires,

and further:

> . . . the pursuit of scientific truth, detached from the practical interests o everyday life, ought to be treated as sacred by every government, and i

is in the highest interests of all that honest servants of truth should be left in peace.

In a discussion about freedom (1940), Einstein states—and we shall have occasion to come back to this point—that fundamental values and aims do not lend themselves to rational discussion; once we agree about certain of them, *then* we can embark on rational argument as to how to achieve them. Here, for instance, are two objectives on which there is general agreement:

1. Those instrumental goods which should serve to maintain the life and health of all human beings should be produced by the least possible labour of all.
2. The satisfaction of physical needs is indeed the indispensable precondition of a satisfactory existence, but in itself it is not enough. In order to be content, men must also have the possibility of developing their intellectual and artistic powers to whatever extent accords with their personal characteristics and abilities.

Einstein hated most things that other men hold dear. 'Comfort and happiness,' he declared in later life, 'have never appeared to me as a goal. I call these ethical bases the ideals of the swineherd. . . . The commonplace goals of human endeavour—possessions, outward success and luxury have always seemed to me despicable, since early youth.'
(G. J. Whitrow, *Einstein: The Man and His Achievement*)

The first of these requires that we should research into the phenomena of nature and society, which presupposes the greatest degree of liberty of expression and communication. It is not enough to have laws guaranteeing this liberty, which we shall call 'external'. It is also vital that through education a spirit of tolerance be cultivated in all of us. Moreover, this liberty cannot be gained in its entirety once and for all; it must be preserved and enlarged by means of a continual struggle. External liberty and the first of the two objectives stated above are, therefore, in a mutual cause-and-effect relationship.

He stresses this point (among others) in his essay 'Education for Independent Thought' (1952), insisting that:

> It is not enough to teach man a specialty. Through it he may become a kind of useful machine, but not a harmoniously developed personality. It is essential that the student acquire an understanding of and a lively feeling for values. He must acquire a vivid sense of the beautiful and of the morally good. Otherwise he—with his specialized knowledge—more closely resembles a well-trained dog than a harmoniously developed person. He must learn to understand the motives of human beings, their illusions, and their sufferings in order to acquire a proper relationship to individual fellow-men and to the community.
>
> These precious things are conveyed to the younger generation through personal contact with those who teach, not—or at least not in the main—through textbooks. It is this that primarily constitutes and preserves culture. This is what I have in mind when I recommend the 'humanities' as important, not just dry specialized knowledge in the fields of history and philosophy.
>
> Overemphasis on the competitive system and premature specialization

on the ground of immediate usefulness kill the spirit on which all cultural life depends, specialized knowledge included.

It is also vital to a valuable education that independent critical thinking be developed in the young human being, a development that is greatly jeopardized by overburdening him with too much and with too varied subjects (point system). Over-burdening necessarily leads to superficiality. Teaching should be such that what is offered is perceived as a valuable gift and not as a hard duty.

Einstein (1948) also deplores the competitive aspects of most educational systems:

> This competitive spirit prevails even in school and, destroying all feeling of human fraternity and cooperation, conceives of achievement not as derived from the love for productive and thoughtful work, but as springing from personal ambition and fear of rejection,

and he makes a strong plea (1952) in favour of reading the classic authors, in whatever field, literary or scientific:

> Somebody who reads only newspapers and at best books of contemporary authors looks to me like an extremely near-sighted person who scorns eyeglasses. He is completely dependent on the prejudices and fashions of his times, since he never gets to see or hear anything else. And what a person thinks on his own without being stimulated by the thoughts and experiences of other people is even in the best case rather paltry and monotonous.
>
> There are only a few enlightened people with a lucid mind and style and with good taste within a century. What has been preserved of their work belongs among the most precious possessions of mankind. We owe it to a few writers of antiquity that the people in the Middle Ages could slowly extricate themselves from the superstitions and ignorance that had darkened life for more than half a millennium.
>
> Nothing is more needed to overcome the modernist's snobbishness.

Einstein was concerned with the corrupting influence that the need to be successful has on the scientist. He frequently discussed this both in print and in conversation. He suggested that it would be a very nice profession for a scientist to be a lighthouse-keeper, for it would not be very demanding intellectually and would leave plenty of time to think about other matters.
(P. Bergmann, in G. J. Whitrow, *Einstein: The Man and His Achievement*)

2

I have been assuming that the reader is already familiar with 'On Education' and I have sought echoes of its ideas in other writings. But some key topics are either completely absent, or, in my opinion, are not given the importance which they certainly had in Einstein's thought. As an old man he made the statement: 'One thing I have learned in a long life: that all our science, measured against reality, is primitive and childlike—and yet it is the most precious thing we have.'

Further (1948):

> By painful experience we have learned that rational thinking does not suffice to solve the problems of our social life.

Many readers, coming to Einstein's writings for the first time, are surprised by the prominent part that ethics plays in them, and thus religion as a way of approaching ethics. Especially important in this respect are his articles 'Science and Religion' (1939) and 'Religion and Science: Irreconcilable?' (1948). It is worthwhile in the present context to summarize them briefly.

In the nineteenth century, and even towards the end of the eighteenth, many people felt that there was an irreconcilable conflict between science and faith, and that one ought to come down on the side of science.

Einstein says (1939):

> According to this conception, the sole function of education was to open the way to thinking and knowing, and the school, as the outstanding organ for the people's education, must serve that end exclusively,

and, continuing:

> It is true that convictions can best be supported with experience and clear thinking. On this point one must agree unreservedly with the extreme rationalist. The weak point of his conception is, however, this, that those convictions which are necessary and determinant for our conduct and judgements cannot be found solely along this solid scientific way.

These 'convictions' exist in a healthy society, they are deeply rooted in it; they are established, however, not by demonstrations but by the revelations vouchsafed to extraordinary people.

The highest of these convictions are to be found in the Judaeo-Christian religious tradition. Freed of their external ritual accretions and considered simply as human values, they enjoin

> . . . free and responsible development of the individual, so that he may place his powers freely and gladly in the service of all mankind,

and they declare that:

> . . . the high destiny of the individual is to serve rather than to rule, or to impose himself in any other way.

Furthermore:

> If one looks at the substance rather than at the form, then one can take these words as expressing also the fundamental democratic position.

There comes, then, the conclusion (1939):

> What, then, in all this, is the function of education and of the school? They should help the young person to grow up in such a spirit that these fundamental principles should be to him as the air which he breathes. Teaching alone cannot do that.

This sheds light on the following exhortation to educators, in 'On Education', and explains its origin: 'The aim must be the training of independently acting and thinking individuals, who, however, see in the service of the community their highest life problem'.

It is not my purpose here to give an exhaustive illustration of Einstein's special brand of religiousness, which denies the existence of a personal God and, as he himself affirmed, takes up a position very like Spinoza's. What I find very important, however, from the educational point of view, is the fact that comparing science and religion he offers certain rather limiting definitions of science, describing it, for example, as 'methodical thinking directed toward finding regulative connections between our sensual experiences' (1948). Thus he denies that science, or rather, rational thought can point out final objectives for our human aspirations or can make ethical judgements. Indeed, he says (1939):

> To make clear these fundamental ends and valuations, and to set them fast in the emotional life of the individual, seems to me precisely the most important function which religion has to perform in the social life of man,

and even more explicitly:

> Fulfilment on the moral and aesthetic side is a goal which lies closer to the preoccupations of art than it does to those of science. Of course *understanding* of our fellow-beings is important. But this understanding becomes fruitful only when it is sustained by sympathetic feeling in joy and in sorrow. The cultivation of this most important spring of moral action is that which is left of religion when it has been purified of the elements of superstition. In this sense, religion forms an important part of education, where it receives far too little consideration, and that little not sufficiently systematic.

It was Einstein's belief that, far from there being a conflict between science and religion, the development of science was a source of enrichment to religion (1941):

> After religious teachers accomplish the refining process indicated they will surely recognize with joy that true religion has been ennobled and made more profound by scientific knowledge.

However, he also expressed the opinion (1953) that:

> . . . in this materialistic age of ours the serious scientific workers are the only profoundly religious people.

To complete the picture of Einstein's thoughts on Education, we must recall his rejection of any sort of military education; this rejection is rooted in his loathing of militarism in general and his antipathy towards any kind of oppression of man by his fellow men.

'In the matter of physics, the first lessons should contain nothing but what is experimental and interesting to see. A pretty experiment is in itself often more valuable than twenty formulae extracted from our minds; it is particularly important that a young mind that has yet to find its way about in the world of phenomena should be spared from formulae altogether. In his physics they play exactly the same weird and fearful part as the figures of dates in Universal History.'

(A.E.)

In the quotation that follows, Einstein is, no doubt, swayed by his personal feelings; but he concludes on an educational note (1931):

> This topic brings me to that worst outcrop of herd life, the military system, which I abhor. That a man can take pleasure in marching in fours to the strains of a band is enough to make me despise him. He has only been given his big brain by mistake; unprotected spinal marrow was all he needed. This plague-spot of civilization ought to be abolished with all possible speed. Heroism on command, senseless violence, and all the loathsome nonsense that goes by the name of patriotism—how passionately I hate them! How vile and despicable seems war to me! I would rather be hacked in pieces than take part in such an abominable business. My opinion of the human race is high enough that I believe this bogey would have disappeared long ago, had the sound sense of the peoples not been systematically corrupted by commercial and political interests acting through the schools and the Press.

And later his concern for the educational side of this question is expressed in unequivocal warnings (1934):

> In the schools, history should be used as a means of interpreting progress in *civilization*, and not for inculcating ideals of imperialistic power and military success. In my opinion, H. G. Wells' *World History* is to be recommended to students for this point of view. Finally, it is at least of indirect importance that in geography, as well as in history, a sympathetic understanding of the characteristics of various peoples be stimulated, and this understanding should include those peoples commonly designated as 'primitive' or 'backward'.

It is to be noted that 'primitive' and 'backward' are placed in quotation marks. In similar contexts Einstein never omits to use them, which fact surely offers a clue to how he viewed the comparison between different cultures and educational systems.

He had further, and drastic, things to say on the topic of military education. Take the following, for instance (1934):

> . . . unless military and aggressively patriotic education is abolished, we can hope for no progress. . . . In addition, the state considers it necessary to educate its citizens for the possibilities of war, an 'education' not only corrupting to the soul and spirit of the young, but also adversely affecting the mentality of adults.

He takes a stand on his principles. However, we should remember that this did not prevent Einstein from encouraging the Belgians to defend their country when attacked by the Nazis, nor from speaking in favour of American intervention in the Second World War. He found nothing inconsistent in all this, nor did many others. It is not my intention to argue the matter here; but I cannot help noting that he seems to have ignored the

One point about Einstein which impressed me perhaps more than any other was this: Einstein was highly critical of his own theories, not only in the sense that he was trying to discover and point out their limitations, but also in the sense that he tried, with respect to every theory he proposed, to find under what conditions he would regard it as refuted by experiment.
(K. Popper in G. J. Whitrow, *Einstein: The Man and His Achievement*)

point that no nation can muster effective defence against an aggressor without having previously undertaken some kind of military training.

Einstein was convinced that there was a vicious circle: military/patriotic education—compulsory military service—war. He saw it as his duty to help break this vicious circle by committing himself as far as possible to the struggle against war, the campaign for conscientious objectors and world government.

Indeed, we must recognize that one of the most important sources of Einstein's educational thought is his concern for improving the destiny of mankind. Speaking to an audience of young people in 1930, he said:

> ... I began by telling you that the fate of the human race was more than ever dependent on its moral strength today. The way to a joyful and happy existence is everywhere through renunciation and self-limitation. Where can the strength for such a process come from? Only from those who have had the chance in their early years to fortify their minds and broaden their outlook through study. Thus we of the older generation look to you and hope that you will strive with all your might and achieve what was denied to us.

And it was above all his concern for educational questions that led him to express himself on society and, more particularly, in favour of socialism. After remarking on a 'crippling of the social consciousness of individuals', he declares:

> This crippling of individuals I consider the worst evil of capitalism. Our whole educational system suffers from this evil. An exaggerated competitive attitude is inculcated into the student, who is trained to worship acquisitive success as a preparation for his future career.
>
> I am convinced there is only *one* way to eliminate these grave evils, namely through the establishment of a socialist economy, accompanied by an educational system which would be oriented toward social goals. In such an economy, the means of production are owned by society itself and are utilized in a planned fashion. A planned economy, which adjusts production to the needs of the community, would distribute the work to be done among all those able to work and would guarantee a livelihood to every man, woman, and child. The education of the individual, in addition to promoting his own innate abilities, would attempt to develop in him a sense of responsibility for his fellow-men in place of the glorification of power and success in our present society.
>
> Nevertheless, it is necessary to remember that a planned economy is not yet socialism.

3

Let us conclude by recalling in what way Einstein was himself involved in the educational process.

The military-type methods employed in the Munich schools he found very unpleasant, both in the elementary schools and in the Luitpold Gymnasium where he felt particularly oppressed by the mechanical and verbalistic way of learning. Incompatibility between the student and his environment gave rise to a very difficult situation in, among other things, the personal relationship between him and his teachers. So at the age of fifteen, cut off and alone, Albert decided to leave the gymnasium and follow his parents to Milan.

At sixteen he sat the entrance examination for the Zürich Polytechnic, but without success. Eventually he was happily surprised to find an atmosphere quite different from that of the Luitpold Gymnasium in the Swiss canton school of Aarau, where he spent a year.

Figure 32

Einstein at 17

The ideas on education I have outlined above sprang from a mature mind bent on a noble aim. These ideas came into being on the hard benches of the Munich schools and Albert became fully aware of them at Aarau. He always recalled the Swiss school with pleasure and gratitude, and a month before his death said: 'It made an unforgettable impression on me, thanks to its liberal spirit and the simple earnestness of its teachers who based themselves on no external authority'. His own description of this school and that of his biographers suggested that it was largely inspired by Pestalozzi's principles; indeed, it seems to have had close affinity with the ideal school implied in 'On Education'. In that congenial atmosphere Einstein acquired a faith in his own intellectual abilities and for the first time in his life found himself in a setting where these abilities were encouraged to develop instead of being suppressed.

Although it is notoriously difficult to acquire Swiss citizenship, it was granted to him. Later when he went to Berlin he was again made a German citizen and many years later, after he had settled at Princeton, New Jersey, American citizenship was conferred upon him by an act of Congress, but these successive nationalities were bestowed upon him almost like honorary degrees. Nevertheless, he retained his Swiss citizenship until the end of his life. In virtue of this, he had a traditionally international neutral status, and he was certainly vividly aware of its significance. In this connection it may be mentioned that the only diploma he had on the walls of his office in Princeton was that of an honorary member of the Bern Society of Sciences.
(H. Mercier, in G. J. Whitrow, *Einstein: The Man and His Achievement*)

At that age a young person views society essentially through the medium of school. So Einstein's decision to apply for Swiss nationality derived, above all, from his comparison of the two educational systems in which he had taken part. His application was soon granted and he retained his Swiss nationality for the rest of his life. His tendencies and his scale of values are clearly reflected in the decision, on entering the Polytechnic at seventeen years of age, to become a teacher rather than an engineer as his family background prompted.

He left the Polytechnic in 1900 at twenty-one with a diploma, but the years spent there were not happy ones. He found the routine oppressive and of the exams he said: 'One had to cram all this stuff into one's mind for the examinations, whether one liked it or not. This coercion had such a deterring effect on me that, after I had passed the final examination, I found the consideration of any scientific problems distasteful to me for an entire year'.

We should bear in mind that however constricting and harmful to the spiritual and intellectual development of pupils the German secondary-school environment undoubtedly was, nonetheless it would be impossible to imagine a university—much less a polytechnic—suited to a student as exceptional as Einstein; he was already irresistibly drawn to the study of pure physics and was to set out on a path of independent research that would enable him in the course of a very few years to achieve the results we all know today.

So Einstein's reaction to the Polytechnic was again one of rejection, with the consequence that none of the professors wanted him as assistant. The best he could do was some temporary supply-teaching at a technical school at Winterthur in 1901. Following that he was taken on as tutor to two boys by a teacher who kept a students' lodging-house at Schaffhausen. Einstein enjoyed his new job and gave himself up to it with enthusiasm. Perhaps a little too much enthusiasm, indeed, for when he found that the methods of the other teachers did not harmonize with his own, he asked that the teaching of the two pupils be entrusted entirely to him. The gymnasium teacher, offended by this and alarmed by Einstein's attitude, dismissed him.

Following the extraordinarily productive period at the Bern patent office, which he entered in 1902, and during which he also taught as *privatdozent* at the local university, Einstein at last officially crossed the threshold of the academic world in 1909 as Associate Professor of Theoretical Physics at the University of Zürich. He was to remain in that world for the rest of his life in various, often very important, positions: in Prague, at the Zürich Polytechnic, at Berlin, and after 1933 at Princeton.

Different opinions have been expressed on Einstein as a teacher. The most vivid and eloquent account comes from his colleague, friend and biographer, Philipp Frank, and what follows here is mainly based on it.

Einstein was never what the majority of students would call a good teacher. For instance, when Kleiner, head of the faculty of Physics at Zürich University, went to Bern to attend a lecture given by the *privatdozent* (preparatory to Einstein's employment at Zürich), he thought Einstein's teaching unsuited to the students. There are excellent reasons for thinking that he was right. Among other things, the subject matter of the lessons was too original and difficult to be explained at the pupils' level; in fact, Einstein's audience was usually limited to a few friends.

At Zürich and after things went along better. Einstein found that collaboration with pupils and colleagues, usually on an individual basis, was not only feasible, but a great pleasure and profit into the bargain, even if his own behaviour did not always arouse favourable responses. For instance, he made no distinction in his way of talking to the rector of the university and to a member of the cleaning staff, and he was very fond of jokes and satire.

His wish to be of use and his artistic sensitivity stood him in good stead in his teaching. Along with these, however, he was afflicted with a sometimes acute aloofness, of which he was himself aware. This made it difficult, even impossible, to form close relationships, affective or cultural, with his fellows.

Utterly devoid of all ambition and vanity, he expounded the matter of his lessons in the simplest way in order to make it comprehensible to everyone concerned. He would illustrate his argument with imaginative comparisons and make it entertaining with a leavening of humour. But he always found it hard to work through the kind of systematic course of lectures which involved merely supplying the students with items of knowledge; he preferred, instead, to talk about what interested him at that particular moment. So his lectures were uneven, not part of an organic whole. But they always contained much of value and left an inextinguishable memory in their audiences.

In his relations with his students, Einstein was lavish with help and advice over problems encountered in the course of study or research; and he was greatly in favour of facing up to difficulties, even if one didn't succeed in solving them. He would have nothing to do, however, with the

In the fall of 1912 I first realized that Einstein's theory of the 'relativity of time' was about to become a world sensation. At that time, in Zürich, I saw in a Viennese daily newspaper the headline: 'The Minute in Danger, a Sensation of Mathematical Science.' In the article a professor of physics explained to an amazed public that by means of an unprecedented mathematical trick a physicist named Einstein had succeeded in proving that under certain conditions time itself could contract or expand, that it could sometimes pass more rapidly and at other times more slowly. This idea changed our entire conception of the relation of man to the universe. Men came and went, generations passed, but the flow of time remained unchanged. Since Einstein this is all ended.

(Philipp Frank, *Einstein: His Life and Times*)

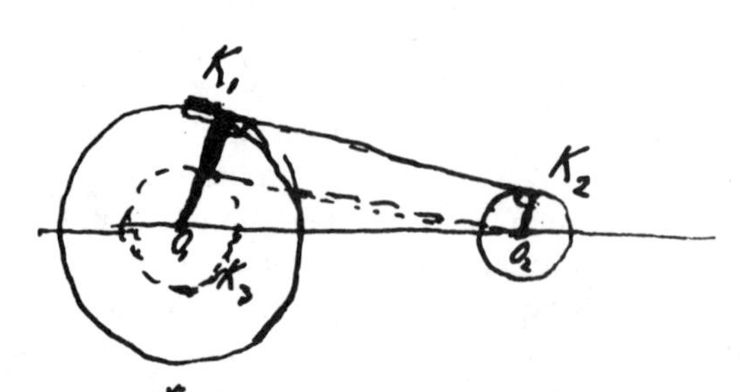

K1
K2
K3
The radius of K3 is the difference r3 = r1 − r2.
The tangent O2 → K3 is || to the tangent on K1 and K2
can be easily constructed. This gives the solution.
A. E.

Figure 33

Einstein's solution to the problem of finding a common tangent to two circles of different radii. The help was requested by a fifteen-year-old schoolgirl

production of 'waste paper' in the form of academic publications, which was already in his time the great blemish of the academic world. Thus it was not as a professor but as a friend that he willingly devoted his abundant free time to helping his students. He seemed to have no difficulty in taking up the thread of his own work after dealing with something quite unconnected with it. Moreover, he showed exceptional interest in the objections and criticisms propounded by so-called 'uninitiates', and when they were mistaken, was incredibly patient in pointing out their mistakes.

Remarkable, too, are some of his statements made *in vivo* on Education and School. For example, in a talk to students undergoing treatment at the Davos sanatorium, he remarks on the physically tonic effect that a certain amount of intellectual activity may have; or when he declares that a book which fascinates the reader by its lively style of argument leads to a knowledge which is not merely stored away in the mind, but actually lived in experience; or when he, a born researcher, claims full dignity for the person who has made teaching his mission (1932):

> It is just as important to make knowledge live and to keep it alive as to solve specific problems.

And in his typically humorous fashion Einstein expresses grave concern over an all too frequent vexation of the teacher's life, one which may even jeopardize the effective performance of his job:

> Now to the salaries of teachers. In a healthy society, every useful activity is compensated in a way to permit of a decent living. The exercise of any socially valuable activity gives inner satisfaction; but it cannot be

considered as part of the salary. The teacher cannot use his inner satisfaction to fill the stomachs of his children.

4

To round out our understanding of Einstein's views in the field of Education, we must also see him as he saw himself:

> I do not believe in human freedom in the philosophical sense. Everybody acts not only under external compulsion, but also in accordance with inner necessity. Schopenhauer's saying, 'A man can do what he wants, but not want what he wants,' has been a very real inspiration to me since my youth.

In Einstein's case this inner compulsion was explicitly linked by him to his Jewishness:

> The pursuit of knowledge for its own sake, an almost fanatical love of justice and the desire for personal independence—these are the features of the Jewish tradition which make me thank my stars that I belong to it.

Just as he was concerned for the emancipation of the American Negro through education and social integration, so he became a convinced Zionist, seeing this movement as preeminently a cultural and educational event of the greatest importance for the Jewish minorities suffering various forms of oppression in so many countries. Einstein's main efforts in the sphere of Zionism were directed towards the creation of a Jewish University.

From the evidence presented in this article, I believe we can conclude that Einstein's interest in education was not merely marginal and episodic, but was something deeply rooted and constant, even though it occupies only a fairly modest part of his speeches and writings.

(Acknowledgement—I should like to acknowledge my indebtedness to my colleagues Carmen Malagodi and Marisa Michelini for constant help and useful discussions. A.L.)

12 Philosophical concepts of space and time

Herbert Hörz

The justification for a physical concept lies exclusively in its clear and unambiguous relation to facts that can be experienced.
(A.E., 'On the Theory of Relativity')

The concepts of space and time are inseparably linked to the theories concerning the structure, motion, and change of physical objects. In the pre-scientific understanding of the world, space was conceived intuitively as the order of objects relative to each other, and time as the sequential order of changes. Later, through mathematical and physical insight into the existing spatial and temporal structures, there developed a better understanding of the essence of space and time, their inner unity, and the relationship between space, time, and motion.

By linking the teaching of philosophical concepts to the history of space and time concepts, one can arrive at a deeper understanding of space–time structures, beginning with the systematization of intuition and proceeding to scientific knowledge. In so doing we must distinguish the pre-scientific philosophical understanding of motion in space and time, which lacked any physical basis, from the space–time concepts of classical physics and their philosophical interpretation, and especially from our revolutionized ideas about space and time due to Einstein's discovery of the special and general theories of relativity. The development of our understanding of the nature of space and time is not yet complete, and certainly other sciences than physics and mathematics will make their contribution. But this remains outside the scope of the present article.

1. WHAT ARE SPACE AND TIME?

Material objects have extension. If we neglect any specific structure of a system, its interactions, and any specific motion, what remains is the abstract extension as the form of an object's existence.

The pure concept of space is nothing other than the sum total of our knowledge about existing extensions. These exist, in an overall sense, as extended regions of matter undergoing changes in which objects interact with each other. One such spatial region we can label as a *system*, whose extent is limited by the range of validity of the relevant theories that are understood to apply to that region. Examples of such spatial regions are atoms, molecules, macroscopic objects, galaxies, and metagalaxies. The interrelation of different systems leads to a hierarchy of spatial relationships. If one postulates the existence of an all-encompassing region in which changes occur, one then comes to the assumption of an absolute space.

Extension exists in an elementary sense as the extension of relatively elementary objects. For objects that do not interact, one is justified in asking about the inner structure. This does not mean, however, that the understanding of space as extension necessarily requires the existence of parts of a whole. It can also be a question here of the interrelationships of elementary objects in which the objects themselves do not have an elementary extension. Extension exists in a structural relationship of objects interacting with each other. If extension in its elementary sense may be postulated as absolute, then one can conceive of the existence of non-material, point-like elements without structure. Spatial relationships exist between relatively global systems and relatively elementary objects through the spatial form of individual objects, through their relative positions, or through the trajectory of a given object. The structures of space, i.e. the forms and properties of existing extensions, are becoming progressively easier to recognize. There exists no extension without matter (absolute space) and there exists no real object without extension (non-material, point-like elements).

Space as extension is understood in science in different ways. First, the concept of space embodies the manifest structural extension of objects. This insight is linked to the change in our conception of the vacuum. With the acceptance of atomism, which assumed the existence of qualitatively identical smallest particles—particles that are impenetrable and closely packed in space and out of which the world is supposedly built—it was assumed that one must necessarily also have empty space in between to account for the movement of these particles. If one observes changes in the position of impenetrable bodies, then these bodies cannot completely occupy the whole of space (the region in which the motion takes place). That was also the idea of Democritus, who used the terms the *plenum* and the Void for the explanation of motion.

As science developed it came to be realized that the hypothesis of an independent empty space was untenable, that everything could be described in terms of qualitatively identical small particles. Modern physics not only shows the dependence of elementary particles on each other, but also their ability to change into each other. With this, the hypothesis of empty space to explain motion becomes unnecessary. Motion is not merely the change in position of bodies; it is change in the most general sense. To it belong change in position, external and internal interactions, the ability to change matter from one form into another, and development. Because material processes can interpenetrate each other, no matter-free space is needed for the existence of motion in its many forms. The investigation of fields surrounding matter shows that the outer form of an object does not represent the limit of its influence. Elementary objects have the *possibility* of exerting an effect in a larger region than they do in fact. If one perceives space as the region in which motion occurs, then one must conclude that this region is filled with matter. Furthermore, the characteristics of a region must follow from the characteristics of the material processes in that region. Therefore, in opposition to Kant's ideas, one can no longer imagine space without matter.

Second, the concept of space includes certain spatial regions that are determined by the existence of relatively closed systems operating under the same laws within the system. Such systems are the domains of quantum-mechanical behaviour, atomic motion, molecular transformation, the Earth, the solar system, galaxies.

'From this hour on, space as such and time as such shall recede to the shadows and only a kind of union of the two retain significance.'
(H. Minkowski)

Third, the concept of space includes the relative positions and trajectories of objects. Later, the problem of understanding motion as a sum of points at rest will be discussed.

In order to define the concept of time, we distinguish between definite real changes, with which we measure time, and pure time. We obtain the latter when we turn our eyes from definite real changes to just the existence of a duration, the length of an event, without taking into account definite content. Thus, the pure concept of time is expressed as pure duration, which is measured by concrete changes. This results in certain consequences for the concept of time. On the one hand, the character of objectively real changes determines the duration of change, and thus the time. Every system, therefore, has its characteristic time (*Eigenzeit*), namely the duration of structural changes as determined by the laws within the system. In this context the question arises whether there is an all-encompassing system by which absolute time could be determined. That is evidently not the case. On the other hand, the relationship between the various characteristic times is always obtainable. Out of them one can then abstract the concept of time. In this respect, for sub-systems of an encompassing system, along with the characteristic times (*Eigenzeiten*) of the sub-systems there exists the time of the whole system. The time relationships are to be determined as inter-relationships between global times of the encompassing system and relatively elementary times, which are constituted out of the irreversibility of elementary changes. While space is pure extension, time is pure duration. Also, it can be made absolute on both a global and an elementary basis, thus leading to absolute time or timeless existence. Both contradict our experience.

2. THE DEVELOPMENT OF OUR UNDERSTANDING OF SPACE AND TIME

With the special and general theories of relativity, the understanding of space and time was revolutionized. These theories showed that systematic intuition was no longer adequate. It even appeared as though the theory of relativity contradicted common sense. However, this resulted from the prevailing concept (until that time) of the absolute independence of space, time, and motion. The careful philosophical analysis of the theory of relativity led to significantly new insights that gave us a deeper understanding of space and time. First, the inner relationship between space and time was discovered, without sacrificing the specific nature of these two aspects of the existence of matter. Second, the inner unity of space–time and matter in motion was shown, which contributed to the re-analysis of the previous understanding of motion. Third, the space–time structure turned out to be the essential determinant of causal structure, because space–time structure is the framework for causal behaviour.

His fame had already taken on a legendary aura, so that not many years ago a schoolgirl from a remote corner of British Columbia wrote him a letter which began with the words: 'I am writing to you to find out whether you really exist.'
(Carl Seelig, *Albert Einstein: A Documentary Biography*)

2.1 *Physical and philosophical space–time theories*

The development of the understanding of space–time took place in different stages. We find an early development of a scientific understanding of space, arising from the needs of surveying, and then later for navigation, etc. Already in 300 BC, in Euclid's *Elements*, a comprehensive theory of spatial relationships was made available to us. Tied to this understanding of space was the assumption of an absolute time that is the same for all spatial systems. In Euclidean geometry, one has a geometry of three-dimensional space, which can be described by straight-line coordinate axes. The essence of this understanding of space is the following:

> Space is three-dimensional. The coordinates of space are rectilinear. The sum of the angles in a triangle is 180°.

Later, classical physics became tied to this conception of space. At this stage physics and geometry are separate. Physics studies the motion of real bodies and geometry the structure of space. The connection between them resides only in the fact that physics uses the theory of the structure of space in order to represent motion. What is philosophically important is, therefore, that space is absolute, i.e. independent of the motion of matter.

An attack on this conception of space occurred in the last century. In philosophy the absoluteness of space was rejected. In contrast to Kant, who saw space and time as forms of perception, Feuerbach and Engels emphasized that space and time are real basic forms of being. However, if space and time are real aspects of the existence of matter, then there must also be determined a connection between space and time as well as the dependence of both on material motion. Also, the invalidity of the axiom of parallel lines and the possibility of non-Euclidean geometry were asserted. The new geometry took over from the old the three-dimensionality and absoluteness of space. Quite new, however, was the discovery that the sum of the angles in a triangle need not be 180° and the coordinate axes may not be rectilinear. At this point, however, the thoughts about non-Euclidean geometry only had the character of a logical extension of the traditional concept of space.

At first there was no application of it in physics. The classical conception of time demanded the one-dimensionality of time (i.e. it did not require a resolution into different components—it is not a vector, but a scalar). Furthermore, the irreversibility of time was emphasized. Time was thus the same in all systems, i.e. independent of the moving matter in them.

At the beginning of our century the special theory of relativity developed by Einstein took into account the inner connection between space and time. The assumption of an absolute time scale is wrong. The basic assumption for this new concept is the constancy of the speed of light. With the Lorentz transformation the dependence of space on time and time on space was mathematically formulated. Minkowski used a four-dimensional schema (a light cone) for the representation of an event. An event required spatial as well as temporal information for its specification. The new space–time concept, although still operating within the classical framework of rectilinear coordinates, gave the philosophically interesting result that the one-dimensionality of time and the three-dimensionality of space remained as before, but the connection between space and time was confirmed. The philosophically weak point was the still remaining independence of space–time from the motion of matter. Further progress had to lie in the proof of a connection between the two. This occurs with the general theory of relativity. Its philosophically important viewpoint is the relationship of space–time to moving matter. In the general theory of relativity the mass distribution determines the equations of motion of bodies. The distribution of the moving matter in turn determines the geometry. However, this geometry *changes* constantly according to the changing mass distribution. The material bodies move according to the geometry that is determined by them. In so doing they change the mass distribution, and hence also the geometry. Thus, we have to deal with the mutual interdependence of space, time and motion, as suspected by philosophy and verified by physics.

By introducing sophisticated mathematical concepts . . . into physics, Einstein not only abandoned the popular principle attributed to Rutherford that 'an alleged scientific discovery has no merit unless it can be explained to a barmaid', but he even outraged many professional scientists.
(G. J. Whitrow, *Einstein: The Man and His Achievement*)

2.2 *Space–time and motion*

The equations of motion of classical mechanics, for example Hamilton's equations, provide the means for determining the time variations of the momentum p and the coordinates q as follows:*

$$\dot{q}=\frac{\partial H}{\partial p}, \qquad \dot{p}=-\frac{\partial H}{\partial q}$$

* Here, H is the Hamiltonian function, equal in the simplest cases to the total energy. E.g., for one-dimensional motion of a particle in a potential $V(x)$,

$$H=\tfrac{1}{2}m\dot{x}^2+V(x)=\frac{p^2}{2m}+V(x)$$

$$\frac{\partial H}{\partial p}=\frac{p}{m}=\dot{x}; \quad -\frac{\partial H}{\partial x}=-\frac{\mathrm{d}V}{\mathrm{d}x}=F=\dot{p}$$

Once one has specified the exact location and momentum (or velocity) of a particle, its dynamical state is precisely determined, and its future path can be exactly predicted.

If we investigate this representation of motion somewhat more closely, we find that the particle must always be found at a definite location, provided the assumptions of classical mechanics are valid. In the same way, the velocity is obtained by measuring the position, because one can determine the time at any given location. For the values of two such positions and corresponding times the velocity of the particle is given by

$$v=\frac{x_2-x_1}{t_2-t_1}$$

So where does this leave Einstein and his exceptional merit? Ah, but however much all was ready and waiting, there was still one more step to be taken and Einstein was the only man able to take this step. And what was so singular about this step? It was primarily methodological. It consisted first in re-examining the generally accepted fundamental ideas, that is, the ideas then current, about time, simultaneity and the ether. This re-examination arose from the fact that Einstein could never tolerate being unable to explain with perfect clarity any question in which he was interested.
(Reverend François Russo, 'From Plurality to Unity', in *Science and Synthesis*)

Note, however, that in order to apply the equations of motion, the position, the velocity at a specified position, and the momentum of the particle must be simultaneously determined. The momentum, of course, is given by the velocity multiplied by the mass. The determination of the velocity at a given place is obtained by letting t_2 approach t_1. In classical mechanics x_2 then automatically approaches x_1. The velocity at a specific position is then given by the derivative of the displacement with respect to time: $v=ds/dt$. Should it happen that with the approach of t_2 to t_1 the transition from x_2 to x_1 does not follow automatically, then the motion of a particle would not be precisely determined from the specification of its location and momentum; rather one would find for v the paradoxical result that the velocity at a specific location would be infinite. This disagrees with the real velocity, which is finite. However, the assumption that the transition $t_2\rightarrow t_1$ also implies $x_2\rightarrow x_1$ is justified only under certain conditions, and leads to a simplification of the concept of motion. The conditions are as follows:

1. At every specified instant, the body must have an exactly defined position. Were this not the case, i.e. if the body had no definite location at a specified instant, then we would obtain an infinitely large velocity for $t_2\rightarrow t_1$.
2. The transition from x_1 to x_2 must be continuous, as otherwise the value of the derivative at the limit cannot be obtained from the ratio of differences. If the position were precisely determined, but the change of position were discontinuous, the result would be an indeterminate velocity.

With this conception of motion in classical mechanics we have understood motion only from one point of view. If, in accordance with condition 1 above, we understand motion to be the observation of a body at a certain position at one instant and at another position at a subsequent instant, we have identified only the result of the motion, not the motion itself. Motion is in these terms just a succession of states of rest.

Now one might think that this limitation could be removed by using condition 2. However, although continuity allows one to consider the notion at a single moment, one must actually break through this continuity, and carry out the construction of the limiting value, if one is to grasp what is meant by motion in terms of our basic concepts.

If one were not to break the continuity, one would obtain the paradoxical esult that the body *is* located at a given position and is *not* located at that position at the same time. Therefore, the two conditions enunciated above are interrelated aspects of motion, and this results in definite consequences for the classical concept of motion. For example, let us assume (in accord- nce with classical physics) that a moving body is at a definite location at every instant; then we really are not dealing with a moving body, but ather, at this instant, with a (relatively) motionless body. Thus, as remarked bove, the motion of a body would be understood as only a series of dis- continuous steps forward. Continuity of motion indeed requires not only hat the body is found just at a specific position, but also that it passes hrough this position.

The theoretical difficulties resulting from this dual requirement were of ourse exposed long ago by Zeno in his paradoxes of the flying arrow and of the contest between Achilles and the tortoise.

If one tries to restore the relationship between continuity and dis- ontinuity by regarding motion as a unity of both, then we obtain results which go beyond classical mechanics. The position at which the moving body is found is an abstraction, as is its instantaneous velocity. There is no theoretical problem if one assumes that action is not quantized, and that mass is constant. With the existence of a quantum of action, our observa- tions of space and time are no longer independent of the motion of matter. This becomes manifest in relativistic effects at high energies and serves as a asis for the construction of theories that assume a change of the space-time tructure in areas not yet investigated.

The philosophical theory of space–time stresses the objectivity of space and time, defines them as forms of existence of matter, shows the relativity of our notions of space and time, and investigates the directions of develop- nent of these notions in the light of new observations and physical ypotheses, such as the existence of the graviton.

Einstein spent his life searching for what is changeless in an incessantly changing world. He searched for unity in multiplicity. In his model of physical reality, space, time, energy, matter are bound together in a single continuum. The crown of his efforts—to find a set of field equations that would unite gravitational and electromagnetic phenomena—may have eluded him. But his achievement is beyond measure or praise.
(James R. Newman, *Science and Sensibility*)

.3 *Space–time and causality*

f, as the first stage in the development of ideas about causality in physics, we take classical mechanics, we find that the framework for possible causal elationships is provided by the law of the conservation of energy and by he necessary relationship between the initial and final states of a process, whereby the states are characterized by position and momentum. The next tage is represented by the development of thermo- and electrodynamics. Here the law of the conservation of energy is maintained, but the form of ausal relationships is stipulated in a different way because of the abandon- nent of action at a distance and its replacement by field laws. In this development the framework for possible causal relationships was maintained, but their form was better understood. There were no manifest contradictions, because in principle it was regarded as possible to reduce the statistical quantities derived from statistical thermodynamics to the motion of classical particles and the determination of their state of motion.

The subsequent development of the theory of relativity necessitated a refinement in the thesis of a universal relationship. Not everything interrelates at the same time and universally; rather, the thesis about objective relationship says: 'There is no material region that is not connected to other regions by material processes'. A limiting velocity, and the rejection of action at a distance, led to a structure of space and time in which the possibilities of causal behaviour were limited to time-like and light-like processes, and space-like processes were not allowed. The structure of space and time turned out to be the basis of causal structure, in the sense that it determined the space–time possibilities of causal behaviour. The development of the general theory of relativity led to a further difficulty concerning the use of the law of conservation of energy for determining possible causal relationships. In general relativity, no global law of the conservation of energy can be formulated. This does not mean that causality in a general sense no longer exists as a means of interrelation. But it must be examined in its specific form as a physical concept. The problem of locality is then interesting, insofar as it leads to connections between statements of philosophy and physics.

People complain that our generation has no philosophers. Quite unjustly: it is merely that today's philosophers sit in another department, their names are Planck and Einstein.
(A. Harnack, quoted in Stanley L. Jaki, *The Relevance of Physics*)

The causal principle in physics is characterized by the special theory of relativity. The Minkowski world (light cone) supplies the framework for possible relationships in physics, whereby the *possible* relationship is a necessary condition for causal relations. Within this space, including the possible influences of events elsewhere on point $P(x, y, z, t)$ and of an event at P on others, the causal relationship has to be determined more precisely. It is determined by the influence from point to point whereby there is no interaction, and hence no cause, that can propagate faster than the speed of light.

The conditions for the use of this specific principle of causality resulting from the special theory of relativity are the following:

1. For the causation of effects there is a critical velocity, namely the speed of light, that limits the universally conceivable connection between all events to an objectively restricted connection. However, there is no material region that cannot in principle be connected with other regions by material processes. The cone of causality includes the region of all possible causations of effects on P or through P.
2. Because of this, the region where causal effects are possible is distinctly separated from the region where such effects are physically impossible. However, if one takes into account the extension of interacting objects then this dividing line is no longer sharp, but becomes somewhat blurred

3. The causation of an effect comes about by the influence of one world point upon others. Causal relationships can be characterized by world-lines, i.e. connecting lines between world points.
4. The localization of events must be considered. An event is characterized only by its space–time relationship and not by its inner structure.
5. No direction of time is distinguished. (In the general theory of relativity, when gravitation is taken into account, the form of light cone changes, and certain asymmetries can be used to define a direction of time.)

The principle of action at close range (as opposed to action at a distance) is essential for the understanding of the connection of causality to locality. Actually, however, we use two different principles of action at close range. On the one hand we are concerned with the direct mediation of the objective relationship between two events as expressed through causality. This direct mediation will not be investigated in all of its aspects or on all existing levels. On the other hand, certain interrelationships do exist that are necessary in general; that is, they are reproducible and essential, and they determine the character of the phenomenon. These interrelationships, also, are based on causal connections, and they determine relationships between the initial and final states of a process, but without involving consideration of the specific and direct mediations in between.

During his Zürich stay the woman doctor, Paulette Brupbacher, asked the whereabouts of his laboratory. With a smile he took a fountain pen out of his breast pocket and said: 'Here.'

(Carl Seelig, *Albert Einstein: A Documentary Biography*)

The connection between these two principles in our understanding is that in the inquiry into causes we always try to investigate the direct relationship between events and thereby try to discover certain laws. Thus, to elucidate the laws concerning a complex system, we are forced to go to a simpler system from which we can obtain the same conclusion. The principle of action at a distance, in its original form, produces relationships between events that physically cannot be connected with each other. Sometimes, however, the second principle of close range interaction may be considered to be a principle of action at a distance, if the description 'close range' is interpreted as being limited to a *direct* mediation of the relationship.

Insofar as it can be regarded as a generally necessary and essential connection, *any* relationship between different events must have as its basis a complex of direct connections which with the passage of time become more and more precisely known. In this sense, the quest for objective laws is at the same time a plea for research to concern itself with basic direct interrelationships, in order to find laws at this basic level. Thus, the principle of close range in its more limited sense is concerned with causal relationships, whereas the principle of close range in its more general sense is concerned with laws existing on the basis of *complexes* of causal relations. A law is an abstraction from a direct interaction, and the generally necessary and essential connection between two events will be emphasized.

Thus, there exist differences between the philosophical conception of causality and the strict physical requirements of *local* causality. First, philosophy regards causality as objectively direct, concrete, and fundamental mediation of the interrelationships between processes, whereby one process produces another. This results in a directionality of content and time. Local causality is a defining and narrowing down of these requirements, depending on the localization of events, on their influence upon each other from point to point, and on the separation of the region of possible causal effects from those which are impossible.

Second, philosophy distinguishes between causality and a law, i.e. a generally necessary and essential connection. Local causality determines the space–time conditions for causal relationships, but its meaning for the laws of physics must be examined. Here the point at issue above all is the localization of events. This can always be accomplished if one assumes a still more elementary level of events than the one under examination.

'In every true searcher of Nature there is a kind of religious reverence; for he finds it impossible to imagine that he is the first to have thought out the exceedingly delicate threads that connect his perceptions. The aspect of knowledge which has not yet been laid bare gives the investigator a feeling akin to that experienced by a child who seeks to grasp the masterly way in which elders manipulate things.'
(A.E.)

Third, the philosophical conception does not require a linear relationship between the initial and final states. Laws can be formulated for this relationship even if the complex of causal relationships underlying these laws is not yet completely investigated. The initial state can be regarded as a definite influence on a system, and the final state can be regarded as having a definite effect upon this influence.

We distinguish, therefore, between the principles of close-range interaction in a narrower and a broader sense. The principle of close range in a narrower sense pertains to the direct, concrete, and fundamental mediation of the relationship. In a broader sense, it relates to the generally necessary and essential relationship between events, that can only exist on the basis of a direct relationship.

Finally one has to consider if there can be a directionality to time, without the possibility of spatially local differentiation of events. From a philosophical point of view the direction of events according to content and time is sufficient. The ability to localize is not required; it is, however, an idealization necessary for a physical understanding that does not take into account the extension of objects or their inner structure. This idealization can be employed in the discovery of physical laws, because the structure of interacting objects can be neglected up to a certain degree. Only if it leads to new effects in the interaction must we give up the idealization of the point-shaped object. As we see, local causality in the physical sense is not identical with the philosophical conception; also, certain non-local concepts would be incompatible with the latter. But the assumption of local causality turns out to be an important means for achieving understanding of physical interactions.

3. PHYSICAL AND MATHEMATICAL SPACE

We have already referred to the development of the concept of space from the assumption of absolute space, independent of material motion, to the evidence that space–time is a form of existence of matter. On the one hand, this was done by proving the unity between physics and geometry, in that the physical importance of non-Euclidean geometry was recognized, and that Kant's thesis of the *a priori* validity of Euclidean geometry was rejected. On the other hand, the concept of space in mathematics became more and more generalized, so that abstract spaces constitute a set of elements with definite representation in functional analysis. Within these abstract spaces one can define the results of measurements. Thus, they do not exist independently of material processes; if they can be used for the representation of the results of physical measurements, they represent objectively real relations. In this context, the controversy that took place concerning the indefinite metric of the Hilbert space in Heisenberg's matrix formulation of quantum theory is not a difference of opinion about the existence of abstract spaces, but a dispute about their use in the theoretical understanding of physical processes. Therefore, the connection between mathematical and physical spaces is interesting also for philosophy, because the mathematical generalization of the concept of space allows for an improved representation of actual physical processes, and this generalized space turns out to be more and more an objectively real structure. The space of ordinary experience with its three dimensions is then separated from the general structure as a special case of abstract spaces.

When Einstein had thought through a problem, he always found it necessary to formulate this subject in as many different ways as possible and to present it so that it would be comprehensible to people accustomed to different modes of thought and with different educational preparations. He liked to formulate his ideas for mathematicians, for experimental physicists, for philosophers, and even for people without much scientific training if they were at all inclined to think independently.
(Philipp Frank, *Einstein: His Life and Times*)

Mathematics deals with possible structures in systems of ideal objects independently from the real properties of these objects. In constructing its theories, mathematics must meet the requirements of logical criteria, such as freedom from contradiction and others, and in certain philosophical approaches it must also provide prescriptions for measurement. But it is not without relationship to objective spaces; their conceivable structures serve as a representation of real or possible structures, and the interpretation of mathematical objects in a mathematically described physical theory is an important task in the understanding of physics. Here one should recall Dirac's theory of holes and its importance for the discovery of positrons. The mathematical space becomes interesting for physics and philosophy as a structure that may be physically interpreted, and as a component of the deeper penetration into the structure of matter. In that sense, those theoreticians are indeed correct who emphasize the difference between mathematical and physical space, but, at the same time, a continual improvement in the representation of objectively real structures in mathematical spaces is in our view essential.

Mathematical theories can certainly develop independently of physics and other sciences, besides originating as mathematical solutions of scientific and practical problems. What is required is the representational character of mathematical theories, which reveals itself in their interpretation. The requirement for new mathematical theories is thus identical with the requirement to find new theoretical ways of understanding the structure of matter in mathematically formulated scientific theories. Thus, we can also see the heuristic value of mathematics.

Since new directions of thought not only permit better theories, but also lead to as yet uninterpreted relationships, the search for the real content of certain mathematical forms, which are used for the representation of well-known facts of physics, has been stimulated. Without attention to logical and internal mathematical criteria for the construction of theories, mathematics could not fulfil this role. New paths of thought are found; one turns away from previous conceptions about existing objects with definite properties, and instead looks at possible relations between abstract objects. Thus, besides the differences between conceivable and actual structures we must also pay attention to the links between them. First, the abstract spaces of mathematics allow us to formulate results of physical measurements. We thereby arrive at a system of statements whose consequences can be examined, and which substitutes the description of observations by essential, functional, qualitative, and quantitative dependences. We can call this the representational function of mathematics, which allows us to describe physical knowledge in a mathematical form. Second, when examining the consequences of a mathematical structure we may find quantities and relations not interpreted by mathematics. These either indicate the inadequacies of the mathematical formalism for the corresponding physical theory, in which case the mathematical theory must be substituted by another one in order to obtain a better representation, or they indicate physical objects and relations that have not yet been found. The difficulty of distinguishing between these possibilities is shown by the example of Shrödinger, who could not bring his relativistic wave equation for the hydrogen atom in tune with observations. He then put it aside for half a year. Meanwhile, the discovery of electron spin proved his equation to be basically correct in non-relativistic form. In order to understand the heuristic function of mathematics, with which we are concerned when mathematical objects and relations first have to be found in their physical content, we need extensive work, discussion and a certain intuition on the part of the theoretician who has to make suggestions for experiments. Third, abstract spaces of mathematics are representations of objectively real structures, if they are to have physical content.

Thus, the constructive and heuristic functions of mathematics are united in its representational function. If abstract spaces are used for the description of physically real relationships resulting from results of measurements and if certain mathematical objects and relationships turn out to be theoretical predictions for objective processes yet to be discovered, then one can regard this mathematical space as a correct representation of the objectively real structure. In this way, in our understanding, objective space and objective structure are not to be separated from mathematical space. Objective space and objective structure exist before mathematical space, even if they are not yet recognized, but mathematical space is a product of human thought; it demonstrates the creativity of the human mind in conceiving possible relationships which are suited for the representation of physical structures, either known or yet to be found. Einstein has suggested this again and again.

When Eddington undertook to verify Einstein's predictions by observations of the eclipse of 1919, Einstein was much less interested in the result than Eddington was. I was reminded of the story about a female admirer of Whistler who told him that she had seen Battersea Bridge looking just as it did in one of his pictures, to which Whistler replied, 'Ah, Nature's coming on!' One felt that Einstein thought the solar system was 'coming on' when it decided to confirm his predictions. It is difficult to turn Einstein's method into a set of textbook maxims for the guidance of students. The recipe would have read as follows: 'First acquire a transcendent genius and an all-embracing imagination, then learn your subject, and then wait for illumination.' It is the first part of this recipe that offers difficulties.

(Bertrand Russell, in *Einstein on Peace*)

Our knowledge about objectively real structures develops in a twofold way. On the one hand, from experiments we receive new data that have to be interpreted. For this we need mathematical spaces. On the other hand, using mathematics we investigate new possible ways of thinking in order to obtain better assumptions for the representation of complicated objective facts in mathematical spaces. This process of understanding began with general conceptions about space in philosophy, and it had its first high point in the construction of Euclidean geometry. As a conception of space this was used for a long time for the description of physical processes, whereby space was regarded as absolutely existent. The criticism of this conception, and the proof that space and time are forms of existence of matter, did not at first require the elimination of general properties of space–time, such as the relation between larger and smaller, before and after, etc. In elementary-particle physics, however, we see the difficulty of separating elementary particles spacewise. At the present time it cannot be determined what 'smaller in size' is supposed to mean, although we can undertake the division of quantum numbers, which leads to the hypothesis of quarks if we divide the elementary charge. Here, space turns out to be in a true sense the structure of material processes, whose general characteristics are either too abstract or can no longer be determined. As the philosophers have already done, when characterizing the development of our conception of space, one can arrive at space as the contiguity of material processes. But we cannot stop at this point. Rather we must analyse this contiguity according to its content by unveiling laws and relationships of material processes. Then objectively real space turns out to be the structure of material processes in which, under certain conditions, we can determine distances, regions and trajectories. In this sense our philosophical conception of space approaches the generality of topological spaces, which for the representation of physical facts have to be furnished with certain properties. In the process the philosophical conception of space as a form of existence of matter must be made more precise.

'It is the moral qualities of its leading personalities that are perhaps of even greater significance for a generation and for the course of history than purely intellectual accomplishments. Even these latter are, to a far greater degree than is commonly credited, dependent on the stature of character.'

(A.E.)

Further Reading

Einstein, A., and Infeld, L., *The Evolution of Physics* (Cambridge: Cambridge University Press, 1938; New York: Simon and Schuster, 1961).

Grünbaum, A., *Philosophical Problems of Space and Time*, 2nd edition (Dordrecht and Boston: D. Reidel, 1973).

Hörz, H., *Materiestruktur* (Berlin, 1971).

Jammer, M., *Concepts of Space* (Cambridge, Mass.: Harvard University Press, 1954).

Reichenbach, H., *The Philosophy of Space and Time* (New York: Dover, 1958).

Smart, J. J. C. (Editor), *Problems of Space and Time* (London and New York: Macmillan, 1964).

Schlegel, R., *Time and the Physical World* (New York: Dover, 1968).

Sklar, L., *Space, Time, and Spacetime* (Berkeley: University of California Press, 1976).

ALBERT EINSTEIN
1879-1955
60 GR POLSKA

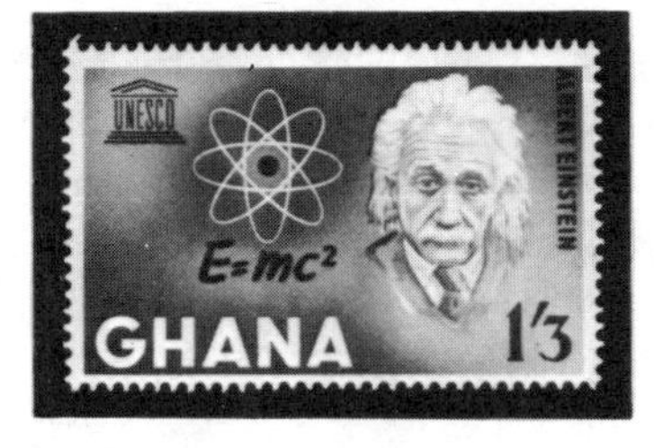
UNESCO
E=mc²
ALBERT EINSTEIN
GHANA
1/3

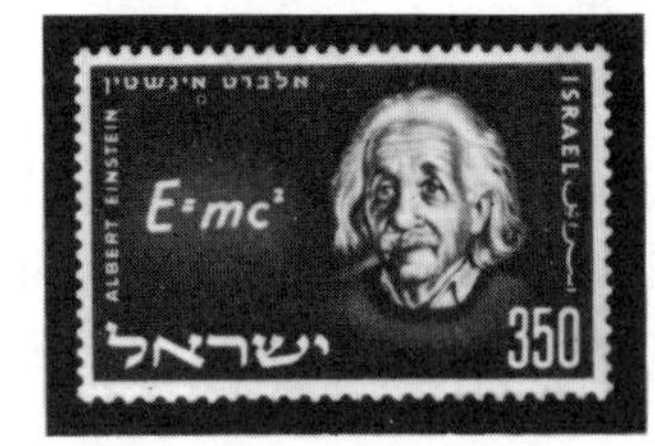
ALBERT EINSTEIN
E=mc²
ISRAEL
ישראל
350

40
1879-1955
HELVETIA

INTRODUCCION DE LA
ELECTRONICA A LA
EXPLOTACION POSTAL
25c
REPUBLICA ARGENTINA
CASA DE MONEDA
"1971"
ALVAREZ BOERO DIB.

REPUBLIQUE DU MALI
e=mc²
1879 1955
Albert Einstein
90F
POSTE AERIENNE 1975

NICARAGUA
E = mc²,
LEY DE
EINSTEIN
CORREO
20
CENTAVOS
LAS 10 FORMULAS MATEMATICAS QUE CAMBIARON LA FAZ DE LA TIERRA

PRIX NOBEL DE PHYSIQUE 1921
100F
POSTE AERIENNE
ALBERT EINSTEIN
REPUBLIQUE DU TCHAD

EDUCATION
L'INSTRUCTION
STRENGTH THROUGH
5
FAIT LA FORCE
POSTES CANADA POSTAGE

E=MC²

United States 8c
EINSTEIN

EINSTEIN THEORIE DE LA RELATIVITE
1976
ETAT COMORIEN
POSTES
10F

CORREO DEL PARAGUAY
E=mc²
ALBERTO
EINSTEIN
14 MARZO 1879
AEREO
G18.15
LAP+9.00

Einstein on postage stamps

E. J. Burge

Israel	1956	S.G. 127	Commemorative.
Poland	1959	S.G. 1128	Famous scientists (one of a set of six with Darwin, Mendeleev, Pasteur, Newton, and Copernicus).
Canada	1962	S.G. 522	Education year.
Ghana	1964	S.G. 355	UNESCO week (one of a set of three with G. Washington Carver).
USA	1965	S.G. 1267	Prominent Americans (one of first series of a set of twenty).
Paraguay	1965	S.G. Appendix	Famous scientists (two identical designs for each of four scientists, Einstein, Newton, Galileo, and Copernicus).
Argentine	1971	S.G. 1356	Electronics in postal development, with phosphorescent scanners.
Nicaragua	1971	S.G. 1765	'The ten mathematical equations that changed the face of the Earth'. The ten stamps show: 1+1=2, and the laws of Newton, Einstein, Tsiolkovsky, Maxwell, Napier, Pythagoras, Boltzmann, de Broglie, and Archimedes.
Switzerland	1972	S.G. 842	Swiss celebrities (one of a set of five with A. Giacometti, C. F. Ramsey, Le Corbusier, and A. Honegger).
Mali	1975	S.G. 492	20th anniversary of Einstein's death.
Chad	1976	S.G. 471	Nobel prizewinners. One of a set of five with R. Koch (medicine, 1905), Anatole France (literature, 1921), Einstein (physics, 1921), Dag Hammarskjøld (peace, 1961), S. Tomonaga (physics, 1965), plus miniature sheet, A. Fleming (medicine, 1945).
Comoro Islands	1976	S.G. 192	American bicentenary and 'Success of Operation Viking'; shows Einstein, Sagan, and Young. Others show Copernicus, Viking orbiting Mars, the Vikings' discovery of America, and the first colour photograph of Martian terrain, plus miniature sheet of Viking on Mars.

13 Approaches to the teaching of special relativity*

Geoffrey Dorling

> 'Every scientist, in working out his own research, gravitates to particular points on the boundary which separates the known from the unknown. and becomes inclined to take his particular perspective from these points. It must not, however, be expected that these individual aspects will form a complete picture, and will indicate the only paths along which science can or will advance.'
> (A.E.)

INTRODUCTION

Einstein's special theory of relativity, first published in *Annalen der Physik* in 1905, has had an overwhelming influence on physical ideas during the twentieth century. However, the conceptual difficulties of its origins and foundations, and the mathematical sophistication often demanded in understanding its consequential influence, have restricted until recently its introduction into an education in physics to the later stages of a university degree course.

Many actively engaged in teaching physics have argued for the inclusion of some work on the special theory at a much earlier stage. J. Rekveld (whose approach to relativity we shall mention later) puts forward several arguments for the inclusion of some relativity in secondary-school courses in his chapter in *Teaching Physics Today*.

During the past ten years there have been several successful attempts at presenting the special theory in a way which, while acceptable to the less mature physicist, still does justice to its concepts and consequences. In particular, all these attempts have implicitly recognized Eric Rogers' dictum in *Physics for the Inquiring Mind*: 'Since relativity is a piece of mathematics, popular accounts that try to explain it without mathematics are almost certain to fail.' Thus one of their problems has been to present the mathematical aspects in an intelligible way. In this they are to be distinguished from the many 'popular' accounts written for a lay audience.

The difficulties which have faced such innovators become clear when we recall Einstein's original assertion of the principle of relativity.

* This article is reprinted, with only minor changes, from *Teaching School Physics*, edited by J. L. Lewis (Harmondsworth: Penguin Books—Unesco, 1972).

Conceptual difficulties abound and yet it remains at the heart of relativity theory.

After mentioning certain apparent anomalies in physical observations made on the electromagnetic field he went on to say:

> Examples of this sort, together with the unsuccessful attempts to discover any motion of the Earth relatively to the 'light medium', suggest that the phenomena of electrodynamics as well as of mechanics possess no properties corresponding to the idea of absolute rest. They suggest rather that, as has already been shown to the first order of small quantities, the laws of electrodynamics and optics will be valid for all frames of reference for which the equations of mechanics hold good. We will raise this conjecture (the purport of which will hereafter be called the 'principle of relativity') to the status of a postulate, and also introduce another postulate, which is only apparently irreconcilable with the former, namely, that light is always propagated in empty space with a definite velocity *c*, which is independent of the state of motion of the emitting body.

This statement, together with his own development, dictated a fairly uniform teaching approach over the succeeding fifty years. The sequence opposite, typical of these presentations, is laid out in diagrammatic form as all the newer presentations have either been closely related to this or have attempted to alleviate the difficulties inherent in it.

It can be seen that the problem of presenting the theory as part of a general education in physics can be classified under two broad headings:

(a) showing that Einstein's principle of relativity is a reasonable description of physical behaviour;
(b) showing the consequences of this principle for our ideas of measurement of mass, length and time.

Recently there have been several different and successful approaches to both these problems. This article will confine itself to just a few noteworthy examples to illustrate some of the different presentations which have evolved over the past ten years. The examples chosen are a sample only and are not to be thought of as a complete survey of the field.

VARIOUS METHODS OF ESTABLISHING EINSTEIN'S PRINCIPLE OF RELATIVITY AND THE INVARIANCE OF THE SPEED OF LIGHT

Recent proposed developments of the theory at a level suitable for introductory courses have fallen very markedly into two camps. There is on the one side a careful simplification of the traditional approach already outlined, which discusses the essence of the ether-drift controversy, leading ultimately to the Michelson–Morley experiment and the invariance of the speed of light. From this experimental fact the principle of relativity is developed.

On the other side are those accounts which feel that the ether-drift controversy is a piece of interesting history quite unnecessary to an understanding of the theory *per se*. We will consider briefly some examples of each approach.

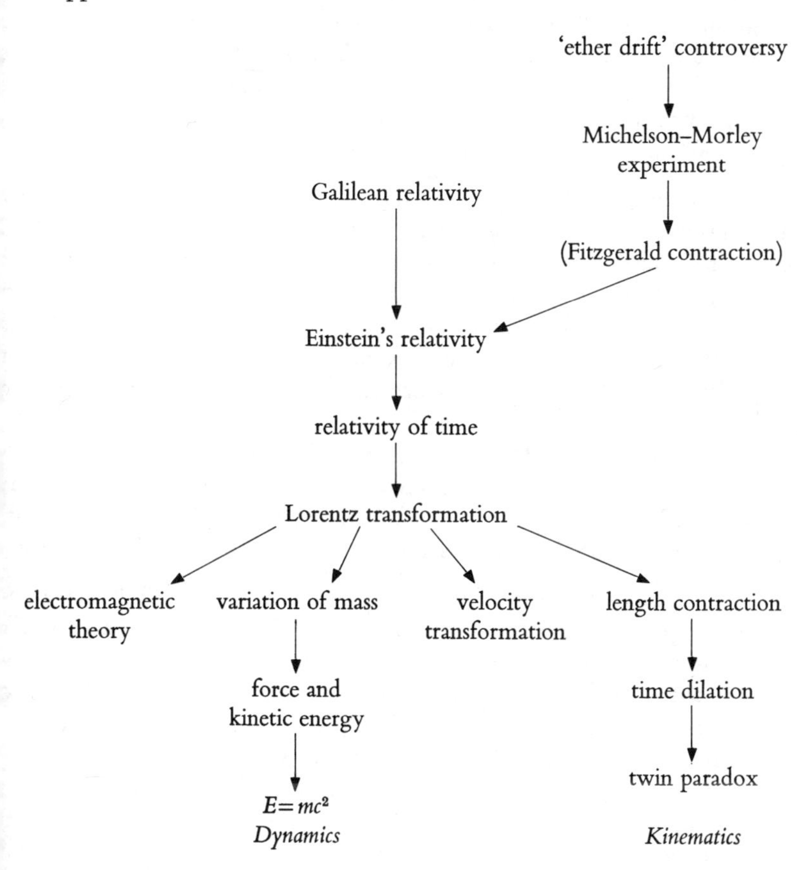

1. *The ether-drift approach*

The PSSC start their account in the *Advanced Topics Supplement* as follows:

> The waves on a coil spring, water waves, sound waves, and the 'starting wave' in a line of cars at a traffic light all propagate in a medium. There is always something, the shape of which moves. It is natural to ask what is the medium in which light waves travel. Or, to put it differently, what is it that is waving in a light wave?
>
> This question puzzled many physicists of the nineteenth century and they devised various experiments in order to prove the presence of the light-carrying medium, the 'ether'. They realized that the ether must be very different from all other wave-carrying media because it apparently is present even in the highest vacuum as well as in transparent materials. It is therefore unlikely that the ether is a form of matter with such

properties as a chemical composition, density and the like. The physicists of the nineteenth century did not look for these material properties but asked themselves the following question:

The ether occupies all space out to the farthest stars. The Earth moves through this space, rotating on its axis and around the Sun. How does the ether move with respect to the Earth? Does the ether follow the Earth's motion, being therefore at rest with respect to the Earth, or is the ether at rest with respect to the Sun and other fixed stars? In the latter case it is obvious that the ether would move with respect to the Earth.

After indicating that experimental evidence exists to contradict any assumption that the ether is at rest with respect to the Earth, the PSSC course sets out to show how physicists attempted to measure an ether-drift velocity. The discussion is centred on the Michelson–Morley experiment. Some of the difficulties inherent in a thorough understanding of this experiment are relieved by the introduction of a *laboratory* experiment utilizing an interference pattern formed by light passing partly through water.

An account of relativity theory in *Senior Science for High School Students* in New South Wales follows a similar introductory pattern. After a short account of Roemer's measurement of the speed of light the 'ether drift' question is introduced with the same problem as that posed by the PSSC.

... the question was asked 'What is it that carries the waves?' Nineteenth-century physicists answered this in what to us now may appear to be a strange way. They suggested that the whole of space, *empty* space, was filled with some 'stuff' they called the ether. They assumed that it had no physical properties by which it could be detected so that it appeared to us precisely as a vacuum. Its only property was that it carried electromagnetic radiation and that the radiation moved through it at the speed of light, *c*. Despite the fact that the ether was assumed to have no detectable properties, it was realized that if it existed at all it should betray its presence in one special circumstance. To understand this it will be necessary to consider two simple analogies.

The two analogies discussed are the time taken for a boat to travel a measured distance up and down a flowing stream and the time for the boat to make a journey of the same distance *across* the flow. This analogy to the Michelson and Morley measurements was also proposed by H. Bondi (1965), and by an arithmetical method it avoids some of the difficult algebra associated with an analysis of this experiment.

Consequent upon the null result for an ether-drift velocity obtained by Michelson and Morley, Einstein's solution is proposed.

In 1905 Einstein suggested that it was absurd to introduce the concept of the ether just because we think that light in a vacuum should behave like sound in the atmosphere. He took as a fundamental experimental fact

It occurred to him that time measurement depends on the idea of simultaneity. Suddenly he was struck by the fact that, although this idea was perfectly clear when two events occur in the same place, it was not equally clear for events in different places. This was the crucial stage in his thinking. For he saw that he had discovered a great gap in the classical treatment of time. It took him roughly ten years to arrive at this point, but from the moment when he came to question the traditional idea of time, only five weeks were needed to write his paper, although he was working all day at the Patent Office.
(G. J. Whitrow, *Einstein: The Man and His Achievement*)

that *the speed of light in a vacuum is always constant, no matter how the observer moves.* On the basis of this 'law of light' he was able to present a completely new and different picture of space and time.

2. *'Linear' approaches*

Writing under the heading 'Relativity' in *Teaching Physics Today*, J. Rekveld took quite a different view of the way relativity theory should be introduced at elementary level. He says:

> Although a short historical sketch has been added to this paper the author does not intend to propagate the opinion that relativity theory should necessarily be preceded by a more or less complete survey of the pertinent struggle in the last century that led to Einstein's theory. To understand what was going on in connection with the theory of an all-pervading ether is in itself a difficult enterprise.
>
> It is true of course that a discussion of the historical process motivates the need for a new and revolutionary theory, but it does not in the least prepare or promote an understanding of relativity theory proper.
>
> Therefore an introductory course on relativity, taught in the second half of the twentieth century with only a limited amount of time at one's disposal, might better be started directly from the fundamentals of Einstein's theory, leaving historical development perhaps to a more advanced course.

He then proposes that an account of the theory should be preceded by a careful look at Galilean relativity which calls

> . . . the attention of our students explicitly to the importance of the notion of 'frame of reference' and makes them familiar with the expression: inertial frame. It leads to a restricted principle of invariancy, namely the laws of mechanics. It teaches students the simple transformations enabling them to go from one inertial system to another and paves the way for the more complicated transformation equations in Einstein's theory.

This introduction leads on to a discussion of how we usually determine velocities relative to different inertial frames, and he continues:

> We know that our Earth is moving round the Sun in a nearly circular path with a velocity of 30 km s^{-1}. It moves in different directions in different parts of the year. We might take this opportunity to investigate the dependence of the velocity of light on the direction of the Earth's motion.
>
> We can refer here to the famous experiment performed by Michelson and Morley for the first time in 1887. Because of the role of an interference effect it does not seem likely that we could explain this experiment on the level we are assuming here. We will have to state the results of the

Einstein had to put the applications for patents, which were frequently vaguely written, into a clearly defined form. He had to be able, above all, to pick out the basic ideas of the inventions from the descriptions. This was frequently not easy and it gave Einstein an opportunity to study thoroughly many ideas that appeared new and interesting. Perhaps it was this work that developed his unusual faculty of immediately grasping the chief consequence of every hypothesis presented, a faculty that has aroused admiration in so many people who have had an opportunity to observe him in scientific discussion.

(Philipp Frank, *Einstein: His Life and Times*)

experiment and point to the constancy of the velocity of light in all inertial frames.

Thus, he continues:

> If light has the same velocity relative to all inertial frames it can be shown by a rather simple thought experiment that the duration of an event observed by different observers does not have an absolute meaning.

Sears and Brehme (1968) take an even curter view of the historical background. They introduce their account with the words:

> The speed of light in a vacuum is 2.9979×10^8 m s^{-1}, or very nearly 3×10^8 m s^{-1}. All experimental evidence leads to the conclusion that this speed is the same for all observers, regardless of their motions relative to each other or to sources of light. This fact is the basis of the theory of relativity.

Though he kept expenses at a minimum, he had to spend money for things from which he derived no pleasure, but which were required by his social position. In order to improve the financial situation, his wife took in students to board. He once said jokingly: 'In my relativity theory I set up a clock at every point in space, but in reality I find it difficult to provide even one clock in my room.'

(Philipp Frank, *Einstein: His Life and Times*)

The remainder of their account is concerned with the impact of this fact on a wide range of physical theory. In their preface to the book, they say:

> This is a text in physics. No attempt is made to discuss the philosophical or metaphysical aspects of relativity. Nor is this text an account of the history of relativity. The famous Michelson–Morley experiment, the first to suggest the invariance of the speed of light, is barely mentioned and the ether appears only in a footnote. These interesting and historically important aspects are not essential to an understanding of the theory.

Both of the last two accounts stress the lack of *necessity* for an historical background. H. Bondi (1965) goes even further and stresses its *irrelevance*. He calls himself a 'traditionist' seeing the theory as a natural growth from classical physics. He introduces his account as follows:

> When the theory of relativity first came out, and for many years afterward, it was looked on as something revolutionary. Attention was focused on the most extraordinary aspects of the theory. With the passage of time, though, the sensational aspects of Albert Einstein's work have ceased to cause wonderment, at least among scientists, and now one begins to see the theory not as a revolution, but as a natural consequence and outgrowth of all the work that has been going on in physics since the days of Isaac Newton and Galileo.

This introduction sets the scene for the rest of the account. In developing the special theory, he tries to show how unextraordinary the *theory* is, but how wrong our original (pre-relativity) concept of time was, due to our limited experience of high relative speeds.

The uniqueness of light is highlighted by dealing briefly with the 'absurdity of the ether concept'. A critical account of the Michelson–Morley experiment concludes with the thought-provoking observation:

'There can be no greater merit in a scientific discovery than that before long it should appear very odd that it ever was considered a discovery.'

A brief but important survey of 'route-dependent' quantities, like the distance of a journey between two given points, introduces the idea that time is a 'route-dependent' quantity as well.

3. *The 'dynamics-first' approach*

Before concluding this account of the various ways that have evolved of establishing the Einstein principle of relativity, we must take note of one other quite different approach to the whole theory. This we might call 'dynamics first'. There are today many experimental observations which can be made on bodies travelling at speeds approaching that of light. This material was not available to Einstein. But there is no reason why such experimental material should not be used to *teach* the special theory. By using data from the acceleration of electrons in a linear accelerator, A. P. French (1968) is enabled to introduce the study by considering why the speeds of the electrons depart so widely from those predicted by Newtonian dynamics. The data are obtained from a filmed experiment* made for the PSSC's Advanced Topics treatment of relativity, only they used it to help develop the dynamical consequences of relativity in its traditional place *after* kinematic considerations.

A relationship between photon energy and momentum is shown to be similar to that for *high-speed* electrons and French goes on:

> This serves to reinforce our belief that the dynamics of photons and of other particles can be brought, for some purposes at least, within the same descriptive framework. Our next step will be to suggest what that framework might be. Our argument will appeal to one's sense of what is plausible; it will not be logically inescapable.

A thought experiment, originated by Einstein, shows that photons of energy E have an effective mass E/c^2. The assumption is made that this link between energy and mass may be universal. As a consequence formulae relating masses and kinetic energies with speed are evolved and the latter is shown to fit the experimental data from the film.

In this way the one 'familiar' result that most people associate with the Einstein theory, namely $E=mc^2$, is derived right at the outset, giving an incentive for a deeper inquiry into the special nature of the speed of light.

When Einstein was in Hollywood in 1931, Charlie Chaplin invited him to dinner in his villa and later to see in his private cinema a performance of the film *City Lights*. During the drive to the town they were recognized by the crowd and enthusiastically cheered. Chaplin calmly remarked to his guest: 'The people are applauding you because none of them understands you and applauding me because everybody understands me.'
(Carl Seelig, *Albert Einstein: A Documentary Biography*)

THE CONSEQUENCES OF THE PRINCIPLE OF RELATIVITY FOR OUR IDEAS OF MASS, LENGTH AND TIME

One of the major difficulties which have always faced students in gaining an insight into these consequences has been the lack of any real observational

* *The Ultimate Speed*, a film produced by the Education Development Center, Newton, Massachusetts.

material concerning the description of events taking place at speeds approaching that of light. It is noteworthy that a body must have a speed of about one-seventh that of light relative to an observer before a one per cent increase in its mass could be detected. This scarcity of observational evidence has hitherto been supplemented by 'thought experiments'. Einstein was probably the originator of these in his 'popular' account of relativity theory in 1916. Of late such thought experiments have been extensively used in developing quantitative results, but as Bondi (1965) points out these have taken on a new realism. Much of the quantitative development in *Relativity and Common Sense* concerns the caperings of astronauts Alfred, Brian, and Charles, and he says:

How can it be that mathematics, being after all a product of human thought which is independent of experience, is so admirably appropriate to the objects of reality? Is human reason, then, without experience, merely by taking thought, able to fathom the properties of real things?

In my opinion the answer to this question is, briefly, this: as far as the propositions of mathematics refer to reality, they are not certain; and as far as they are certain, they do not refer to reality.
(A.E., 'Geometry and Experience')

> When Einstein in 1916 wrote a book on relativity for the general public, he could think of no better example to illustrate his ideas than to imagine indefinitely long trains running past indefinitely long embankments at speeds approaching the velocity of light! . . . Far fetched as they were, those trains afforded about the only possible images that would fall within the layman's understanding and not be dismissed as Jules Verne absurdities . . .
>
> Today, all this has changed. We send rockets to the moon and the vicinity of Venus. The most stubborn sceptic no longer doubts that space stations of some sort will exist within the lifetime of the youngest readers of these pages. Russian and American astronauts circle the Earth at speeds approaching 20 000 miles an hour, and while our Brian's 71 000 miles per second is a far stretch from that figure, yet we can think quite realistically of speeds beyond the ken of our fathers and grandfathers. Every day experimenters at the great accelerating machines (the 'atom smashers') work with speeds nine-tenths of the velocity of light; relativistic effects are the regular order of their business. In a very few years special relativity has come down from the clouds of phantasy or philosophic speculation to its rightful foothold on the solid ground of the public domain.
>
> It is in the nature of the human mind that learning is easier when a demonstrable *need* to learn exists. Our fathers had no actual need to understand relativity, but we have, and we can address ourselves to the adventures of Alfred, Brian, and Charles without the emotional misgivings that, forty years ago, upset passengers on Einstein's indefinitely long trains. Alfred, Brian, and Charles are no less fictional but their manoeuvrings in space are representative of situations which, in more complex details and refined form, command the attention and challenge the laboratory skills of today's scientists and engineers.

The work of 'today's scientists and engineers' has also been put to good effect. We have already seen how French has used a filmed experiment on high-speed electrons to introduce his account. The PSSC, which originally introduced this film, use another on the half-life of muons* to give reality to their work on time dilation. French makes use of this film in his own account as well.

Certainly these real and imagined experiments are an essential prop to understanding, but the mathematics of the Lorentz transformations are unavoidable in accounts that seek to be, in the words of Sears and Brehme, 'a text in physics'. The traditional sequence has already been outlined. An account of the relativity of simultaneity precedes the Lorentz transformations. These in turn are applied first of all to *kinematic* problems (length contraction, time dilation, composition of velocities) and then to *dynamic* problems. Many recent authors have departed from this order. The PSSC develops the law of composition of velocities first of all; Bondi deduces the time-dilation formula at the outset; Sears and Brehme *begin* their account with the Lorentz transforms. French, as we have already seen, begins with $E=mc^2$.

In developing the essential mathematics, approaches have been numerous and overlapping, but at the risk of over-simplification three trends can be distinguished:

(a) Simplification of the traditional algebraic approach.
(b) The *k*-calculus.
(c) Geometrical approaches.

We will look briefly at each of these.

Just as it is the pride of many people never to have any time, so it has been Einstein's always to have time. I recall a visit I once paid him on which we decided to visit the astrophysical observatory at Potsdam together. We agreed to meet on a certain bridge in Potsdam, but since I was a good deal of a stranger in Berlin, I said I could not promise to be there at the appointed time. 'Oh,' said Einstein, 'that makes no difference; then I will wait on the bridge.' I suggested that that might waste too much of his time. 'Oh no,' was the rejoinder, 'the kind of work I do can be done anywhere. Why should I be less capable of reflecting about my problems on the Potsdam bridge than at home?'
(Philipp Frank, *Einstein: His Life and Times*)

Simplification of the traditional algebraic treatment

An example of this is the PSSC's treatment (PSSC, 1966). To give reality to the new formulae they describe Fizeau's experiment on the passage of light through moving water. The shift in interference pattern is not used (as he used it) to measure the Fresnel 'drag coefficient', but to show the need for a new law of combination of velocities. Their derived result

$$w=\frac{u+v}{1+uv/c^2}$$

is shown to fit the experimental results and they say:

> Velocity, by definition, is displacement divided by time. If large velocities do not behave the way we know low velocities to behave, then we become suspicious that our notions about length and time may not be adequate. We shall have to examine carefully what we really mean when we measure intervals of length and time in a frame of reference which is moving with respect to us.

The need to synchronize two clocks for these measurements soon leads to the realization that clocks synchronized for one observer will not be

* *Time Dilation—An Experiment with Mu-Mesons*, a film produced by the Education Development Center, Newton, Massachusetts.

synchronized for another moving with respect to the first. By assuming that the relative velocity of two observers (frames of reference) is much less than c, some simple (first-order) transformations are worked out and the relativistic law for velocity combination is shown to follow.

The Lorentz transformations are developed by introducing a necessary second-order correction of $\sqrt{(1-v^2/c^2)}$ to make the transformations of distance and time between the two frames symmetrical. The more detailed treatment of length contraction and time dilation which follows is enhanced by the filmed experiment on the half-life of muons.

Eric Rogers in the section on 'Relativity and Mathematics' in *Physics for the Inquiring Mind* deals with the difficulties of mathematics in a different way. Acknowledging its complexity, he says:

Einstein always began with the simplest possible ideas and then, by describing how he saw the problem, he put it into the appropriate context. This intuitive approach was almost like painting a picture. It was an experience that taught me the difference between knowledge and understanding.
(E. H. Hutten, in G. J. Whitrow, *Einstein: The Man and His Achievement*)

> To understand relativity you should either follow its algebra through in standard texts, or, as here, examine the origins and final results, taking the mathematical machine work on trust.

To prepare for this approach, four pages of the chapter have previously been devoted to obtaining a proper perspective of the place of mathematics in physics. Here mathematics is shown to be a 'language' and a 'clever servant'. After a detailed discussion of two attempts to measure the Earth's speed relative to the 'ether' the mathematical analysis needed to resolve the contradiction resulting from these is assumed to be incorporated in a logic machine. This is the 'clever servant' which can work out the answers to given questions when told all the information you have and assumptions you wish to make.

The k-calculus

This approach to the mathematics of relativity was developed by H. Bondi in *Relativity and Common Sense* and has been used in its essentials in *Senior Science for High School Students*. We will outline Bondi's own approach here.

He tackles the mathematical development of the theory of relativity by an immediate consideration of the way different inertial observers will measure time intervals. This is achieved by comparing the rate at which a succession of light signals are sent out by one observer with the rate they are received by another in relative motion with respect to the first. In order to discuss the matter in concrete terms an earth-bound observer, Alfred, is imagined to be sending a regular succession of light signals to a space station manned by David at rest relative to Alfred. A third observer, Brian, is travelling from Alfred to David and intercepts these signals, sending out one of his own immediately he receives one of Alfred's. Bondi concludes:

> . . . if Alfred flashed his light at intervals h, then David would have seen these flashes at intervals h, each flash taking the same time to reach him. Brian would have seen them at some interval kh by his watch so

that k is the ratio of the interval of reception to the interval of transmission. If Brian flashes his torch at intervals kh by his watch, then these flashes, travelling in company with those emitted by Alfred, would be seen at intervals h, giving the reciprocal ratio $1/k$ between Brian and David.

After noting that the relationship between two inertial observers is completely specified by the value of k, he says:

> Note that the principle of relativity, by insisting on the equivalence of all inertial observers, makes it quite clear that the ratio k must be the same whichever of a pair of inertial observers does the transmitting. It is through this rule that our work on light differs so sharply from the work in sound where, it will be remembered, the speed of transmitter and receiver relative to the air had also to be taken into account.

Once this is established it can be shown that different inertial observers will disagree about the length of apparently corresponding time intervals. The extent of their disagreement is of course the essential part of relativity theory. Bondi goes on to show how k is related to the relative speeds of the two observers, and the law of velocity composition and the Lorentz transformations are then evolved in terms of k.

Geometrical approaches: Minkowski diagrams

In 1908 H. Minkowski described a geometrical interpretation of the Lorentz transformations. In this an event is described by *four* coordinates x, y, z, and t in any one frame of reference. The complete kinematic history of any point is represented by a line in four-dimensional space with axes x, y, z, and t. This line is called a 'world line'. Since relativity theory is usually concerned with two frames of reference in uniform motion with respect to each other, the direction of this motion is made the x-axis and problems connected with the description of a succession of events are limited to a consideration of the two dimensions x and t.

Not very many recent textbook accounts of relativity make much use of these diagrams. However, Rekveld (1965) and French (1968) use them extensively. Rekveld says:

> The kinematical results of the theory on relativity . . . can also be derived by a geometrical approach, in which the so-called Minkowski diagrams are used as visual aids. A geometrical presentation of the theory may be applied either as an independent method or as a way of supporting the algebraic discussion. In some cases the geometrical approach has advantages, especially for the teaching of concepts which ask a great deal of the imaginative ability of the students.

He prepares the ground for their use later in developing the Lorentz transformations by considering a geometrical representation of the Galilean transformation (Diagram 13.1a). An event E is described by coordinates (x_1, t_1) in frame (x, t) and by (x_1', t_1') in frame (x', t').

He later shows, as French does, that in order to describe correctly the passage of a light signal in the two frames the x-axes should not be coincident (Diagram 13.1b).

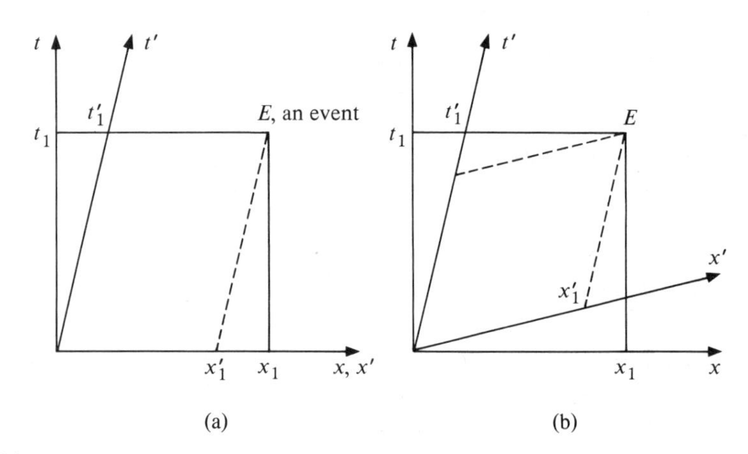

Diagram 13.1

Once Diagram 13.1b has been accepted it becomes an exercise in geometry to deal with the Lorentz transformations, and length contraction and time dilation effects can be visualized as direct outcomes from changes in $(\Delta E')_{t \text{ constant}}$ and $(\Delta E)_{x \text{ constant}}$.

The use of Brehme diagrams

The Minkowski diagrams are not the only way to make a geometrical approach to relativity theory. To illustrate this we will consider one other geometrical approach from the many which are possible. This is the approach devised by R. W. Brehme and used extensively in *Introduction to the Theory of Relativity*.

Sears and Brehme also introduce their geometrical representation by considering first the Galilean transformation. An event E is represented by coordinates (x_1, t_1) in frame (x, t) and by (x_1', t_1') in frame (x', t') (Diagram 13.2a). As they state in their book: 'The coordinates of events are found by dropping perpendiculars to the axes, even though the axes are not orthogonal'.

In this case it is the t-, t'-axes which are coincident. One special difficulty in using the Minkowski diagrams is that the scales on the x- and x'- and on the t- and t'-axes are not identical, i.e. unit time interval is not the same graph length on both t and t'. As a result lengths may *look longer* in (x, t) than in (x', t'), but may *in fact* be *shorter*. In the Brehme diagrams the graph scales are identical.

In order to describe correctly the passage of a light signal in the two frames of reference, it can be shown that the t- and t'-axes cannot be coincident (Diagram 13.2b). The angle ϕ between the t- and t'-axes is only the same as the angle between the x- and the x'-axes if there is a scale factor c (the speed of light) between x- and t- (and thus x'- and t'-) axes.

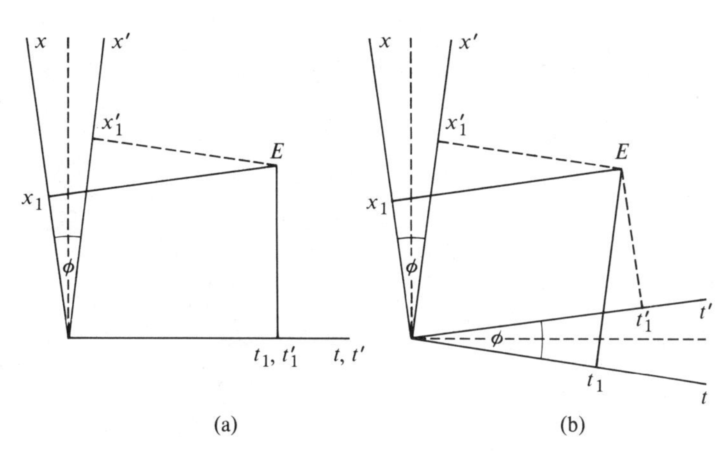

Diagram 13.2

Again, once these diagrams have been understood, the consequences of the principle of relativity are very easy to visualize and calculate. Another feature of these particular diagrams is that owing to their symmetry neither the (x, t) nor the (x', t') frame of reference seems specially preferred.

CONCLUSION

The aim of this article has been to show how numerous and various have been the recent approaches to elementary or introductory treatments of the theory of relativity. It must be emphasized again that the examples used have only been chosen in order to illustrate this variety. It will, it is hoped, have been seen that no one approach is unique, and yet all have unique features. It is thus clear that if twenty different approaches were described, a twenty-first could be devised by using particularly favoured features from each one.

References

Bondi, H., *Relativity and Common Sense* (London: Heinemann, 1965).
Einstein, A., 'On the electrodynamics of moving bodies', from *Principle of Relativity*, translated by W. Perrett and G. B. Jeffery (New York: Dover, 1923).
French, A. P., *Special Relativity* (London: Nelson, 1968).
PSSC, *Advanced Topics Supplement* (Farnborough: Heath, 1966).
Rekveld, J., 'Relativity', Chapter 10 of *Teaching Physics Today*, OECD.

Rogers, E. M., *Physics for the Inquiring Mind* (New Jersey: Princeton University Press, 1960).

Sears, F. W., and Brehme, R. W., *Introduction to the Theory of Relativity* (New York: Addison-Wesley, 1968).

Sikjaer, S. (Editor), *Seminar on the Teaching of Physics in Schools* (Copenhagen: Gyldendal, 1971).

Some other books on relativity theory

Bergmann, P. G., *Introduction to the Theory of Relativity* (New Jersey: Prentice-Hall, 1942).

Bohm, D., *The Special Theory of Relativity* (New York: Benjamin, 1965).

Cortini, G., *La Relativitá Ristretta*, with a historical note by S. Bergia (Turin: Loescher Editore, 1978).

Nevanlinna, R., *Space, Time and Relativity* (New York: Addison-Wesley, 1968).

Ney, E. P., *Electromagnetism and Relativity* (New York: Harper and Row, 1962).

Rosser, W. G. V., *Relativity and High Energy Physics* (London: Wykeham Publications, 1969).

Sexl, R., and Schmidt, H. K., *Relativitätstheorie* (Vieweg, 1978).

Taylor, E. F., and Wheeler, J. A., *Spacetime Physics* (New York: Freeman, 1966).

Figure 34

Cartoon by Wim van Wieringen, 1950. The caption reads 'Our learned professors studying an Einstein problem'

APPENDIX

Since the foregoing article was written, there have indeed been various further contributions to the teaching of relativity. It is a subject that seems to hold an inexhaustible attraction for teachers of physics, as well as for students and the lay public.

Several of these later developments attest to a growing concern with pedagogic strategies as well as with a preferred selection and sequence of topics. Full descriptions of some of these newer courses are to be found in *Seminar on the Teaching of Physics in Schools* (edited by S. Sikjaer). Here, we simply describe a few of them in brief outline.

Haber-Schaim (1971), in the volume cited above, gives a detailed description of a development which, like that of French (1968) uses as its starting-point the *Ultimate Speed* filmed experiment, but then goes on to explore the possible form of the velocity-addition law in the way that this problem is treated in the PSSC Advanced Topics (1966). This development closely parallels what is done in detail (but in reverse order) in the PSSC *College Physics* (1968), chapters 30 and 32.

Messel (1971), also in Sikjaer, *op. cit.*, gives an outline of an approach that stems from Bondi's *k*-calculus, but casts its net wide, focussing at first on the fundamentals of time and time scales, but later extending to cosmological questions (Olbers' paradox) and to the consideration of magnetism as an essentially relativistic phenomenon.

An article by Swartz (1971), again in Sikjaer, *op. cit.*, also concerns itself with 'the relativistic relationship between electricity and magnetism'. In addition, Swartz points to the desirability of developing individualized study materials for this and other subjects.

Angotti *et al.* (1977) describe a programme tried out at Sao Paulo and another Brazilian institution that is organized completely around a pedagogical structure of 'attitudinal objectives'. Beginning with the *Ultimate Speed* film, the students are invited to try to develop their own versions of possible connections between energy, momentum and speed. Similarly, they are asked to consider the implications of the *Time Dilation* filmed experiment. The Lorentz transformations *per se* are not emphasized. In fact, wealth of content is deliberately sacrificed to what is felt to be a valuable learning experience through trial-and-error, class discussion and guided speculation.

Kagan and Mendoza (*The Physics Teacher*, 1978, **16**, 225–7) describe their successful experience with a number of twelfth-grade high-school students in Israel, using a deliberately non-historical approach based upon experiments in relativistic particle dynamics (along much the same lines as PSSC and Haber–Schaim).

Various teachers, especially at university level, have taken steps to draw upon the enormous wealth of bubble-chamber photographs, accumulated by high-energy particle physics research groups, to provide real data for students to work on. By measuring track curvatures and ranges, etc., students can see the workings of relativistic dynamics at first hand.

Duboc (*Bulletin de l'Union des Physiciens*, 1974, **569**, 139–71; 1975, **577**, 43–73) describes in detail a programme along these lines that was successfully conducted with high-school students at a number of lycées in France, using bubble-chamber photographs obtained from C.E.R.N. These papers describe the experimental arrangements, show a number of photographs of different kinds of events, and give examples of the results of their analysis by students.

A particularly interesting and original approach has been developed by R. Sexl (1976) around the existence of atomic clocks, which provide an accuracy of time measurement quite unattainable previously. Synchronization of widely separated clocks (perhaps on different continents) by exchange of radio signals has become a reality instead of a hypothetical procedure, and gives direct verification that the time of transit is the same with or against the motion of the Earth. Time dilation has been directly measured for caesium clocks carried on aircraft at 550 km/h. Dynamical effects (mass–energy equivalence, variation of mass with speed) are then inferred from the kinematic results. This approach is embodied in *Relativitätstheorie* by Sexl and Schmidt (1978).

Although the above remarks are concerned with relatively recent developments, it is appropriate to end this epilogue with a salute to Max Born's superb semipopular book, *Einstein's Theory of Relativity* (1924, reprinted Dover Publications, 1962). Although not a textbook, this is an authoritative presentation at a level corresponding closely to that of introductory courses at secondary or early university level. In fact one can find in Born's book the essential groundwork for most of the textbook treatments that were developed many decades later, and it can still be highly recommended as one of the best expositions of relativity theory at an elementary level.

A. P. F.

EINSTEIN'S LETTERS

Introduction

Einstein's correspondence was very voluminous and extraordinarily wide-ranging. Amongst the people with whom he exchanged letters were Anatole France, Thomas Mann, Albert Schweitzer, H. G. Wells, and Martin Buber. His correspondence with major figures in politics and world government included Mahatma Gandhi, Dag Hammarskjøld, Thomas Masaryk and Adlai Stevenson. Much of this correspondence has been published elsewhere, and all of it will no doubt be published in due course.

In this book we limit ourselves to selections from Einstein's correspondence with a few of his closest acquaintances, mostly on matters relating to physics.

Einstein writes to his best friend

P. Speziali

THE FRIENDSHIP

On 23 June 1918 Albert Einstein wrote a letter from Berlin to his friend, Michele Angelo Besso. It began with these words: 'When I see your handwriting, I am always glad in a special way, for nobody else is so close to me, nobody knows me so well, nobody is so kindly disposed towards me as you are.' These words, born of affection, receive clear confirmation in the long-continuing epistolary dialogue that the two friends maintained in a regular and assiduous way until their lives ended, only a few weeks apart.

The 110 letters by Einstein and the 119 by Besso that remain to us from this voluminous correspondence* deal with a great variety of subjects: science, philosophy, religion, literature, politics, economics, personal matters, current events, and many other things. Thanks to these letters, several aspects of Einstein's life, work, and also his character come through to us more clearly and sometimes even in new perspectives. Thanks to them, also, we learn about the important role played by Besso at several points in the life of his distinguished friend and the services that he thereby rendered to science. One of these services was already known in another way, long before the publication of the letters in question. At the end of the 1905 paper, 'On the Electrodynamics of Moving Bodies', in the *Annalen der Physik* in which Einstein first published his special theory of relativity, one reads: 'In closing, I wish to say that my friend and colleague, M. Besso, has constantly lent his valuable advice while I was working on this problem, and that I am indebted to him for many interesting suggestions.'

Michele Besso was six years older than Einstein. The first of five children of Joseph Besso and Erminia Cantoni, he was born near Zürich on 25 May 1873. His family came originally from Trieste, where they were actively engaged in the insurance business. In 1879 Joseph Besso obtained Swiss nationality for himself and his children. Michele first studied at Trieste, and then at the University of Rome where he took courses in mathematics and physics. On the advice of his uncle David, who taught mathematics at the University of Modena (Italy), he left for Zürich and enrolled, in October 1891, in the mechanics section of the Federal Polytechnic School (ETH).

* *Albert Einstein–Michele Besso Correspondance 1903–1955* (edited by P. Speziali) (Paris: Hermann, 1972).

There he took courses from Frobenius, Hurwitz, Fiedler, and Stodola (among others). After four years of brilliant study he obtained his diploma in mechanical engineering and, soon afterwards, a position in an electrical-machinery factory in Zürich.

Michele Besso came frequently to Zürich to attend musical soirées—he played the violin—and there he first met Einstein. The latter, then seventeen years old, was lodging (with his sister Maja) with the Winteler family and was about to enter the Polytechnic School. The two young men met frequently and became friends, drawn together by the same tastes and the same thirst for knowledge. Thanks to Einstein, Besso made the acquaintance of the Winteler family—to such an extent that in 1898 he married the older daughter, Anna, by whom he had a son, Vero. (We may also note that in 1910 Maja Einstein married Paul Winteler, the younger son in the family.)

We next find the Besso family at Milan and Trieste, but Michele, through the efforts of his friend Albert, obtained a position in Bern, at the Federal Patent Office, where he became Albert's colleague from 1904 to 1908. Thereafter, however, their paths separated. Einstein remained a further year at Bern before being appointed Professor of Theoretical Physics at the University of Zürich; after that he went to Prague, then back to Zürich (this time to the Polytechnic), and thence to Berlin, at the Kaiser Wilhelm Institute. The rest—such as the travels that took him to South America and to Japan—is well known. Through all this he kept in close touch with his friend, who himself worked for a while as a technical consultant at Gorizia, near Trieste, then came back to Switzerland to work on patents—from 1916 to 1938 he was a *privatdozent* at the Polytechnic in this field—and finally returned to the Federal Patent Office, where he remained until his retirement. From 1939 until his death on 15 March 1955 he lived at Geneva with his son's family. Einstein, before he finally settled at Princeton, frequently returned to Switzerland on family business; each time he would, without fail, have a reunion with his friend, who was, as we have said, his closest and perhaps his only true confidant.

We may remark that Besso published a score of articles on topics in economics, industrial organization, patents, the biological theory of heredity, the physics of fluids, and the geometric structure of electronic shells. Einstein, in his letter of condolence to Besso's son and sister (sent from Princeton on 21 March 1955), dwelt in moving terms on his relationship with Michele: 'The circle of his interests seemed truly without limits,' and, further on, 'And now he has been ahead of me once again, in leaving this strange world.'

2. THE LETTERS

The chief purpose of this article is to present a selection of typical extracts from Einstein's letters to Besso, together with one letter in its entirety. This choice, quite arbitrarily made, though exclusively limited to scientific matters, should in our opinion suffice to give an idea of the richness of this correspondence. Here, to begin with, a sentence from a letter early in March 1914, sent from Zürich to Gorizia: 'I have been working like a madman—and, which is what matters, with great success.' What was this about?

> I have succeeded in demonstrating, by a simple calculation, that the equations of gravitation are valid for every reference system obeying these conditions [a set of four third-order equations for the basic tensor components of general relativity theory]. From this it follows that there are many different kinds of transformation for acceleration, which transforms the equations into themselves (through a rotation, for example) in such a way that the equivalence principle is preserved in its most basic form, indeed to an unsuspected and far-reaching extent.
>
> I believe that, at the time of your visit, I had already demonstrated the rigorous equivalence between inertial and gravitational mass, and also that of the gravitational field. Now I am completely convinced, and I no longer doubt the validity of the whole system, whether or not the solar eclipse observations succeed. The logic of the whole thing is just too clear.

Other equally important results are announced on 15 February 1915:

> Gravitation. Displacement of spectral lines towards the red. The members of a spectroscopic binary star have the same mean speed along the line of sight. The masses of the stars are obtained from the periodic Doppler shifts. The heavier member of the binary must show a larger *average* red-shift than the lighter member. *This is confirmed.* Since one can estimate the radii of the stars (from their spectral type) there follows an approximate quantitative verification of the theory, which is thus shown to be satisfactory.

Let us jump ahead to 1 March 1931. Einstein has just returned from the United States, thrilled by his visit to the Mount Wilson Observatory:

> The trip to America was very interesting, though also very tiring. The people at the Mount Wilson Observatory are outstanding. They have recently found that the spiral nebulae are distributed approximately uniformly in space, and that they show a strong Doppler effect, proportional to their distances, that one can readily deduce from general relativity theory (without the 'cosmological' term). The snag, however, is the fact that the expansion of the universe extrapolates back to an origin in time, 10^{10} or 10^{11} years ago. Since any other explanation of these Doppler shifts leads to grave difficulties, the situation is truly exciting.

One very interesting passage concerns the influence of Ernst Mach on the evolution of Einstein's thinking. It comes from a letter sent from Princeton on 6 January 1948:

> As to Mach, I must make a distinction between his general influence and the effect he had on me personally. Mach achieved some important results (for example, the discovery of shock waves, which is based on a

truly ingenious optical method). However, let us not talk about that, but about his influence on the general attitude to the fundamentals of physics. His great merit was to soften the dogmatism that reigned in these matters during the 18th and 19th centuries. He tried to show, especially in mechanics and the theory of heat, that the concepts are born of experience. He convincingly defended the point of view according to which these concepts—even the most fundamental—derive their justification only from experience and that they are not in any way necessary from a *logical* standpoint. His approach was particularly valuable when he showed clearly that the most important problems in physics are not mathematical deductive in nature, but are those that relate to basic principles. His weakness, as I see it, lies in the fact that he believed, more or less strongly, that science consists only of putting experimental results in order; that is, he does not recognise the free constructive element in the creation of a theory. He thought that theories are somehow the result of a *discovery* and not of an *invention*. He even went so far as to consider 'sensations' to be not only a kind of substance, but also, in a certain sense as the building blocks of the real world; he believed that in this way he could bridge the gulf between psychology and physics. If he had been truly consistent, he would have had to reject not only atomism but also the whole idea of physical reality.

As for the influence of Mach on my own thinking, it has certainly been very great. I remember very well how, during my early years as a student you directed my attention to his treatise on mechanics and to his theory of heat, and how these two works made a deep impression on me. Frankly, however, I cannot clearly see to what extent they affected my own work. So far as I can recall, David Hume had a greater direct influence on me; I read him at Bern in the company of Conrad Habich and Solovine. However, as I have just said, I am in no position to analyse what is buried in my subconscious. For the rest, it is interesting to note that Mach vehemently rejected the special theory of relativity. (He did not live to see the general theory.) The theory seemed to him excessively speculative. He did not realise that this speculative character applies also to Newtonian mechanics and, in general, to every conceivable theory. There is only a difference of degree between the theories, in the extent to which the paths of thought from basic principles to experimentally verifiable consequences are different in length and complexity.

On 28 February 1952, after re-reading the 'Autobiographical Notes' in *Albert Einstein: Philosopher-Scientist* (1949), Besso asked for an explanation of a passage in which Einstein had written (*inter alia*) these words:* 'A theory can be tested by experience, but there is no way from experience to the construction of a theory. Equations of such complexity as those of the gravitational field can be found only through the discovery of a logically

* See also the letter to Maurice Solovine on page 271.

imple mathematical condition that determines the equations completely or almost completely. Once one has those sufficiently strong formal conditions, one needs only a small amount of factual knowledge for the construction of a theory.'

In the following month, in a reply dated 20 March 1952, one finds the following amplification:

> The remark on p. 88 means this: A vast collection of facts is essential for the establishment of any theory that is to have a chance of success. But this material does not of itself constitute a starting point for a deductive theory. However, with the help of this material, one may succeed in finding a general principle that can be the starting point for a logical (deductive) theory. But there is no *logical* path leading from the empirical material to the general principle on which the logical deduction will then rest.
>
> Thus, I do not believe in Mill's path to knowledge via induction—at least, not as a logical method. For instance, I do not think that there is any experience from which one can deduce the concept of number.
>
> The further theory progresses, the clearer it becomes that one cannot discover fundamental laws by induction, starting from facts of experience (for example, the field equations of gravitation or Schrödinger's equation in quantum mechanics). In general one can say: the path that leads from the particular to the general is an intuitive one; that which leads from the general to the particular is logical.

And now, to conclude, the letter by Einstein that exceeds all the preceding ones in length. It is also the last letter that he sent to his old friend, and it provides us with a veritable scientific testament. It was written in reply to letter that Besso (still intellectually very alive at eighty-one) wrote to the seventy-five-year-old Einstein, asking for comments on a brief statement of his (Besso's) view of the essential content of general relativity.

(Princeton) August 10th, 1954.

Dear Michele:

Your picture of the general theory of relativity characterizes its genetic aspects very well. However, it is also valuable to analyse the whole thing from the standpoint of formal logic. For if mathematical difficulties make the empirical content of the theory temporarily inaccessible, logical simplicity (even though it is not by itself sufficient) becomes the only criterion of the value of the theory.

The special theory of relativity is basically nothing more than a grafting of the idea of an inertial system onto the firm conviction, supported by experiment, of the constancy of the speed of light for every inertial system. The theory cannot dispense with the concept of an inertial system, a concept that cannot be supported from the viewpoint of the theory of knowledge. (The inconsistency of this concept was illuminated very clearly by Mach, but it had already been recognizec with less clarity by Huygens and Leibniz.)

The core of this objection with respect to Newtonian principles is bes clarified by a comparison with the 'centre of the universe' in Aristotelear physics: there exists a centre towards which all heavy bodies tend. In thi way, for example, one accounts for the spherical shape of the earth. Th objectionable feature is that the centre of the universe acts on all the rest but that the rest (i.e., material bodies) does not react on the centre (one-way causal connection).

This is what happens with the inertial system. It determines the inertia properties of bodies without itself being influenced. (Basically, it woul be better to speak of the totality of inertial frames, but this is not essential. The essence of the general theory of relativity is to go beyond th inertial system. (At the time when general relativity was being created this was not so clear, but it was recognised later, principally by Levi Civita.) In constructing the theory, I chose as a starting point the sym metric tensor g_{ik}. This provided the possibility of defining the 'displace ment field' Γ^{l}_{ik} which, for each vector at a point P, determines a vecto at an infinitesimally distant point P' ($\delta A^{\nu} = -\Gamma^{\nu}_{\sigma\tau} A^{\sigma} \mathrm{d}x_{\tau}$).

This concept of a displacement field is in itself independent of th existence of a metric field g_{ik}; the fact that it was initially introduced onl in connection with the metric field came about because Riemann starte from the Gaussian theory of the curvature of surfaces, by which surface acquires a metric through being immersed in Euclidean space.

But why is it the displacement field that allows one to be freed fror the obstacle of the inertial system? If, in an inertial system, one has two points P and Q, separated by an arbitrary distance, two vectors wit the same components, there is an objective (invariant) connection: th vectors are equal and parallel. It follows that, in an inertial system, th differentiation of a tensor with respect to the coordinates yields anoth tensor, and that, for example, the wave equation represents an objectiv proposition in inertial systems. The displacement field allows one t construct such tensors by differentiation with respect to an arbitrar coordinate system. It is thus the invariant substitute for the inertial syste —and, thereby, the foundation for every relativistic field theory.

If one introduces the displacement field as the fundamental fie magnitude, it determines a curvature tensor through the invariant act displacing a vector along the boundary of an infinitesimal element of t surface. Thus there belong to the field Γ (which itself does not have t character of a tensor) the curvature tensors R^{i}_{klm} and R_{kl}.*

To obtain the field equations, it is best to use the variational metho as this always yields 4 identities between the field equations, which a

* See the article by Hermann Bondi on page 113.

necessary for the compatibility of a relativistic system of equations. To construct the scalar magnitude that one needs for the variational integral, one must have a tensor g_{kl} (or g^{kl}) which with the R_{kl} gives the scalar $g^{kl}R_{kl}$. This is the formal reason why one needs a tensor in addition to the Γ^l_{ik}.

The theory of the pure gravitational field is obtained in this way, if one chooses Γ^l_{ik} (in its lower indices) and also g_{ik} to be symmetric, which is significant from the point of view of invariance.

On the other hand, it is evident from the definition of infinitesimal displacement that there is no particular reason to choose the Γ-field to be symmetric in its lower indices. The more general condition would then imply a need to make the g_{ik}-field non-symmetric. This then leads to the theory of the non-symmetrical field without any choice.

The reason why I do not know if this theory corresponds to physical reality resides solely in the fact that it has not been possible to say anything about either the existence or the structure of solutions, everywhere free of singularities, of such systems of non-linear equations.

However, one must not think that this theory would be determined only by the requirements of relativity. In the usual theory of gravitation one has, for example, a right-hand side that represents the field-producing and the field-influenced masses. Field-theoretically one would be introducing a second supplementary invariant of the field.

Such a procedure would necessitate the introduction of new kinds of field, independent of the quantities Γ. Moreover, the sign of the additional invariant could be chosen arbitrarily, so that, for example, one would never know why gravitating masses all have the same sign. In brief, one would be combining expressions that have no logical connection with one another. I am sufficiently optimistic to be convinced that our universe is not patched together in this fashion.

In this sense, the theory is quite sufficiently and uniquely determined by the requirements of relativity. I concede, however, that it is quite possible that physics cannot be founded on the concept of field—that is to say, on continuous elements. But then, out of my whole castle in the air—including the theory of gravitation, but also most of current physics—there would remain almost *nothing*.

With warm regards,

Your A.E.

Mit der erkenntnis-theoretischen Sache haben Sie mich gründlich missverstanden; wahrscheinlich habe ich mich schlecht ausgedrückt. Ich sehe die Sache schematisch so

A, System der Axiome

S S' S'' gefolgerte Sätze

E — Mannigfaltigkeit der unmittelbaren (Sinnen-) Erlebnisse

(1) Die E (Erlebnisse) sind uns gegeben.

(2) A sind die Axiome, aus denen wir Folgerungen ziehen. Psychologisch beruhen die A auf E. Es gibt aber keinen logischen Weg von den E zu A, sondern nur einen intuitiven (psychologischen) Zusammenhang, der immer „auf Widerruf" ist.

(3) Aus A werden auf logischem Wege Einzel-Aussagen S abgeleitet, welche Ableitungen den Anspruch auf Richtigkeit erheben können.

(4) Die S werden mit den E in Beziehung gebracht (Prüfung an der Erfahrung). Diese Prozedur gehört genau betrachtet ebenfalls der extra-logischen (intuitiven) Sphäre an, weil die Beziehung der in den S auftretenden Begriffe zu den Erlebnissen E nicht logischer Natur sind.

Diese Beziehung der S zu den E ist aber (pragmatisch) viel weniger unsicher als die Beziehung der A zu den E. (Beispiel der Begriff Hund und die entsprechenden Erlebnisse). Wäre solches Entsprechen nicht mit grosser Sicherheit erzielbar, (obwohl nicht logisch fassbar), so wäre die logische Maschinerie für das „Begreifen der Wirklichkeit" völlig wertlos (Beispiel Theologie). —

Die Quintessenz ist der ewig problematische Zusammenhang alles Gedanklichen mit dem Erlebbaren (Sinnen-Erlebnisse).

Die Arbeit für den De Broglie-Jubiläums-Band wird von den dortigen Kollegen selbst ins Französische übersetzt. Der Inhalt ist für die Leute aber schlimmste Ketzerei. Ich kann es Ihnen erst schicken, wenn es gedruckt ist.

Mir geht es allen recht ordentlich. Mein Arbeitsvermögen hat aber schon empfindlich abgenommen. Nun, es hat auch sein Gutes. Ende alles Herzliche

Ihr A. E.

Der Seelig scheint ein sehr lieber Mensch zu sein, das merke ich aus seinem ganzen Verhalten, ohne ihn persönlich zu kennen.

Figure 35 Facsimile of part of Einstein's letter to Maurice Solovine, 7 May 1952

Letter to Maurice Solovine, 7 May 1952

Dear Solo:

In your letter you reproach me for having committed two offences. The first is to have taken an uncritical attitude towards the plan for world government. It is not that you consider this as *undesirable*, but rather as unattainable in the near future. You give good reasons for its impracticality. You might well have added, as a good reason for apprehension, that world government might be intolerable and, in particular, more unjust than the present state of anarchy. . . . On the other hand there exists the danger of self-annihilation by mankind, a matter that should be of heavy concern to us all. Therefore one should at least (even if with some hesitancy) retract the 'undesirable'.

As for the 'impossible', one can say this: it changes into 'possible' if people *really want it*, if only through fear of an intolerable state of insecurity. We need to put out all our strength to bring about such a condition voluntarily. This effort is useful, even if the goal is not attained, for it will certainly have a good educational effect, provided that it is directed against stupid and pernicious nationalism.

Now, you say that one should first educate youth to examine historical events objectively. Only thus could one hope to achieve something in the political realm. But this priority is like that of the hen and the egg; that is to say, we are in a vicious circle. The hen is the political order, and the egg is rational education. Since there is no free end to this tangle, from which we could unravel it, we must try everything and not lose courage in the process.

But if all our efforts are in vain, and mankind does go down to self destruction, the universe will shed no tears for us. . . .

With regard to the epistemological question, you have radically misunderstood me; I probably expressed myself badly. I see the matter schematically like this:

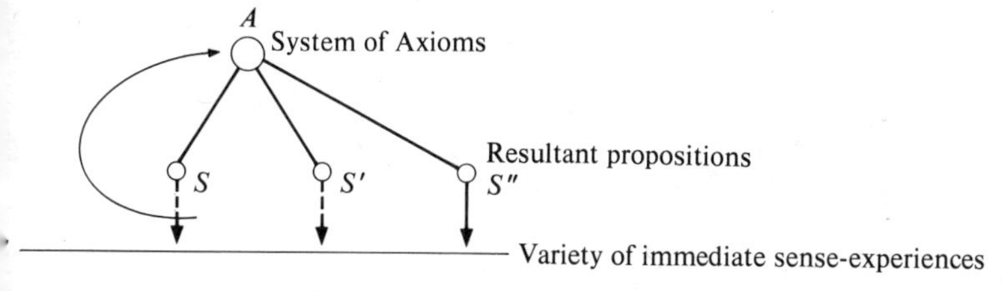

(1) The E (direct experiences) are given to us.
(2) A are the axioms, from which we draw consequences. Psycho-logically the A rest on the E. But there exists no logical path leading from the E to A, only an intuitive (psychological) connection, whicl is always merely 'until further notice'.
(3) From the A are deduced, *by a logical path*, particular assertions S tha can claim to be exact.
(4) The S are brought into relation with the E (testing by experience) This procedure belongs also to the extra-logical (intuitive) sphere because the connection[s] between the concepts appearing in S an the immediate experiences E are not of a logical nature.
But this relationship between the S and the E is (pragmatically) muc less uncertain than the relation of the A to the E. (Example: the con cept 'dog' and the corresponding direct experiences.) If such a cor respondence could not be achieved with great certainty (even thoug it is not logically 'grasped' a bit better), the logical machiner would be of no value at all for the 'comprehension of reality' (fo example, theology).

The quintessence is the always problematical connection between th world of ideas and that which can be experienced (direct sense-experiences

We are all quite well. But my capacity for work has perceptibl decreased. Ah well, that also has its good side.

Best regards to you.
Yours, A.E.

Letter from Bohr to Einstein

Copenhagen
11 November, 1922

Dear Professor Einstein:

I should like to convey my most heartfelt congratulations to you on he award of the Nobel Prize. The external recognition cannot mean much o you, but the associated funds will perhaps bring about an easing of your working conditions.

It was for me the greatest honour and pleasure that external circum-tances resulted in my being considered for the award of the prize at the ame time as you.* I know how little I have deserved this, but I did want o say that I have felt it as great good fortune that—quite apart from your great contribution to the world of human thought in general—the funda-mental contribution that you made to the more special field in which I am working should have been publicly recognised, as were those of Rutherford nd Planck, before I myself was considered for such an honour.

With most heartfelt greetings to you and your wife from my wife and myself.

Your devoted
N. Bohr

* This no doubt refers to the fact that the choice of Einstein to receive the Nobel Prize for 921 was not announced until a year later, along with the announcement of Bohr's omination for the 1922 prize.

Letter from Einstein to Bohr

near Singapore
11 January, 1923

Dear (or rather beloved) Bohr!

Your cordial letter reached me shortly before my departure from Japan I can say without exaggeration that it gave me as much pleasure as the Nobel Prize. I find especially charming your concern lest you might have received the prize before I did—that is truly 'Bohrish'. Your new investigations on the atom have accompanied me on my travels and have further increased my regard for your intellect. I believe that I have finally understood the connection between electricity and gravitation. Eddington ha come closer to the truth than Weyl.

The trip is splendid. I am charmed by Japan and the Japanese and am sure that you would be too. Moreover, a sea voyage like this is a delightfu existence for a dreamer—it is like a cloister. . . .

Hearty greetings. I look forward to seeing you again, at the latest in Stockholm.

Yours in admiration,
A. Einstein

Letters to Max Born

Max Born was a very distinguished German physicist who held the chair of theoretical physics at Göttingen, and later the Tait Chair of Natural Philosophy at the University of Edinburgh. He was the originator of the now accepted interpretation of the wave function in Schrödinger's wave mechanics. He was a close acquaintance of Einstein in Germany, and they kept in touch by letter throughout their lives.

9 SEPTEMBER 1920

'Don't be too hard on me. Everyone has to sacrifice at the altar of stupidity from time to time, to please the Deity and the human race.'

29 APRIL 1924

'. . . I should not want to be forced into abandoning strict causality without defending it more strongly than I have so far. I find the idea quite intolerable that an electron exposed to radiation should choose *of its own free will*, not only its moment to jump off, but also its direction. In that case, I would rather be a cobbler, or even an employee in a gaming-house, than a physicist. Certainly my attempts to give tangible form to the quanta have foundered again and again, but I am far from giving up hope. And even if it never works there is always that consolation that this lack of success is entirely mine.'

4 DECEMBER 1926

'Quantum mechanics is certainly imposing. But an inner voice tells me that it is not yet the real thing. The theory says a lot, but does not really bring us any closer to the secret of the "old one". I, at any rate, am convinced that *He* is not playing at dice.'

15 JANUARY 1927

'What applies to jokes, I suppose, also applies to pictures and to plays. I think they should not smell of logical scheme, but of a delicious fragment of life, scintillating with various colours according to the position of the beholder. If one wants to get away from this vagueness one must take up mathematics. And even then one reaches one's aim only by becoming completely insubstantial under the dissecting knife of clarity. Living matter and clarity are opposites—they run away from one another. We are now experiencing this rather tragically in physics.'

7 SEPTEMBER 1944

'We have become Antipodean in our scientific expectations. You believe in the God who plays dice, and I in complete law and order in a world which objectively exists, and which I, in a wildly speculative way, am trying to capture. I firmly *believe*, but I hope that someone will discover a more realistic way, or rather a more tangible basis than it has been my lot to find. Even the great initial success of the quantum theory does not make me believe in the fundamental dice-game, although I am well aware that our younger colleagues interpret this as a consequence of senility. No doubt the day will come when we will see whose instinctive attitude was the correct one.'

12 APRIL 1949

'. . . you ask me what my attitude is towards the simple life. I simply enjoy giving more than receiving in every respect, do not take myself nor the doings of the masses seriously, am not ashamed of my weaknesses and vices, and naturally take things as they come with equanimity and humour. Many people are like this, and I really cannot understand why I have been made into a kind of idol. I suppose it is just as incomprehensible as why an avalanche should be triggered off by *one particular* particle of dust, and why it should take a certain course.'

15 SEPTEMBER 1950

'There is nothing analogous in relativity to what I call incompleteness o description in the quantum theory. Briefly it is because the ψ-function is incapable of describing certain qualities of an individual system, whose "reality" we none doubt (such as a macroscopic parameter).

Take a (macroscopic) body which can rotate freely about an axis. It state is fully determined by an angle. Let the initial conditions (angle and angular momentum) be defined as precisely as the quantum theory allows The Schroedinger equation then gives the ψ-function for any subsequen time interval. If this is sufficiently large, all angles become (in practice) equally probable. But if an observation is made (e.g. by flashing a torch) a definite angle is found (with sufficient accuracy). This does not prove that the angle had a definite value before it was observed—but we believe this to be the case, because we are committed to the requirements o reality on the macroscopic scale. Thus, the ψ-function does not expres the real state of affairs perfectly in this case. This is what I call "incomplete description".

So far, you may not object. But you will probably take the position that a complete description would be useless because there is no mathe matical relationship for such a case. I do not say that I am able to disprove this view. But my instinct tells me that a complete formulation of th relationships is tied up with complete description of its factual state. I am convinced of this although, up to now, *success* is against it. I also believ that the current formulation is true in the same sense as e.g. thermo dynamics, i.e. as far as the concepts used are adequate. I do not expect to convince you, or anybody else—I just want you to understand the way I think.

I see from the last paragraph of your letter that you, too, take the quantum theoretical description as incomplete (referring to an ensemble). But you are after all convinced that no (complete) laws exist for a complete description, according to the positivistic maxim *esse est percipi*. Well, this is a programmatic attitude, not knowledge. This is where our attitudes really differ. For the time being, I am alone in my views—as Leibniz was with respect to the absolute space of Newton's theory.'

MAX BORN'S COMMENTS

This is probably the clearest presentation of Einstein's philosophy of reality. The last but one paragraph is particularly revealing. He calls my way of describing the physical world 'incomplete'; in his eyes this is a flaw which he hopes to see removed, while I am prepared to put up with it. I have in fact always regarded it as a step forward, because an exact description of the state of a physical system presupposes that one can make statements of infinite precision about it, and this seems absurd to me.

28 NOVEMBER 1954 (To Einstein from Max Born)

'I read in the paper recently that you are supposed to have said: "If I were born a second time, I would become not a physicist, but an artisan". These words were a great comfort to me, for similar thoughts are going around in my mind as well, in view of the evil which our once so beautiful science has brought upon the world. . . .'

17 JANUARY 1955

'. . . What I wanted to say was just this: In the present circumstances, the only profession I would choose would be one where earning a living had nothing to do with the search for knowledge.'

EINSTEIN'S WRITINGS

Introduction

This section of the book contains a selection of Einstein's writings, an is intended to illustrate Einstein's articulateness both as a commentator o human problems and as an expositor of his scientific discoveries.

The selection is arranged chronologically, beginning with the first par of his most famous paper. The paper concerning a new derivation of $E=mc$ illustrates Einstein's ability to construct a simple pedagogic argument.

Since Einstein has been so widely regarded as a man who lived in a abstruse theoretical world of his own making, it is particularly interestin to read his essay on the very down-to-earth problem of the formation o meanders in rivers.

'On the Method of Theoretical Physics' should be read in conjunctio with the memoir by Abraham Pais, on page 35.

'On Education' should be read in conjunction with 'Einstein an Education' by Arturo Loria, on page 215.

On the electrodynamics of moving bodies

Einstein's most famous paper—his 1905 paper introducing the special theory of relativity—appeared in the Annalen der Physik *under the title* 'Zur Elektrodynamik bewegter Körper' (*On the Electrodynamics of Moving Bodies*). *It consists of a brief introduction followed by two main parts: one developing the new relativistic transformations for space, time and velocity (including what later became famous as the 'twin paradox'), the other a specific application of the theory to electrodynamics.*

The whole paper is a model of simplicity and clarity, and contains surprisingly little in the way of complicated mathematics. We reproduce here a translation of the introduction and Part I (kinematical part).

It is known that Maxwell's electrodynamics—as usually understood at the present time—when applied to moving bodies, leads to asymmetries which do not appear to be inherent in the phenomena. Take, for example, the reciprocal electrodynamic action of a magnet and a conductor. The observable phenomenon here depends only on the relative motion of the conductor and the magnet, whereas the customary view draws a sharp distinction between the two cases in which either the one or the other of these bodies is in motion. For if the magnet is in motion and the conductor at rest, there arises in the neighbourhood of the magnet an electric field with a certain definite energy, producing a current at the places where parts of the conductor are situated. But if the magnet is stationary and the conductor in motion, no electric field arises in the neighbourhood of the magnet. In the conductor, however, we find an electromotive force, to which in itself there is no corresponding energy, but which gives rise—assuming equality of relative motion in the two cases discussed—to electric currents of the same path and intensity as those produced by the electric forces in the former case.

Examples of this sort, together with the unsuccessful attempts to discover any motion of the earth relatively to the 'light medium', suggest that the phenomena of electrodynamics as well as of mechanics possess no properties corresponding to the idea of absolute rest. They suggest rather that, as has already been shown to the first order of small quantities, the same laws of electrodynamics and optics will be valid for all frames of reference for which the equations of mechanics hold good. We will raise this conjecture (the purport of which will hereafter be called the 'Principle of Relativity') to the status of a postulate, and also introduce another postulate, which is only apparently irreconcilable with the former, namely, that light is always propagated in empty space with a definite speed c which is independent of the state of motion of the emitting body. These two postulates suffice for the attainment of a simple and consistent theory of the electrodynamics of moving bodies based on Maxwell's theory for stationary bodies. The introduction of a 'luminiferous ether' will prove to be superfluous inasmuch as the view here to be developed will not require an 'absolutely stationary space' provided with special properties, nor assign a velocity-vector to a point of the empty space in which electromagnetic processes take place.

The theory to be developed is based—like all electrodynamics—on the kinematics of the rigid body, since the assertions of any such theory have to do with the relationships between rigid bodies (systems of coordinates), clocks, and electromagnetic processes. Insufficient consideration of this circumstance lies at the root of the difficulties which the electrodynamics of moving bodies at present encounters.

KINEMATICAL PART

1. *Definition of simultaneity*

Let us take a system of coordinates in which the equations of Newtonian mechanics hold good.* In order to render our presentation more precise and to distinguish this system of coordinates verbally from others which will be introduced hereafter, we call it the 'stationary system'.

If a material point is at rest relatively to this system of coordinates, its position can be defined relatively thereto by the employment of rigid standards of measurement and the methods of Euclidean geometry, and can be expressed in Cartesian coordinates.

If we wish to describe the *motion* of a material point, we give the values of its coordinates as functions of the time. Now we must bear carefully in mind that a mathematical description of this kind has no physical meaning unless we are quite clear as to what we understand by 'time'. We have to take into account that all our judgements in which time plays a part are always judgements of *simultaneous events*. If, for instance, I say, 'That train arrives here at 7 o'clock,' I mean something like this: 'The pointing of the small hand of my watch to 7 and the arrival of the train are simultaneous events.'†

It might appear possible to overcome all the difficulties attending the

* i.e. to the first approximation.

† We shall not here discuss the inexactitude which lurks in the concept of simultaneity of two events at approximately the same place, which can only be removed by an abstraction.

definition of 'time' by substituting 'the position of the small hand of my watch' for 'time'. And in fact such a definition is satisfactory when we are concerned with defining a time exclusively for the place where the watch is located; but it is no longer satisfactory when we have to connect in time a series of events occurring at different places, or—what comes to the same thing—to evaluate the times of events occurring at places remote from the watch.

We might, of course, content ourselves with time values determined by an observer stationed together with the watch at the origin of the coordinates, and coordinating the corresponding positions of the hands with light signals, given out by every event to be timed, and reaching him through empty space. But this coordination has the disadvantage that it is not independent of the standpoint of the observer with the watch or clock, as we know from experience. We arrive at a much more practical determination along the following line of thought.

If at the point A of space there is a clock, an observer at A can determine the time values of events in the immediate proximity of A by finding the positions of the hands which are simultaneous with these events. If there is at the point B of space another clock in all respects resembling the one at A, it is possible for an observer at B to determine the time values of events in the immediate neighbourhood of B. But it is not possible without further assumption to compare, in respect of time, an event at A with an event at B. We have so far defined only an 'A time' and a 'B time'. We have not defined a common 'time' for A and B; the latter time can now be defined in establishing *by definition* that the 'time' required by light to travel from A to B equals the 'time' it requires to travel from B to A. Let a ray of light start at the 'A time' t_A from A towards B, let it at the 'B time' t_B be reflected at B in the direction of A, and arrive again at A at the 'A time' t_A'.

In accordance with definition the two clocks synchronize if

$$t_B - t_A = t_A' - t_B$$

We assume that this definition of synchronism is free from contradictions, and possible for any number of points; and that the following relations are universally valid:

1. If the clock at B synchronizes with the clock at A, the clock at A synchronizes with the clock at B.
2. If the clock at A synchronizes with the clock at B and also with the clock at C, the clocks at B and C also synchronize with each other.

Thus with the help of certain imaginary physical experiments we have settled what is to be understood by synchronous stationary clocks located at different places, and have evidently obtained a definition of 'simultaneous', or 'synchronous', and of 'time'. The 'time' of an event is that which is given simultaneously with the event by a stationary clock located at the place of the event, this clock being synchronous, and indeed synchronous for all determinations, with a specified stationary clock.

In agreement with experience we further assume the quantity

$$\frac{2\mathrm{AB}}{t_{\mathrm{A}}'-t_{\mathrm{A}}}=c$$

to be a universal constant—the speed of light in empty space.

It is essential to have time defined by means of stationary clocks in the stationary system, and the time now defined being appropriate to the stationary system we call it 'the time of the stationary system'.

2. *On the relativity of lengths and times*

The following reflections are based on the principle of relativity and on the principle of the constancy of the speed of light. These two principles we define as follows:

1. The laws by which the states of physical systems undergo change are not affected, whether these changes of state be referred to the one or the other of two systems of coordinates in uniform translatory motion.
2. Any ray of light moves in the 'stationary' system of coordinates with the determined speed c, whether the ray be emitted by a stationary or by a moving body. Hence

$$\text{speed}=\frac{\text{light path}}{\text{time interval}}$$

where time interval is to be taken in the sense of the definition in Section 1.

Let there be given a stationary rigid rod; and let its length be l as measured by a measuring-rod which is also stationary. We now imagine the axis of the rod lying along the axis of x of the stationary system of coordinates, and that a uniform motion of parallel translation with velocity v along the axis of x in the direction of increasing x is then imparted to the rod. We now inquire as to the length of the moving rod, and imagine its length to be ascertained by the following two operations:

(*a*) The observer moves together with the given measuring-rod and the rod to be measured, and measures the length of the rod directly by superposing the measuring-rod, in just the same way as if all three were at rest.

(*b*) By means of stationary clocks set up in the stationary system and synchronizing in accordance with Section 1, the observer ascertains at what points of the stationary system the two ends of the rod to be measured are located at a definite time. The distance between these two points, measured by the measuring-rod already employed, which in this case is at rest, is also a length which may be designated 'the length of the rod'.

In accordance with the principle of relativity the length to be discovered by the operation (*a*)—we will call it 'the length of the rod in the moving system'—must be equal to the length l of the stationary rod.

The length to be discovered by the operation (*b*) we will call 'the length of the (moving) rod in the stationary system'. This we shall determine on the basis of our two principles, and we shall find that it differs from l.

Current kinematics tacitly assumes that the lengths determined by these two operations are precisely equal, or in other words, that a moving rigid body at the epoch t may in geometrical respects be perfectly represented by *the same* body *at rest* in a definite position.

We imagine further that at the two ends A and B of the rod, clocks are placed which synchronize with the clocks of the stationary system, that is to say that their indications correspond at any instant to the 'time of the stationary system' at the places where they happen to be. These clocks are therefore 'synchronous in the stationary system'.

We imagine further that with each clock there is a moving observer, and that these observers apply to both clocks the criterion established in Section 1 for the synchronization of two clocks. Let a ray of light depart from A at the time* t_A, let it be reflected at B at the time t_B, and reach A again at the time t_A'. Taking into consideration the principle of the constancy of the speed of light we find that

$$t_B - t_A = \frac{r_{AB}}{c-v} \text{ and } t_A' - t_B = \frac{r_{AB}}{c+v}$$

where r_{AB} denotes the length of the moving rod—measured in the stationary system. Observers moving with the moving rod would thus find that two clocks were not synchronous, while observers in the stationary system would declare the clocks to be synchronous.

So we see that we cannot attach any *absolute* significance to the concept of simultaneity, but that two events which, viewed from a system of coordinates, are simultaneous, can no longer be looked upon as simultaneous events when envisaged from a system which is in motion relatively to that system.

3. *Theory of the transformation of coordinates and times from a stationary system to another system in uniform motion of translation relatively to the former.* Let us in 'stationary' space take two systems of coordinates, i.e. two systems, each of three rigid material lines, perpendicular to one another, and issuing from a point. Let the axes of X of the two systems coincide, and their axes of Y and Z respectively be parallel. Let each system be provided with a rigid measuring-rod and a number of clocks, and let the

* 'Time' here denotes 'time of the stationary system' and also 'position of hands of the moving clock situated at the place under discussion'.

two measuring-rods, and likewise all the clocks of the two systems, be in all respects alike.

Now to the origin of one of the two systems (k) let a constant velocity v be imparted in the direction of the increasing x of the other stationary system (K), and let this velocity be communicated to the axes of the coordinates, the relevant measuring-rod, and the clocks. To any time of the stationary system K there will then correspond a definite position of the axes of the moving system, and from reasons of symmetry we are entitled to assume that the motion of k may be such that the axes of the moving system are at the time t (this 't' always denotes a time of the stationary system) parallel to the axes of the stationary system.

We now imagine space to be measured from the stationary system K by means of the stationary measuring-rod, and also from the moving system k by means of the measuring-rod moving with it; and that we thus obtain the coordinates x, y, z, and ξ, η, ζ respectively. Further, let the time t of the stationary system be determined for all points thereof at which there are clocks by means of light signals in the manner indicated in Section 1; similarly let the time τ of the moving system be determined for all points of the moving system at which there are clocks at rest relatively to that system by applying the method, given in Section 1, of light signals between the points at which the latter clocks are located.

To any system of values x, y, z, t, which completely defines the place and time of an event in the stationary system, there belongs a system of values ξ, η, ζ, τ, determining that event relatively to the system k, and our task is now to find the system of equations connecting these quantities.

In the first place it is clear that the equations must be *linear* on account of the properties of homogeneity which we attribute to space and time.

If we place $x'=x-vt$, it is clear that a point at rest in the system k must have a system of values x', y, z, independent of time. We first define τ as a function of x', y, z, and t. To do this we have to express in equations that τ is nothing else than the summary of the data of clocks at rest in system k, which have been synchronized according to the rule given in Section 1.

From the origin of system k let a ray be emitted at the time τ_0 along the X-axis to x', and at the time τ_1 be reflected thence to the origin of the coordinates, arriving there at the time τ_2; we then must have $\frac{1}{2}(\tau_0+\tau_2)=\tau_1$, or, by inserting the arguments of the function τ and applying the principle of the constancy of the speed of light in the stationary system:

$$\frac{1}{2}\left[\tau(0,0,0,t)+\tau\left(0,0,0,t+\frac{x'}{c-v}+\frac{x'}{c+v}\right)\right]=\tau\left(x',0,0,t+\frac{x'}{c-v}\right)$$

Hence, if x' be chosen infinitesimally small,

$$\frac{1}{2}\left(\frac{1}{c-v}+\frac{1}{c+v}\right)\frac{\partial\tau}{\partial t}=\frac{\partial\tau}{\partial x'}+\frac{1}{c-v}\frac{\partial\tau}{\partial t}$$

or

$$\frac{\partial\tau}{\partial x'}+\frac{v}{c^2-v^2}\frac{\partial\tau}{\partial t}=0$$

It is to be noted that instead of the origin of the coordinates we might have chosen any other point for the point of origin of the ray, and the equation just obtained is therefore valid for all values of x', y, z.

An analogous consideration—applied to the axes of Y and Z—it being borne in mind that light is always propagated along these axes, when viewed from the stationary system, with the speed $\sqrt{(c^2-v^2)}$, gives us

$$\frac{\partial\tau}{\partial y}=0, \qquad \frac{\partial\tau}{\partial z}=0$$

Since τ is a *linear* function, it follows from these equations that

$$\tau=a\left(t-\frac{v}{c^2-v^2}x'\right)$$

where a is a function $\phi(v)$ at present unknown, and where for brevity it is assumed that at the origin of k, $\tau=0$ when $t=0$.

With the help of this result we easily determine the quantities ξ, η, ζ by expressing in equations that light (as required by the principle of the constancy of the speed of light, in combination with the principle of relativity) is also propagated with speed c when measured in the moving system. For a ray of light emitted at the time $\tau=0$ in the direction of the increasing ξ

$$\xi=c\tau \text{ or } \xi=ac\left(t-\frac{v}{c^2-v^2}x'\right)$$

But the ray moves relatively to the initial point of k, when measured in the stationary system, with the speed $c-v$, so that

$$\frac{x'}{c-v}=t$$

If we insert this value of t in the equation for ξ, we obtain

$$\xi=a\frac{c^2}{c^2-v^2}x'$$

In an analogous manner we find, by considering rays moving along the two other axes, that

$$\eta=c\tau=ac\left(t-\frac{v}{c^2-v^2}x'\right)$$

when

$$\frac{y}{\sqrt{(c^2-v^2)}}=t, \qquad x'=0$$

Wobei a eine vorläufig unbekannte Funktion $\varphi(v)$ ist und der Kürze halber angenommen ist, dass im Anfangspunkt von k für $\tau = 0$ $t = 0$ sei.

Mit Hilfe dieser Resultate ist es leicht, die Grössen ξ, η, ζ, zu ermitteln, indem man durch Gleichungen ausdrückt, dass sich das Licht (wie das Prinzip der Konstanz der Lichtgeschwindigkeit in Verbindung mit dem Relativitätsprinzip verlangt) auch im bewegten System gemessen mit der Geschwindigkeit V fortpflanzt. Für einen zur Zeit $\tau = 0$ in Richtung der wachsenden ξ ausgesandten Lichtstrahl gilt

$$\xi = V\tau$$

oder

$$\xi = aV\left(t - \frac{v}{V^2 - v^2} x'\right)$$

Nun bewegt sich aber der Lichtstrahl relativ zum Anfangspunkt von k im ruhenden System gemessen mit der Geschwindigkeit $V - v$, so dass gilt

$$\frac{x'}{V - v} = t$$

Setzen wir diesen Wert von t in die Gleichung für ξ ein, so erhalten wir:

$$\xi = a \frac{V^2}{V^2 - v^2} x'.$$

Auf analoge Weise erhalten wir durch Betrachtung von längs den beiden andern Axen bewegten Lichtstrahlen:

$$\eta = V\tau = aV\left(t - \frac{v}{V^2 - v^2} x'\right),$$

wobei

$$\frac{y}{\sqrt{V^2 - v^2}} = t\,; \quad x' = 0\,;$$

also

$$\eta = a \frac{V}{\sqrt{V^2 - v^2}} y$$

und

$$\zeta = a \frac{V}{\sqrt{V^2 - v^2}} z.$$

Setzen wir für x' seinen Wert ein, so erhalten wir

$$\tau = \varphi(v)\,\beta\left(t - \frac{v}{V^2} x\right)$$
$$\xi = \varphi(v)\,\beta\,(x - vt)$$
$$\eta = \varphi(v)\, y$$
$$\zeta = \varphi(v)\, z$$

wobei $\beta = \dfrac{1}{\sqrt{1 - \frac{v^2}{V^2}}}$

Figure 36 One page of a copy, handwritten by Einstein in November 1943, of his first relativity paper

Thus

$$\eta = a\frac{c}{\sqrt{(c^2-v^2)}}y \text{ and } \zeta = a\frac{c}{\sqrt{(c^2-v^2)}}z$$

Substituting for x' its value, we obtain

$$\begin{aligned}\tau &= \phi(v)\beta(t - vx/c^2)\\ \xi &= \phi(v)\beta(x - vt)\\ \eta &= \phi(v)y\\ \zeta &= \phi(v)z\end{aligned}$$

where

$$\beta = \frac{1}{\sqrt{(1 - v^2/c^2)}}$$

and ϕ is an as yet unknown function of v. If no assumption whatever be made as to the initial position of the moving system and as to the zero point of τ, an additive constant is to be placed on the right side of each of these equations.

We now have to prove that any ray of light, measured in the moving system, is propagated with the speed c, if, as we have assumed, this is the case in the stationary system; for we have not as yet furnished the proof that the principle of the constancy of the speed of light is compatible with the principle of relativity.

At the time $t=\tau=0$, when the origin of the coordinates is common to the two systems, let a spherical wave be emitted therefrom, and be propagated with the speed c in system K. If (x, y, z) be a point just attained by this wave, then

$$x^2+y^2+z^2=c^2t^2$$

Transforming this equation with the aid of our equations of transformation we obtain after a simple calculation

$$\xi^2+\eta^2+\zeta^2=c^2\tau^2$$

The wave under consideration is therefore no less a spherical wave with speed of propagation c when viewed in the moving system. This shows that our two fundamental principles are compatible.*

In the equations of transformation which have been developed there enters an unknown function ϕ of v, which we will now determine.

For this purpose we introduce a third system of coordinates K', which relatively to the system k is in a state of parallel translatory motion parallel to the axis of X, such that the origin of coordinates of system k moves with velocity $-v$ on the axis of X. At the time $t=0$ let all three origins coincide, and when $t=x=y=z=0$ let the time t' of the system K' be zero.

* The equations of the Lorentz transformation may be more simply deduced directly from the condition that in virtue of those equations the relation $x^2+y^2+z^2=c^2t^2$ shall have as its consequence the second relation $\xi^2+\eta^2+\zeta^2=c^2\tau^2$.

We call the coordinates, measured in the system K', x', y', z', and by a twofold application of our equations of transformation we obtain

$$\begin{aligned} t' &= \phi(-v)\beta(-v)(\tau + v\xi/c^2) = \phi(v)\phi(-v)t \\ x' &= \phi(-v)\beta(-v)(\xi + v\tau) \quad = \phi(v)\phi(-v)x \\ y' &= \phi(-v)\eta \qquad\qquad\qquad\quad = \phi(v)\phi(-v)y \\ z' &= \phi(-v)\zeta \qquad\qquad\qquad\quad = \phi(v)\phi(-v)z \end{aligned}$$

Since the relations between x', y', z' and x, y, z do not contain the time t, the systems K and K' are at rest with respect to one another, and it is clear that the transformation from K to K' must be the identical transformation. Thus

$$\phi(v)\phi(-v) = 1$$

We now inquire into the signification of $\phi(v)$. We give our attention to that part of the axis of Y of system k which lies between $\xi=0$, $\eta=0$, $\zeta=0$ and $\xi=0$, $\eta=l$, $\zeta=0$. This part of the axis of Y is a rod moving perpendicularly to its axis with velocity v relatively to system K. Its ends possess in K the coordinates

$$x_1 = vt, \qquad y_1 = \frac{l}{\phi(v)}, \qquad z_1 = 0$$

and

$$x_2 = vt, \quad y_2 = 0, \quad z_2 = 0$$

The length of the rod measured in K is therefore $l/\phi(v)$; and this gives us the meaning of the function $\phi(v)$. From reasons of symmetry it is now evident that the length of a given rod moving perpendicularly to its axis, measured in the stationary system, must depend only on the velocity and not on the direction and the sense of the motion. The length of the moving rod measured in the stationary system does not change, therefore, if v and $-v$ are interchanged. Hence it follows that $l/\phi(v) = l/\phi(-v)$, or

$$\phi(v) = \phi(-v)$$

It follows from this relation and the one previously found that $\phi(v) = 1$, so that the transformation equations which have been found become

$$\begin{aligned} \tau &= \beta(t - vx/c^2) \\ \xi &= \beta(x - vt) \\ \eta &= y \\ \zeta &= z \end{aligned}$$

where

$$\beta = 1/\sqrt{(1 - v^2/c^2)}$$

4. *Physical meaning of the equations obtained in respect to moving rigid bodies and moving clocks*

We envisage a rigid sphere* of radius R, at rest relatively to the moving system k, and with its centre at the origin of coordinates of k. The equation of the surface of this sphere moving relatively to the system K with velocity v is

$$\xi^2+\eta^2+\zeta^2=R^2$$

The equation of this surface expressed in x, y, z at the time $t=0$ is

$$\frac{x^2}{(\sqrt{(1-v^2/c^2)})^2}+y^2+z^2=R^2$$

A rigid body which, measured in a state of rest, has the form of a sphere, therefore has in a state of motion—viewed from the stationary system—the form of an ellipsoid of revolution with the axes

$$R\sqrt{(1-v^2/c^2)},\ R,\ R$$

Thus, whereas the Y and Z dimensions of the sphere (and therefore of every rigid body of no matter what form) do not appear modified by the motion, the X dimension appears shortened in the ratio $1:\sqrt{(1-v^2/c^2)}$, i.e. the greater the value of v, the greater the shortening. For $v=c$ all moving objects—viewed from the 'stationary' system—shrivel up into plane figures. For velocities greater than that of light our deliberations become meaningless; we shall, however, find in what follows that the velocity of light in our theory plays the part, physically, of an infinitely great velocity.

It is clear that the same results hold good of bodies at rest in the 'stationary' system, viewed from a system in uniform motion.

Further, we imagine one of the clocks which are qualified to mark the time t when at rest relatively to the stationary system, and the time τ when at rest relatively to the moving system, to be located at the origin of the coordinates of k, and so adjusted that it marks the time τ. What is the rate of this clock, when viewed from the stationary system?

Between the quantities x, t, and τ, which refer to the position of the clock, we have, evidently, $x=vt$ and

$$\tau=\frac{1}{\sqrt{(1-v^2/c^2)}}(t-vx/c^2)$$

Therefore,

$$\tau=t\sqrt{(1-v^2/c^2)}=t-(1-\sqrt{(1-v^2/c^2)})t$$

whence it follows that the time marked by the clock (viewed in the stationary system) is slow by $1-\sqrt{(1-v^2/c^2)}$ seconds per second, or—neglecting magnitudes of fourth and higher order—by $\frac{1}{2}v^2/c^2$.

From this there ensues the following peculiar consequence. If at the points

* That is, a body possessing spherical form when examined at rest.

A and B of K there are stationary clocks which, viewed in the stationary system, are synchronous; and if the clock at A is moved with the velocity v along the line AB to B, then on its arrival at B the two clocks no longer synchronize, but the clock moved from A to B lags behind the other which has remained at B by $\frac{1}{2}tv^2/c^2$ (up to magnitudes of fourth and higher order), t being the time occupied in the journey from A to B.

It is at once apparent that this result still holds good if the clock moves from A to B in any polygonal line, and also when the points A and B coincide.

If we assume that the result proved for a polygonal line is also valid for a continuously curved line, we arrive at this result: If one of two synchronous clocks at A is moved in a closed curve with constant speed until it returns to A, the journey lasting t seconds, then by the clock which has remained at rest the travelled clock on its arrival at A will be $\frac{1}{2}tv^2/c^2$ seconds slow. Thence we conclude that a balance-clock* at the equator must go more slowly, by a very small amount, than a precisely similar clock situated at one of the poles under otherwise identical conditions.

5. *The composition of velocities*

In the system k moving along the axis of X of the system K with velocity v, let a point move in accordance with the equations

$$\xi = w_\xi \tau,\ \eta = w_\eta \tau,\ \zeta = 0,$$

where w_ξ and w_η denote constants.

Required: the motion of the point relatively to the system K. If with the help of the equations of transformation developed in Section 3 we introduce the quantities x, y, z, t into the equations of motion of the point, we obtain

$$x = \frac{w_\xi + v}{1 + v w_\xi / c^2} t$$

$$y = \frac{\sqrt{(1 - v^2/c^2)}}{1 + v w_\xi / c^2} w_\eta t$$

$$z = 0$$

Thus the law of the parallelogram of velocities is valid according to our theory only to a first approximation. We set

$$V^2 = \left(\frac{dx}{dt}\right)^2 + \left(\frac{dy}{dt}\right)^2$$

$$w^2 = w_\xi^2 + w_\eta^2$$

$$\alpha = \tan^{-1}(w_y / w_x)$$

* Not a pendulum-clock, which is physically a system to which the Earth belongs. This case had to be excluded.

α is then to be looked upon as the angle between the velocities v and w. After a simple calculation we obtain

$$V = \frac{\sqrt{[(v^2 + w^2 + 2vw \cos \alpha) - (vw \sin \alpha/c^2)^2]}}{1 + vw \cos \alpha/c^2}$$

It is worth remarking that v and w enter into the expression for the resultant velocity in a symmetrical manner. If w also has the direction of the axis of X, we get

$$V = \frac{v + w}{1 + vw/c^2}$$

It follows from this equation that from a composition of two velocities which are less than c, there always results a velocity less than c. For if we set $v = c - \kappa$, $w = c - \lambda$, κ and λ being positive and less than c, then

$$V = c\frac{2c - \kappa - \lambda}{2c - \kappa - \lambda + \kappa\lambda/c} < c$$

It follows, further, that the velocity of light c cannot be altered by composition with a velocity less than that of light. For this case we obtain

$$V = \frac{c + w}{1 + w/c} = c$$

We might also have obtained the formula for V, for the case when v and w have the same direction, by compounding two transformations in accordance with Section 3. If in addition to the systems K and k figuring in Section 3 we introduce still another system of coordinates k' moving parallel to k, its origin moving on the axis of X with the velocity w, we obtain equations between the quantities x, y, z, t and the corresponding quantities of k', which differ from the equations found in Section 3 only in that the place of 'v' is taken by the quantity

$$\frac{v + w}{1 + vw/c^2}$$

from which we see that such parallel transformations—necessarily—form a group.

We have now deduced the requisite laws of the theory of kinematics corresponding to our two principles, and we proceed to show their application to electrodynamics. . . .

Geometry and experience

Can we visualize a three-dimensional universe which is finite, yet unbounded?

The usual answer to this question is 'No', but that is not the right answer. The purpose of the following remarks is to show that the answer should be 'Yes'. I want to show that without any extraordinary difficulty we can illustrate the theory of a finite universe by means of a mental picture to which, with some practice, we shall soon grow accustomed.

What do we wish to express when we say that our space is infinite? Nothing more than that we might lay any number of bodies of equal sizes side by side without ever filling space. Suppose that we are provided with a great many cubic boxes all of the same size. In accordance with Euclidean geometry we can place them above, beside, and behind one another so as to fill an arbitrarily large part of space; but this construction would never be finished: we could go on adding more and more cubes without ever finding that there was no more room. That is what we wish to express when we say that space is infinite. It would be better to say that space is infinite in relation to practically-rigid bodies, assuming that the laws of disposition for these bodies are given by Euclidean geometry.

Another example of an infinite continuum is the plane. On a plane surface we may lay squares of cardboard so that each side of any square has the side of another square adjacent to it. The construction is never finished; we can always go on laying squares—if their laws of disposition correspond to those of plane figures of Euclidean geometry. The plane is therefore infinite in relation to the cardboard squares. Accordingly we say that the plane is an infinite continuum of two dimensions, and space an infinite continuum of three dimensions. What is here meant by the number of dimensions, I think I may assume to be known.

Now we take an example of a two-dimensional continuum which is finite, but unbounded. We imagine the surface of a large globe and a quantity of small paper discs, all of the same size. We place one of the discs anywhere on the surface of the globe. If we move the disc about, anywhere we like, on the surface of the globe, we do not come upon a boundary anywhere on the journey. Therefore we say that the spherical surface of the globe is an unbounded continuum. Moreover, the spherical surface is a finite continuum. For if we stick the paper discs on the globe, so that no disc overlaps another, the surface of the globe will finally become so full that there is no room for another disc. This means exactly that the spherical surface of the globe is finite in relation to the paper discs. Further, the spherical surface is a non-Euclidean continuum of two dimensions, that is to say, the laws of disposition for the rigid figures lying in it do not agree with those of the Euclidean plane. This can be shown in the following way. Take a disc and surround it in a circle by six more discs, each of which is to be surrounded in turn by six discs, and so on. If this construction is made on a plane surface, we obtain an uninterrupted arrangement in which there are six discs touching every disc except those which lie on the outside. On the spherical surface the construction also seems to promise success at the outset, and the smaller the radius of the disc in proportion to that of the

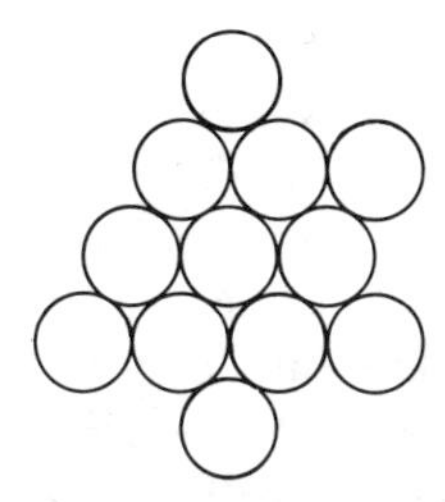

sphere, the more promising it seems. But as the construction progresses it becomes more and more patent that the arrangement of the discs in the manner indicated, without interruption, is not possible, as it should be possible by the Euclidean geometry of the plane. In this way creatures which cannot leave the spherical surface, and cannot even peep out from the spherical surface into three-dimensional space, might discover, merely by experimenting with discs, that their two-dimensional 'space' is not Euclidean, but spherical space.

From the latest results of the theory of relativity it is probable that our three-dimensional space is also approximately spherical, that is, that the laws of disposition of rigid bodies in it are not given by Euclidean geometry, but approximately by spherical geometry, if only we consider parts of space which are sufficiently extended. Now this is the place where the reader's imagination boggles. 'Nobody can imagine this thing,' he cries indignantly. 'It can be said, but cannot be thought. I can imagine a spherical surface well enough, but nothing analogous to it in three dimensions.'

We must try to surmount this barrier in the mind, and the patient reader will see that it is by no means a particularly difficult task. For this purpose we will first give our attention once more to the geometry of two-dimensional spherical surfaces. In the adjoining figure let K be the spherical surface, touched at S by a plane E, which, for facility of presentation, is shown in the drawing as a bounded surface. Let L be a disc on the spherical surface. Now let us imagine that at the point N of the spherical surface, diametrically opposite to S, there is a luminous point, throwing a shadow L' of the disc L upon the plane E. Every point on the sphere has its shadow on the plane. If the disc on the sphere K is moved, its shadow L' on the plane E also moves. When the disc L is at S, it almost exactly coincides with its shadow. If it moves on the spherical surface away from S upwards, the disc

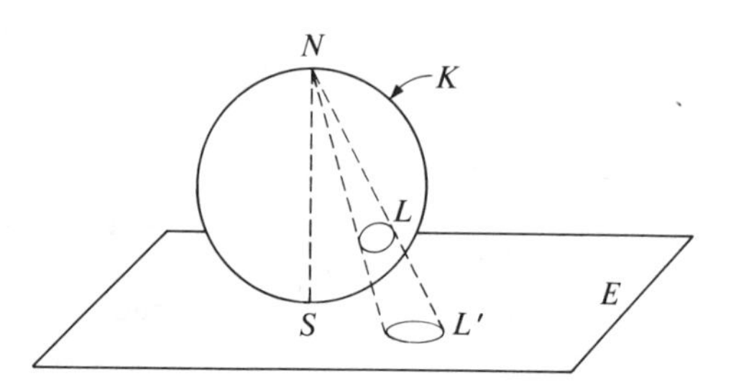

shadow L' on the plane also moves away from S on the plane outwards, growing bigger and bigger. As the disc L approaches the luminous point N, the shadow moves off to infinity, and becomes infinitely great.

Now we put the question: what are the laws of disposition of the disc-shadows L' on the plane E? Evidently they are exactly the same as the laws of disposition of the discs L on the spherical surface. For to each original figure on K there is a corresponding shadow figure on E. If two discs on K are touching, their shadows on E also touch. The shadow-geometry on the plane agrees with the disc-geometry on the sphere. If we call the disc-shadows rigid figures, then spherical geometry holds good on the plane E with respect to these rigid figures. In particular, the plane is finite with respect to the disc-shadows, since only a finite number of the shadows can find room on the plane.

At this point somebody will say, 'That is nonsense. The disc-shadows are *not* rigid figures. We have only to move a two-foot rule about on the plane E to convince ourselves that the shadows constantly increase in size as they move away from S on the plane toward infinity.' But what if the two-foot rule were to behave on the plane E in the same way as the disc-shadows L'? It would then be impossible to show that the shadows increase in size as they move away from S; such an assertion would then no longer have any meaning whatever. In fact the only objective assertion that can be made about the disc-shadows is just this, that they are related in exactly the same way as are the rigid discs on the spherical surface in the sense of Euclidean geometry.

We must carefully bear in mind that our statement as to the growth of the disc-shadows, as they move away from S toward infinity, has in itself no objective meaning, as long as we are unable to compare the disc-shadows with Euclidean rigid bodies which can be moved about on the plane E. In respect of the laws of disposition of the shadows L', the point S has no special privileges on the plane any more than on the spherical surface.

The representation given above of spherical geometry on the plane is important for us, because it readily allows itself to be transferred to the three-dimensional case.

Let us imagine a point S of our space, and a great number of small spheres, L', which can all be brought to coincide with one another. But these spheres are not to be rigid in the sense of Euclidean geometry; their radius is to increase (in the sense of Euclidean geometry) when they are moved away from S toward infinity; it is to increase according to the same law as the radii of the disc-shadows L' on the plane.

After having gained a vivid mental image of the geometrical behaviour of our L' spheres, let us assume that in our space there are no rigid bodies at all in the sense of Euclidean geometry, but only bodies having the behaviour of our L' spheres. Then we shall have a clear picture of three-dimensional spherical space, or, rather of three-dimensional spherical geometry. Here our spheres must be called 'rigid' spheres. Their increase in size as they depart from S is not to be detected by measuring with measuring-rods, any more than in the case of the disc-shadows on E, because the standards of measurement behave in the same way as the spheres. Space is homogeneous, that is to say, the same spherical configurations are possible in the neighbourhood of every point.* Our space is finite, because, in consequence of the 'growth' of the spheres, only a finite number of them can find room in space.

In this way, by using as a crutch the practice in thinking and visualization which Euclidean geometry gives us, we have acquired a mental picture of spherical geometry. We may without difficulty impart more depth and vigour to these ideas by carrying out special imaginary constructions. Nor would it be difficult to represent the case of what is called elliptical geometry in an analogous manner. My only aim today has been to show that the human faculty of visualization is by no means bound to capitulate to non-Euclidean geometry.

(Excerpts from the lecture 'Geometry and Experience'
delivered to the Prussian Academy of Sciences, 1921)

* This is intelligible without calculation—but only for the two-dimensional case—if we revert once more to the case of the disc on the surface of the sphere.

The cause of the formation of meanders in the courses of rivers and the so-called Baer's Law

Read before the Prussian Academy, 7 January 1926. Published in the German periodical, Die Naturwissenschaften, *Vol. 14, 1926.*

It is common knowledge that streams tend to curve in serpentine shapes instead of following the line of the maximum declivity of the ground. It is also well known to geographers that the rivers of the northern hemisphere tend to erode chiefly on the right side. The rivers of the southern hemisphere behave in the opposite manner (Baer's law). Many attempts have been made to explain this phenomenon, and I am not sure whether anything I say in the following pages will be new to the expert; some of my considerations are certainly known. Nevertheless, having found nobody who was thoroughly familiar with the causal relations involved, I think it is appropriate to give a short qualitative exposition of them.

First of all, it is clear that the erosion must be stronger the greater the velocity of the current where it touches the bank in question, or rather the more steeply it falls to zero at any particular point of the confining wall. This is equally true under all circumstances, whether the erosion depends on mechanical or on physico-chemical factors (decomposition of the ground). We must then concentrate our attention on the circumstances which affect the steepness of the velocity gradient at the wall.

In both cases the asymmetry as regards the fall in velocity in question is indirectly due to the formation of a circular motion to which we will next direct our attention.

I begin with a little experiment which anybody can easily repeat. Imagine a flat-bottomed cup full of tea. At the bottom there are some tea leaves, which stay there because they are rather heavier than the liquid they have displaced. If the liquid is made to rotate by a spoon, the leaves will soon collect in the centre of the bottom of the cup. The explanation of this phenomenon is as follows: the rotation of the liquid causes a centrifugal force to act on it. This in itself would give rise to no change in the flow of the liquid if the latter rotated like a solid body. But in the neighbourhood of the walls of the cup the liquid is restrained by friction, so that the angular velocity with which it rotates is less there than in other places nearer the centre. In particular, the angular velocity of rotation, and therefore the centrifugal force, will be smaller near the bottom than higher up. The result of this will be a circular movement of the liquid of the type illustrated in Figure 1 which goes on increasing until, under the influence of ground friction, it becomes stationary. The tea leaves are swept into the centre by the circular movement and act as proof of its existence.

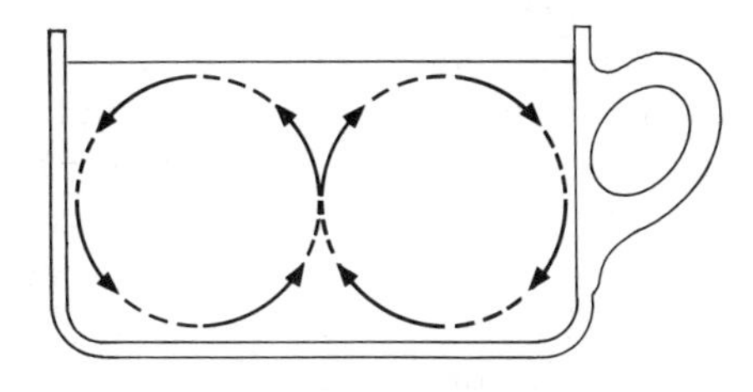

FIG. 1

The same sort of thing happens with a curving stream (Figure 2). At every cross-section of its course, where it is bent, a centrifugal force operates in the direction of the outside of the curve (from A to B). This force is less near the bottom, where the speed of the current is reduced by friction, than higher above the bottom. This causes a circular movement of the kind illustrated in the diagram. Even where there is no bend in the river, a circular movement of the kind shown in Figure 2 will take place, if only on a small scale, as a result of the Earth's rotation. The latter produces a Coriolis-force, acting transversely to the direction of the current, whose right-hand horizontal component amounts to $2v\Omega \sin \phi$ per unit of mass of the liquid, where v is the velocity of the current, Ω the speed of the Earth's rotation, and ϕ the geographical latitude. As ground friction causes a diminution of this force toward the bottom, this force also gives rise to a circular movement of the type indicated in Figure 2.

After this preliminary discussion we come back to the question of the distribution of velocities over the cross-section of the stream, which is the

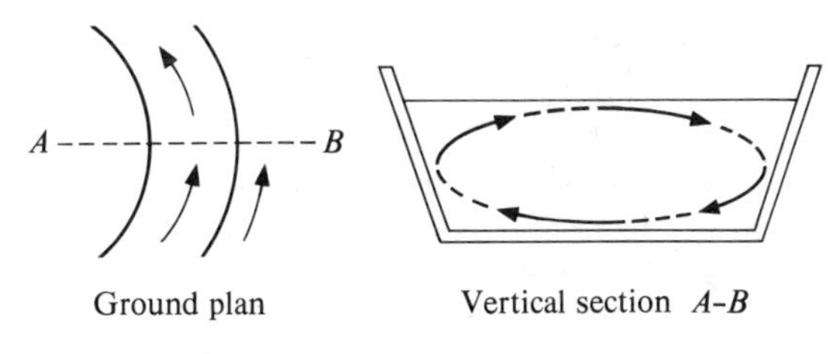

FIG. 2

controlling factor in erosion. For this purpose we must first realize how the (turbulent) distribution of velocities develops and is maintained. If the water which was previously at rest were suddenly set in motion by the action of a uniformly distributed accelerating force, the distribution of velocities over the cross-section would at first be uniform. A distribution of velocities gradually increasing from the confining walls toward the centre of the cross-section would only establish itself after a time, under the influence of friction at the walls. A disturbance of the (roughly speaking) stationary distribution of velocities over the cross-section would only gradually set in again under the influence of fluid friction.

Hydrodynamics pictures the process by which this stationary distribution of velocities is established in the following way. In a plane (potential) flow all the vortex-filaments are concentrated at the walls. They detach themselves and slowly move toward the centre of the cross-section of the stream, distributing themselves over a layer of increasing thickness. The velocity gradient at the walls thereby gradually diminishes. Under the action of the internal friction of the liquid the vortex filaments in the interior of the cross-section are gradually absorbed, their place being taken by new ones which form at the wall. A quasi-stationary distribution of velocities is thus produced. The important thing for us is that the attainment of the stationary distribution of velocities is a slow process. That is why relatively insignificant, constantly operative causes are able to exert a considerable influence on the distribution of velocities over the cross-section. Let us now consider what sort of influence the circular motion due to a bend in the river or the Coriolis-force, as illustrated in Figure 2, is bound to exert on the distribution of velocities over the cross-section of the river. The particles or liquid in most rapid motion will be farthest away from the walls, that is to say, in the upper part above the centre of the bottom. These most rapid parts of the water will be driven by the circulation toward the right-hand wall, while the left-hand wall gets the water which comes from the region near the bottom and has a specially low velocity. Hence in the case depicted in Figure 2 the erosion is necessarily stronger on the right side than on the left. It should be noted that this explanation is essentially based on the fact that the slow circulating movement of the water exerts a considerable influence on the distribution of velocities, because the adjustment of velocities by internal friction which counteracts this consequence of the circulating movement is also a slow process.

We have now revealed the causes of the formation of meanders. Certain details can, however, also be deduced without difficulty from these facts. Erosion will be comparatively extensive not merely on the right-hand wall but also on the right half of the bottom, so that there will be a tendency to assume a profile as illustrated in Figure 3.

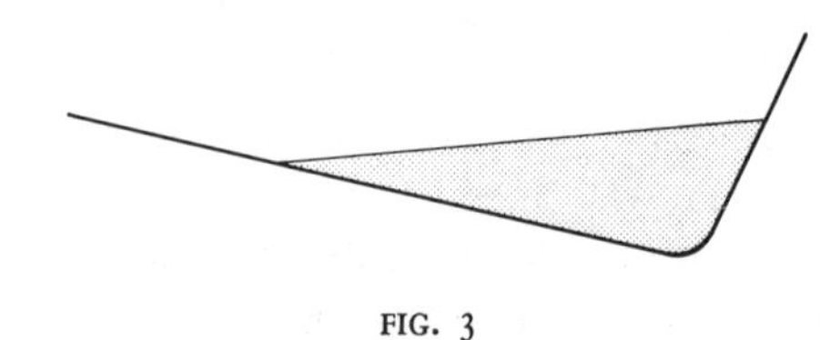

FIG. 3

Moreover, the water at the surface will come from the left-hand wall, and will therefore, on the left-hand side especially, be moving less rapidly than the water rather lower down. This has, in fact, been observed. It should further be noted that the circular motion possesses inertia. The circulation will therefore only achieve its maximum beyond the place of the greater asymmetry of the erosion. Hence in the course of the erosion an advance of the wave-line of the meander-formation is bound to take place in the direction of the current. Finally, the larger the cross-section of the river, the more slowly will the circular movement be absorbed by friction; the wave-line of the meander-formation will therefore increase with the cross-section of the river.

Excerpts from *Ideas and Opinions*

The cult of individuals is always, in my view, unjustified. To be sure
nature distributes her gifts unevenly among her children. But there ar
plenty of the well-endowed, thank God, and I am firmly convinced tha
most of them live quiet, unobtrusive lives. It strikes me as unfair, and eve
in bad taste, to select a few of them for boundless admiration, attributin
super-human powers of mind and character to them. This has been my fate
and the contrast between the popular estimate of my powers and achieve
ments and the reality is simply grotesque. The awareness of this strang
state of affairs would be unbearable but for one pleasing consolation: it i
a welcome symptom in an age which is commonly denounced as material
istic, that it makes heroes of men whose goals lie wholly in the intellectua
and moral sphere. This proves that knowledge and justice are ranke
above wealth and power by a large section of the human race.

(From 'My First Impression of the USA', 1921

How strange is the lot of us mortals! Each of us is here for a brief sojourn
for what purpose he knows not, though he senses it. But without deepe
reflection one knows from daily life that one exists for other people—firs
of all for those upon whose smiles and well-being our own happiness i
wholly dependent, and then for the many, unknown to us, to whos
destinies we are bound by the ties of sympathy.

I do not at all believe in human freedom in the philosophical sense
Everybody acts not only under external compulsion but also in accordanc
with inner necessity. Schöpenhauer's saying, 'A man can do what he wants
but not want what he wants', has been a continual consolation in the fac
of life's hardships, my own and others', and an unfailing well-spring o
tolerance. This realization mercifully mitigates the easily paralysing sens
of responsibility and prevents us from taking ourselves and other peopl
all too seriously; it is conducive to a view of life which, in particula
gives humour its due.

To inquire after the meaning or object of one's own existence or that c
all creatures has always seemed to me absurd from an objective point o
view. And yet everybody has certain ideals which determine the directio

of his endeavours and his judgements. In this sense I have never looked upon ease and happiness as ends in themselves—this ethical basis I call the ideal of a pigsty. The ideals which have lighted my way, and time after time have given me new courage to face life cheerfully, have been Kindness, Beauty, and Truth. Without the sense of kinship with men of like mind, without the occupation with the objective world, the eternally unattainable in the field of art and scientific endeavours, life would have seemed to me empty.

Figure 37

Albert Einstein with Paul Ehrenfest

The trite objects of human efforts—possessions, outward success, luxury—have always seemed to me contemptible.

My passionate sense of social justice and social responsibility has always contrasted oddly with my pronounced lack of need for direct contact with other human beings and human communities. I am truly a 'lone traveller' and have never belonged to my country, my home, my friends, or even my immediate family, with my whole heart; in the face of all these ties, I have never lost a sense of distance and a need for solitude—feelings which increase with the years. One becomes sharply aware, but without regret, of the limits of mutual understanding and consonance with other people. No doubt, such a person loses some of his innocence and unconcern; on the other hand, he is largely independent of the opinions, habits, and judgements of his fellows and avoids the temptation to build his inner equilibrium upon such insecure foundations.

My political ideal is democracy. Let every man be respected as an individual and no man idolized. It is an irony of fate that I myself have been the recipient of excessive admiration and reverence from my fellow-beings, through no fault, and no merit, of my own. The cause of this may well be the desire, unattainable for many, to understand the few ideas to which I have with my feeble powers attained through ceaseless struggle. I am quite aware that it is necessary for the achievement of the objective of an organization that one man should do the thinking and directing and generally bear the responsibility. But the led must not be coerced, they must be able to choose their leader. An autocratic system of coercion, in my opinion, soon degenerates. For force always attracts men of low morality, and I believe it to be an invariable rule that tyrants of genius are succeeded by scoundrels.

The most beautiful experience we can have is the mysterious. It is the fundamental emotion which stands at the cradle of true art and true science. Whoever does not know it and can no longer wonder, no longer marvel, is as good as dead, and his eyes are dimmed. It was the experience of mystery—even if mixed with fear—that engendered religion. A knowledge of the existence of something we cannot penetrate, our perceptions of the profoundest reason and the most radiant beauty, which only in their most primitive forms are accessible to our minds—it is this knowledge and this emotion that constitute true religiosity; in this sense, and in this alone, I am a deeply religious man. I cannot conceive of a God who rewards and punishes his creatures, or has a will of the kind that we experience in ourselves. Neither can I nor would I want to conceive of an individual that survives his physical death; let feeble souls, from fear or absurd egoism, cherish such thoughts. I am satisfied with the mystery of the eternity of life and with the awareness and a glimpse of the marvellous structure of the existing world, together with the devoted striving to comprehend a portion, be it ever so tiny, of the Reason that manifests itself in nature.

(From 'The World as I See It', 1934)

The existence and validity of human rights are not written in the stars. The ideals concerning the conduct of men toward each other and the desirable structure of the community have been conceived and taught by enlightened individuals in the course of history. Those ideals and convictions which resulted from historical experience, from the cravings for beauty and harmony, have been readily accepted in theory by man—and, at all times, have been trampled upon by the same people under the pressure of their animal instincts. A large part of history is therefore replete with the struggle for those human rights, an eternal struggle in which a final victory can never be won. But to tire in that struggle would mean the ruin of society.

(From 'Human Rights', 1954)

You will hardly find one among the profounder sort of scientific minds without a religious feeling of his own. But it is different from the religiosity of the naive man. For the latter, God is a being from whose care one hopes to benefit and whose punishment one fears; a sublimation of a feeling similar to that of a child for its father, a being to whom one stands, so to speak, in a personal relation, however deeply it may be tinged with awe.

But the scientist is possessed by the sense of universal causation. The future, to him, is every whit as necessary and determined as the past. There is nothing divine about morality; it is a purely human affair. His religious feeling takes the form of a rapturous amazement at the harmony of natural law, which reveals an intelligence of such superiority that, compared with it, all the systematic thinking and acting of human beings is an utterly insignificant reflection. This feeling is the guiding principle of his life and work, in so far as he succeeds in keeping himself from the shackles of selfish desire. It is beyond question closely akin to that which has possessed the religious geniuses of all ages.

(From 'The Religious Spirit of Science', 1934)

It was my good fortune to be linked with Mme Curie through twenty years of sublime and unclouded friendship. I came to admire her human grandeur to an ever-growing degree. Her strength, her purity of will, her austerity toward herself, her objectivity, her incorruptible judgement—all these were of a kind seldom found joined in a single individual. She felt herself at every moment to be a servant of society, and her profound modesty never left any room for complacency. She was oppressed by an abiding sense for the asperities and inequities of society. This is what gave her that severe outward aspect, so easily misinterpreted by those who were not close to her—a curious severity unrelieved by any artistic strain. Once she had recognized a certain way as the right one, she pursued it without compromise and with extreme tenacity.

The greatest scientific deed of her life—proving the existence of radioactive elements and isolating them—owes its accomplishment not merely to bold intuition but to a devotion and tenacity in execution under the most extreme hardships imaginable, such as the history of experimental science has not often witnessed.

(From 'Marie Curie In Memoriam', 1935)

Notes on the origin of the general theory of relativity

I gladly accede to the request that I should say something about the history of my own scientific work. Not that I have an exaggerated notion of the importance of my own efforts, but to write the history of other men's work demands a degree of absorption in other people's ideas which is much more in the line of the trained historian: to throw light on one's own earlier thinking appears incomparably easier. Here one has an immense advantage over everybody else, and one ought not to leave the opportunity unused out of modesty.

When by the special theory of relativity I had arrived at the equivalence of all so-called inertial systems for the formulation of natural laws (1905), the question whether there was not a further equivalence of coordinate systems followed naturally, to say the least of it. To put it in another way, if only a relative meaning can be attached to the concept of velocity, ought we nevertheless to persevere in treating acceleration as an absolute concept?

From the purely kinematic point of view there was no doubt about the relativity of all motions whatever; but physically speaking, the inertial system seemed to occupy a privileged position, which made the use of coordinate systems moving in other ways appear artificial.

I was of course acquainted with Mach's view, according to which it appeared conceivable that what inertial resistance counteracts is not acceleration as such but acceleration with respect to the masses of the other bodies existing in the world. There was something fascinating about this idea to me, but it provided no workable basis for a new theory.

I first came a step nearer to the solution of the problem when I attempted to deal with the law of gravity within the framework of the special theory of relativity. Like most writers at the time, I tried to frame a *field-law* for gravitation, since it was no longer possible, at least in any natural way, to introduce direct action at a distance owing to the abolition of the notion of absolute simultaneity.

These investigations, however, led to a result which raised my strong suspicions. According to classical mechanics, the vertical acceleration of a body in the vertical gravitational field is independent of the horizontal component of its velocity. Hence in such a gravitational field the vertical acceleration of a mechanical system or of its centre of gravity works out independently of its internal kinetic energy. But in the theory I advanced, the acceleration of a falling body was not independent of its horizontal velocity or the internal energy of a system.

This did not fit in with the old experimental fact that all bodies have the same acceleration in a gravitational field. This law, which may also be formulated as the law of the equality of inertial and gravitational mass, was now brought home to me in all its significance. I was in the highest degree amazed at its existence and guessed that in it must lie the key to a deeper understanding of inertia and gravitation. I had no serious doubts about its strict validity even without knowing the results of the admirable experiments of Eötvös, which—if my memory is right—I only came to know later. I now abandoned as inadequate the attempt to treat the problem of gravitation, in the manner outlined above, within the framework of the special theory of relativity. It clearly failed to do justice to the most fundamental property of gravitation. The principle of the equality of inertial and gravitational mass could now be formulated quite clearly as follows: In a homogeneous gravitational field all motions take place in the same way as in the absence of a gravitational field in relation to a uniformly accelerated coordinate system. If this principle held good for any events whatever (the 'principle of equivalence'), this was an indication that the principle of relativity needed to be extended to coordinate systems in non-uniform motion with respect to each other, if we were to reach a natural theory of the gravitational fields. Such reflections kept me busy from 1908 to 1911, and I attempted to draw special conclusions from them, of which I do not propose to speak here. For the moment the one important thing was the discovery that a reasonable theory of gravitation could only be hoped for from an extension of the principle of relativity.

What was needed, therefore, was to frame a theory whose equations kept their form in the case of non-linear transformations of the coordinates. Whether this was to apply to arbitrary (continuous) transformations of coordinates or only to certain ones, I could not for the moment say.

I soon saw that the inclusion of non-linear transformations, as the principle of equivalence demanded, was inevitably fatal to the simple physical interpretation of the coordinates—i.e. that it could no longer be required that coordinate differences should signify direct results of measurement with ideal scales or clocks. I was much bothered by this piece of knowledge, for it took me a long time to see what coordinates at all meant in physics. I did not find the way out of this dilemma until 1912, and then it came to me as a result of the following consideration:

A new formulation of the law of inertia had to be found which in case of the absence of a 'real gravitational field' passed over into Galileo's formulation for the principle of inertia if an inertial system was used as coordinate system. Galileo's formulation amounts to this: A material point, which is acted on by no force, will be represented in four-dimensional space by a straight line, that is to say, by a shortest line, or more correctly, a extremal line. This concept presupposes that of the length of a line element that is to say, a metric. In the special theory of relativity, as Minkowsk had shown, this metric was a quasi-Euclidean one, i.e. the square of th 'length' ds of a line element was a certain quadratic function of th differentials of the coordinates.

If other coordinates are introduced by means of a non-linear transform ation, ds^2 remains a homogeneous function of the differentials of the co ordinates, but the coefficients of this function ($g_{\mu\nu}$) cease to be constant an become certain functions of the coordinates. In mathematical terms th means that physical (four-dimensional) space has a Riemannian metric. Th timelike extremal lines of this metric furnish the law of motion of a materi point which is acted on by no force apart from the forces of gravity. Th coefficients ($g_{\mu\nu}$) of this metric at the same time describe the gravitationa field with reference to the coordinate system selected. A natural formulatio of the principle of equivalence had thus been found, the extension of whic to any gravitational field whatever formed a perfectly natural hypothesis.

The solution of the above-mentioned dilemma was therefore as follows A physical significance attaches not to the differentials of the coordinate but only to the Riemannian metric corresponding to them. A workabl basis had now been found for the general theory of relativity. Two furthe problems remained to be solved, however.

(1) If a field-law is expressed in terms of the special theory of relativity how can it be transferred to the case of a Riemannian metric?

(2) What are the differential laws which determine the Riemannia metric (i.e. $g_{\mu\nu}$) itself?

I worked on these problems from 1912 to 1914 together with my frien Grossmann. We found that the mathematical methods for solving problen (1) lay ready in our hands in the absolute differential calculus of Ricci an Levi-Civita.

As for problem (2), its solution obviously required the constructio (from the $g_{\mu\nu}$) of the differential invariants of the second order. We soo saw that these had already been established by Riemann (the tensor o curvature). We had already considered the right field-equations for gravi tation two years before the publication of the general theory of relativity but we were unable to see how they could be used in physics. On th contrary, I felt sure that they could not do justice to experience. Moreove I believed that I could show on general considerations that a law of gravi tation invariant with respect to arbitrary transformations of coordinate was inconsistent with the principle of causality. These were errors of though which cost me two years of excessively hard work, until I finally recognize them as such at the end of 1915, and after having ruefully returned to th Riemannian curvature, succeeded in linking the theory with the facts o astronomical experience.

In the light of knowledge attained, the happy achievement seems almos a matter of course, and any intelligent student can grasp it without too much trouble. But the years of anxious searching in the dark, with their intense longing, their alternations of confidence and exhaustion and the final emergence into the light—only those who have experienced it can understand that.

(From 'Notes on the Origin of the General Theory of Relativity', in *Mein Weltbild* (Amsterdam: Querido Verlag, 1934).)

On the method of theoretical physics

If you want to find out anything from the theoretical physicists about the methods they use, I advise you to stick closely to one principle: don't listen to their words, fix your attention on their deeds. To him who is a discoverer in this field, the products of his imagination appear so necessary and natural that he regards them, and would like to have them regarded by others, not as creations of thought but as given realities.

These words sound like an invitation to you to walk out of this lecture. You will say to yourselves, the fellow's a working physicist himself and ought therefore to leave all questions of the structure of theoretical science to the epistemologists.

Against such criticism I can defend myself from the personal point of view by assuring you that it is not at my own instance but at the kind invitation of others that I have mounted this rostrum, which serves to commemorate a man who fought hard all his life for the unity of knowledge. Objectively, however, my enterprise can be justified on the ground that it may, after all, be of interest to know how one who has spent a lifetime in striving with all his might to clear up and rectify its fundamentals looks upon his own branch of science. The way in which he regards its past and present may depend too much on what he hopes for the future and aims at in the present; but that is the inevitable fate of anybody who has occupied himself intensively with a world of ideas. The same thing happens to him as to the historian, who in the same way, even though perhaps unconsciously, groups actual events round ideals which he has formed for himself on the subject of human society.

Let us now cast an eye over the development of the theoretical system, paying special attention to the relations between the content of the theory and the totality of empirical fact. We are concerned with the eternal antithesis between the two inseparable components of our knowledge, the empirical and the rational, in our department.

We reverence ancient Greece as the cradle of western science. Here for the first time the world witnessed the miracle of a logical system which proceeded from step to step with such precision that every single one of its propositions was absolutely indubitable—I refer to Euclid's geometry. This admirable triumph of reasoning gave the human intellect the necessary confidence in itself for its subsequent achievements. If Euclid failed to kindle your youthful enthusiasm, then you were not born to be a scientific thinker.

But before mankind could be ripe for a science which takes in the whole of reality, a second fundamental truth was needed, which only became common property among philosophers with the advent of Kepler and Galileo. Pure logical thinking cannot yield us any knowledge of the empirical world; all knowledge of reality starts from experience and ends in it. Propositions arrived at by purely logical means are completely empty as regards reality. Because Galileo saw this, and particularly because he drummed it into the scientific world, he is the father of modern physics—indeed, of modern science altogether.

If, then, experience is the alpha and the omega of all our knowledge of reality, what is the function of pure reason in science?

A complete system of theoretical physics is made up of concepts, fundamental laws which are supposed to be valid for those concepts and conclusions to be reached by logical deduction. It is these conclusions which must correspond with our separate experiences; in any theoretical treatise their logical deduction occupies almost the whole book.

This is exactly what happens in Euclid's geometry, except that there the fundamental laws are called axioms and there is no question of the conclusions having to correspond to any sort of experience. If, however, one regards Euclidean geometry as the science of the possible mutual relations of practically rigid bodies in space, that is to say, treats it as a physical science, without abstracting from its original empirical content, the logical homogeneity of geometry and theoretical physics becomes complete.

We have thus assigned to pure reason and experience their places in a theoretical system of physics. The structure of the system is the work of reason; the empirical contents and their mutual relations must find their representation in the conclusions of the theory. In the possibility of such a representation lie the sole value and justification of the whole system, and especially of the concepts and fundamental principles which underlie it. Apart from that, these latter are free inventions of the human intellect, which cannot be justified either by the nature of that intellect or in any other fashion *a priori*.

These fundamental concepts and postulates, which cannot be further reduced logically, form the essential part of a theory, which reason cannot touch. It is the grand object of all theory to make these irreducible elements as simple and as few in number as possible, without having to renounce the adequate representation of any empirical content whatever.

The view I have just outlined of the purely fictitious character of the fundamentals of a scientific theory was by no means the prevailing one in the eighteenth and nineteenth centuries. But it is steadily gaining ground from the fact that the distance in thought between the fundamental concepts and laws on one side and, on the other, the conclusions which have to be brought into relation with our experience grows larger and larger, the simpler the logical structure becomes—that is to say, the smaller the number of logically independent conceptual elements which are found necessary to support the structure.

Newton, the first creator of a comprehensive, workable system of theoretical physics, still believed that the basic concepts and laws of his system could be derived from experience. This is no doubt the meaning of his saying, *hypotheses non fingo*.

Actually the concepts of time and space appeared at that time to present no difficulties. The concepts of mass, inertia, and force, and the laws connecting them, seemed to be drawn directly from experience. Once this basis is accepted, the expression for the force of gravitation appears derivable from experience, and it was reasonable to expect the same in regard to other forces.

We can indeed see from Newton's formulation of it that the concept of absolute space, which comprised that of absolute rest, made him feel uncomfortable; he realized that there seemed to be nothing in experience corresponding to this last concept. He was also not quite comfortable about the introduction of forces operating at a distance. But the tremendous practical success of his doctrines may well have prevented him and the physicists of the eighteenth and nineteenth centuries from recognizing the fictitious character of the foundations of his system.

The natural philosophers of those days were, on the contrary, most of them possessed with the idea that the fundamental concepts and postulates of physics were not in the logical sense free inventions of the human mind but could be deduced from experience by 'abstraction'—that is to say, by logical means. A clear recognition of the erroneousness of this notion really only came with the general theory of relativity, which showed that one could take account of a wider range of empirical facts, and that, too, in a more satisfactory and complete manner, on a foundation quite different from the Newtonian. But quite apart from the question of the superiority of one or the other, the fictitious character of fundamental principles is perfectly evident from the fact that we can point to two essentially different principles, both of which correspond with experience to a large extent; this proves at the same time that every attempt at a logical deduction of the basic concepts and postulates of mechanics from elementary experiences is doomed to failure.

If, then, it is true that the axiomatic basis of theoretical physics cannot be extracted from experience but must be freely invented, can we ever hope to find the right way? Nay, more, has this right way any existence outside our illusions? Can we hope to be guided safely by experience at all when there exist theories (such as classical mechanics) which to a large extent do justice to experience, without getting to the root of the matter? I answer without hesitation that there is, in my opinion, a right way, and that we are capable of finding it. Our experience hitherto justifies us in believing that nature is the realization of the simplest conceivable mathematical ideas. I am convinced that we can discover by means of purely mathematical constructions the concepts and the laws connecting them with each other, which furnish the key to the understanding of natural phenomena. Experience may suggest the appropriate mathematical concepts, but they most certainly cannot be deduced from it. Experience remains, of course, the sole criterion of the physical utility of a mathematical construction. But the creative principle resides in mathematics. In a certain sense, therefore, I hold it true that pure thought can grasp reality, as the ancients dreamed.

In order to justify this confidence, I am compelled to make use of a mathematical concept. The physical world is represented as a four-dimensional continuum. If I assume a Riemannian metric in it and ask what are the simplest laws which such a metric can satisfy, I arrive at the relativistic theory of gravitation in empty space. If in that space I assume a vector-field or an anti-symmetrical tensor-field which can be derived from it, and ask what are the simplest laws which such a field can satisfy, I arrive at Maxwell's equations for empty space.

At this point we still lack a theory for those parts of space in which electrical charge density does not disappear. De Broglie conjectured the existence of a wave field, which served to explain certain quantum properties of matter. Dirac found in the spinors field-magnitudes of a new sort, whose simplest equations enable one to a large extent to deduce the properties of the electron. Subsequently I discovered, in conjunction with my colleague, Dr Walther Mayer, that these spinors form a special case of a new sort of field, mathematically connected with the four-dimensional system, which we called 'semivectors'. The simplest equations which such semivectors can satisfy furnish a key to the understanding of the existence of two sorts of elementary particles, of different ponderable mass and equal but opposite electrical charge. These semivectors are, after ordinary vectors, the simplest mathematical fields that are possible in a metrical continuum of four dimensions, and it looks as if they describe, in a natural way, certain essential properties of electrical particles.

The important point for us to observe is that all these constructions and the laws connecting them can be arrived at by the principle of looking for the mathematically simplest concepts and the link between them. In the limited number of the mathematically existent simple field types, and the simple equations possible between them, lies the theorist's hope of grasping the real in all its depth.

Meanwhile the great stumbling-block for a field-theory of this kind lies in the conception of the atomic structure of matter and energy. For the theory is fundamentally non-atomic in so far as it operates exclusively with continuous functions of space, in contrast to classical mechanics, whose most important element, the material point, in itself does justice to the atomic structure of matter.

The modern quantum theory in the form associated with the names of de Broglie, Schrödinger, and Dirac, which operates with continuous functions, has overcome these difficulties by a bold piece of interpretation which was first given a clear form by Max Born. According to this, the spatial functions which appear in the equations make no claim to be a mathematical model of the atomic structure. Those functions are only supposed to determine the mathematical probabilities to find such structures, if measurements are taken, at a particular spot or in a certain state of motion. This notion is logically unobjectionable and has important successes to its credit. Unfortunately, however, it compels one to use a continuum the number of whose dimensions is not that ascribed to space by physics hitherto (four) but rises indefinitely with the number of the particles constituting the system under consideration. I cannot but confess that I attach only a transitory importance to this interpretation. I still believe in the possibility of a model of reality—that is to say, of a theory which represents things themselves and not merely the probability of their occurrence.

On the other hand, it seems to me certain that we must give up the idea of a complete localization of the particles in a theoretical model. This seems to me to be the permanent upshot of Heisenberg's principle of uncertainty. But an atomic theory in the true sense of the word (not merely on the basis of an interpretation) without localization of particles in a mathematical model is perfectly thinkable. For instance, to account for the atomic character of electricity, the field equations need only lead to the following conclusions: A region of three-dimensional space at whose boundary electrical density vanishes everywhere always contains a total electrical charge whose size is represented by a whole number. In a continuum-theory atomic characteristics would be satisfactorily expressed by integral laws without localization of the entities which constitute the atomic structure.

Not until the atomic structure has been successfully represented in such a manner would I consider the quantum-riddle solved.

On education

A day of celebration generally is in the first place dedicated to retrospect, especially to the memory of personages who have gained special distinction for the development of the cultural life. This friendly service for our predecessors must indeed not be neglected, particularly as such a memory of the best of the past is proper to stimulate the well-disposed of today to a courageous effort. But this should be done by someone who, from his youth, has been connected with this State and is familiar with its past, not by one who like a gypsy has wandered about and gathered his experiences in all kinds of countries.

Thus, there is nothing else left for me but to speak about such questions as, independently of space and time, always have been and will be connected with educational matters. In this attempt I cannot lay any claim to being an authority, especially as intelligent and well-meaning men of all times have dealt with educational problems and have certainly repeatedly expressed their views clearly about these matters. From what source shall I, as a partial layman in the realm of pedagogy, derive courage to expound opinions with no foundations except personal experience and personal conviction? If it were really a scientific matter, one would probably be tempted to silence by such considerations.

However, with the affairs of active human beings it is different. Here knowledge of truth alone does not suffice; on the contrary this knowledge must continually be renewed by ceaseless effort, if it is not to be lost. It resembles a statue of marble which stands in the desert and is continuously threatened with burial by the shifting sand. The hands of service must ever be at work, in order that the marble continue lastingly to shine in the Sun. To these serving hands mine also shall belong.

The school has always been the most important means of transferring the wealth of tradition from one generation to the next. This applies today in an even higher degree than in former times for, through modern development of the economic life, the family as bearer of tradition and education has been weakened. The continuance and health of human society is therefore in a still higher degree dependent on the school than formerly.

Sometimes one sees in the school the instrument for transferring a certain maximum quantity of knowledge to the growing generation. But that is not right. Knowledge is dead; the school, however, serves the living. It should develop in the young individuals those qualities and capabilities which are of value for the welfare of the commonwealth. But that does not mean that individuality should be destroyed and the individual become a mere tool of the community, like a bee or an ant. For a community of standardized individuals without personal originality and personal aims would be a poor community without possibilities for development. On the contrary, the aim must be the training of independently acting and thinking individuals, who, however, see in the service of the community their highest life problem. As far as I can judge, the English school system comes nearest to the realization of this ideal.

But how shall one try to attain this ideal? Should one perhaps try to realize this aim by moralizing? Not at all. Words are and remain an empty sound, and the road to perdition has ever been accompanied by lip service to an ideal. But personalities are not formed by what is heard and said, but by labour and activity.

The most important method of education accordingly always has consisted of that in which the pupil was urged to actual performance. This applies as well to the first attempts at writing of the primary boy as to the doctor's thesis on graduation from the university, or as to the mere memorizing of a poem, the writing of a composition, the interpretation and translation of a text, the solving of a mathematical problem or the practice of physical sport.

But behind every achievement exists the motivation which is at the foundation of it and which in turn is strengthened and nourished by the accomplishment of the undertaking. Here there are the greatest differences and they are of greatest importance to the educational value of the school. The same work may owe its origin to fear and compulsion, ambitious desire for authority and distinction, or loving interest in the object and a desire for truth and understanding, and thus to that divine curiosity which every healthy child possesses, but which so often early is weakened. The educational influence which is exercised upon the pupil by the accomplishment of one and the same work may be widely different, depending upon whether fear of hurt, egoistic passion or desire for pleasure and satisfaction are at the bottom of this work. And nobody will maintain that the administration of the school and the attitude of the teachers does not have an influence upon the moulding of the psychological foundation for pupils.

To me the worst thing seems to be for a school principally to work with methods of fear, force and artificial authority. Such treatment destroys the sound sentiments, the sincerity and the self-confidence of the pupil. It produces the submissive subject. It is no wonder that such schools are the rule in Germany and Russia. I know that the schools in this country are free from this worst evil; this also is so in Switzerland and probably in all democratically governed countries. It is comparatively simple to keep the school free from this worst of all evils. Give into the power of the teacher the fewest possible coercive measures, so that the only source of the pupil's respect for the teacher is the human and intellectual qualities of the latter.

The second-named motive, ambition or, in milder terms, the aiming at recognition and consideration, lies firmly fixed in human nature. With absence of mental stimulus of this kind, human cooperation would be entirely impossible; the desire for the approval of one's fellowman certainly is one of the most important binding powers of society. In this complex of feelings, constructive and destructive forces lie closely together. Desire for approval and recognition is a healthy motive; but the desire to be acknowledged as better, stronger or more intelligent than a fellow being or fellow scholar easily leads to an excessively egoistic psychological adjustment, which may become injurious for the individual and for the community. Therefore the school and the teacher must guard against employing the easy method of creating individual ambition, in order to induce the pupils to diligent work.

Darwin's theory of the struggle for existence and the selectivity connected with it has by many people been cited as authorization of the encouragement of the spirit of competition. Some people also in such a way have tried to prove pseudo-scientifically the necessity of the destructive economic struggle of competition between individuals. But this is wrong, because man owes his strength in the struggle for existence to the fact that he is a socially living animal. As little as a battle between single ants of an ant hill is essential for survival, just so little is this the case with the individual members of a human community.

Therefore one should guard against preaching to the young man success in the customary sense as the aim of life. For a successful man is he who receives a great deal from his fellowmen, usually incomparably more than corresponds to his service to them. The value of a man, however, should be seen in what he gives and not in what he is able to receive.

The most important motive for work in the school and in life is the pleasure in work, pleasure in its result and the knowledge of the value of the result to the community. In the awakening and strengthening of these psychological forces in the young man, I see the most important task given by the school. Such a psychological foundation alone leads to a joyous desire for the highest possessions of men, knowledge and artistlike workmanship.

The awakening of these productive psychological powers is certainly less easy than the practice of force or the awakening of individual ambition but is the more valuable for it. The point to develop is the childlike inclination for play and the childlike desire for recognition and to guide the child over to important fields for society; it is that education which in the main is founded upon the desire for successful activity and acknowledgement. If the school succeeds in working successfully from such points of view, it will be highly honoured by the rising generation and the tasks given by the school will be submitted to as a sort of gift. I have known children who preferred schooltime to vacation.

Such a school demands from the teacher that he be a kind of artist in his province. What can be done that this spirit be gained in the school? For this there is just as little a universal remedy as there is for an individual to remain well. But there are certain necessary conditions which can be met. First, teachers should grow up in such schools. Second, the teacher should be given extensive liberty in the selection of the material to be taught and the methods of teaching employed by him. For it is true also of him that pleasure in the shaping of his work is killed by force and exterior pressure.

If you have followed attentively my meditations up to this point, you will probably wonder about one thing. I have spoken fully about in what spirit, according to my opinion, youth should be instructed. But I have said nothing yet about the choice of subjects for instruction, nor about the method of teaching. Should language predominate or technical education in science?

To this I answer: In my opinion all this is of secondary importance. If a young man has trained his muscles and physical endurance by gymnastics and walking, he will later be fitted for every physical work. This is also analogous to the training of the mind and the exercising of the mental and manual skill. Thus the wit was not wrong who defined education in this way: 'Education is that which remains, if one has forgotten everything he learned in school.' For this reason I am not at all anxious to take sides in the struggle between the followers of the classical philologic-historical education and the education more devoted to natural science.

On the other hand, I want to oppose the idea that the school has to teach directly that special knowledge and those accomplishments which one has to use later directly in life. The demands of life are much too manifold to let such a specialized training in school appear possible. Apart from that, it seems to me, moreover, objectionable to treat the individual like a dead tool. The school should always have as its aim that the young man leave it as a harmonious personality, not as a specialist. This in my opinion is true in a certain sense even for technical schools, whose students will devote themselves to a quite definite profession. The development of general ability for independent thinking and judgement should always be placed foremost, not the acquisition of special knowledge. If a person masters the fundamentals of his subject and has learned to think and work independently, he will surely find his way and besides will better be able to adapt himself to progress and changes than the person whose training principally consists in the acquiring of detailed knowledge.

Finally, I wish to emphasize once more that what has been said here in a somewhat categorical form does not claim to mean more than the personal opinion of a man, which is founded upon *nothing but* his own personal experience, which he has gathered as a student and as a teacher.

(From *Out of My Later Years* (New York: Philosophical Library, 1950).)

An elementary derivation of the equivalence of mass and energy

The following derivation of the law of equivalence, which has not been published before, has two advantages. Although it makes use of the principle of special relativity, it does not presume the formal machinery of the theory but uses only three previously known laws:

(1) The law of the conservation of momentum.
(2) The expression for the pressure of radiation; that is, the momentum of a complex of radiation moving in a fixed direction.
(3) The well known expression for the aberration of light (influence of the motion of the earth on the apparent location of the fixed stars—Bradley).

We now consider the following system. Let the body B rest freely in

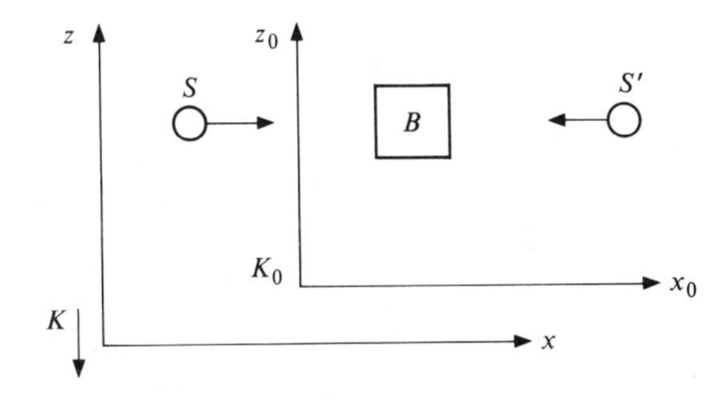

space with respect to the system K_0. Two complexes of radiation S, S' each of energy $E/2$ move in the positive and negative x_0 direction respectively and are eventually absorbed by B. With this absorption the energy of B increases by E. The body B stays at rest with respect to K_0 by reasons of symmetry.

$\frac{E}{2}$ → B ← $\frac{E}{2}$

Now we consider this same process with respect to the system K, which moves with respect to K_0 with the constant velocity v in the negative Z_0 direction. With respect to K the description of the process is as follows:

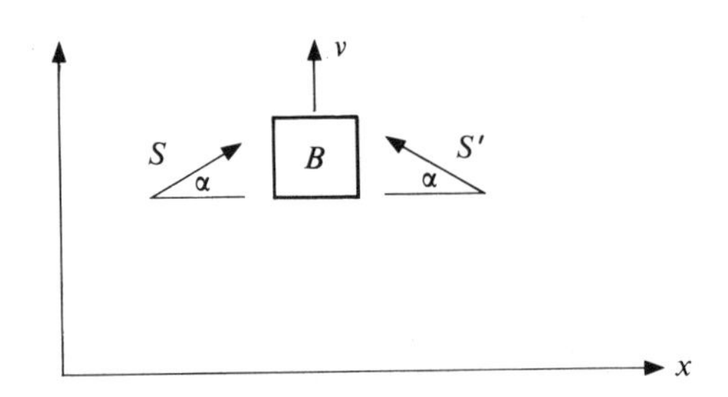

The body B moves in the positive Z direction with velocity v. The two complexes of radiation now have directions with respect to K which make an angle α with the x axis. The law of aberration states that in the first approximation $\alpha = v/c$, where c is the velocity of light. From the consideration with respect to K_0 we know that the velocity v of B remains unchanged by the absorption of S and S'.

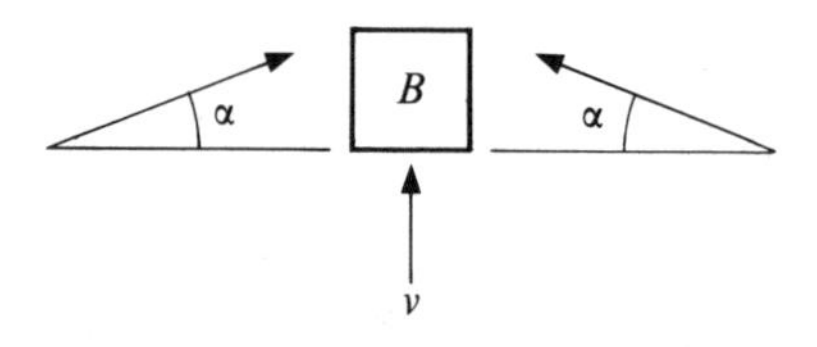

Now we apply the law of conservation of momentum with respect to the z direction to our system in the coordinate-frame K.

1. *Before the absorption* let M be the mass of B; Mv is then the expression of the momentum of B (according to classical mechanics). Each of the complexes has the energy $E/2$ and hence, by a well known conclusion of Maxwell's theory, it has the momentum $E/2c$. Rigorously speaking this is the momentum of S with respect to K_0. However, when v is small with respect to c, the momentum with respect to K is the same except for a quantity of second order of magnitude (v^2/c^2 compared to 1). The z-component of this momentum is $(E/2c) \sin \alpha$ or with sufficient accuracy (except for quantities of higher order of magnitude) $(E/2c)\alpha$ or $(E/2)(v/c^2)$. S and S' together therefore have a momentum Ev/c^2 in the z direction. The total momentum of the system before absorption is therefore

$$Mv + \frac{E}{c^2}v$$

2. *After the absorption* let M' be the mass of B. We anticipate here the possibility that the mass increased with the absorption of the energy E (this

is necessary so that the final result of our consideration be consistent). The momentum of the system after absorption is then

$$M'v$$

We now assume the law of the conservation of momentum and apply it with respect to the z direction. This gives the equation

$$Mv+\frac{E}{c^2}v=M'v$$

or

$$M'-M=\frac{E}{c^2}$$

This equation expresses the law of the equivalence of energy and mass. The energy increase E is connected with the mass increase E/c^2. Since energy according to the usual definition leaves an additive constant free, we may so choose the latter that

$$E=Mc^2$$

(From *Out of My Later Years*)

 Figure 38 Cartoon by Herblock

A Short Selected Bibliography of Einstein's Writings

Essays on Science (New York: Philosophical Library, 1934)

The World as I See It (London: John Lane, 1935)

The Evolution of Physics (with L. Infeld) (Cambridge: Cambridge University Press, 1938; 1961)

'Autobiographical Notes' in *Albert Einstein: Philosopher-Scientist* (Evanston, Ill.: The Library of Living Philosophers, 1949)

Out of My Later Years (New York: Philosophical Library, 1950)

The Meaning of Relativity (Princeton: Princeton University Press, 1950)

The Principle of Relativity (Principal papers by Einstein; also by Lorentz, Minkowski and Weyl) (New York: Dover Publications, 1952)

Ideas and Opinions (New York: Crown Publishing Co., 1954; Dell Publishing Co., 1973)

Investigations on the Theory of the Brownian Movement (Papers by Einstein, edited by R. Fürth) (New York: Dover Publications, 1956)

Relativity: The Special and General Theory (New York: Crown Publishing Co., 1961)

Subject index

Index

Index of names